The ASQ Certified Supplier Quality Professional Handbook

The ASQ Certified Supplier Quality Professional Handbook

Second Edition

Mark Allen Durivage and
Stephanie Parker, Editors

ASQExcellence
Milwaukee, Wisconsin

The ASQ Certified Supplier Quality Professional Handbook, Second Edition
Published by ASQExcellence
Produced and distributed by Quality Press, ASQ, Milwaukee, WI

Publisher's Cataloging-in-Publication data

Names: Durivage, Mark Allen, editor. | Parker, Stephanie L., editor.
Title: The ASQ certified supplier quality professional handbook , second edition / Mark Allen Durivage and Stephanie Parker, editors.
Description: Includes bibliographical references and index. | Milwaukee, WI: ASQExcellence, 2023.
Identifiers: LCCN: 2023945602 | ISBN: 978-1-63694-119-6 (hardcover) | 978-1-63694-120-2 (PDF) | 978-1-63694-121-9 (epub)
Subjects: LCSH Industrial procurement--Quality control. | Materials management--Quality control. | BISAC BUSINESS & ECONOMICS / Logistics & Supply Chain | BUSINESS & ECONOMICS / Quality Control
Classification: LCC HD39.5 .A77 2023 | DDC 658.7--dc23

ASQ and ASQExcellence advance individual, organizational, and community excellence worldwide through learning, quality improvement, and knowledge exchange.

Bookstores, wholesalers, schools, libraries, businesses, and organizations: Quality Press and ASQExcellence books are available at quantity discounts for bulk purchases for business, trade, or educational uses. For more information, please contact Quality Press at 800-248-1946 or books@asq.org.

To place orders or browse the selection of all Quality Press and ASQExcellence titles, visit our website at: http://www.asq.org/quality-press.

27 26 25 24 23 LS 6 5 4 3 2 1

ASQ
Quality Press
600 N. Plankinton Ave.
Milwaukee, WI 53203-2914
Email: books@asq.org
Excellence Through Quality™

ASQE™
Inspire Excellence

Table of Contents

List of Figures and Tables

Part IV

Part V

Part VI

Part VII

Preface

The Certified Supplier Quality Professional (CSQP) certification provides valuable credentials to quality professionals in the growing field of supplier quality engineering. Due to globalization of the supply chain, supplier quality engineers are becoming more important in a broad spectrum of industries, including the manufacturing and service industries.

The purpose of this handbook is to prepare individuals for the CSQP examination and provide a reference for the practitioner. Throughout this handbook, examples are provided based on the collective experience and knowledge of the authors and co-editors. However, these examples are not explicitly specified in regulations, leaving decisions to the company, as well as the burden of justifying practices using sound scientific principles that provide the context of the rationale.

The *ASQ Certified Supplier Quality Professional Handbook, Second Edition* follows the updated Body of Knowledge (BoK) and provides additional materials to support the exam as well as providing techniques that can be readily applied to develop a new supplier quality program or enhance the existing system.

Preface

Acknowledgments

The ASQ Certified Supplier Quality Professional Handbook, Second Edition, is dedicated to the individuals around the world who work tirelessly to ensure that the goods and services purchased and then received by their companies adhere to specifications, requirements, and regulations.

The following individuals are to be recognized as contributing chapter authors for the second edition of the handbook: Ed Cook, Chris Riegel, Mariel David-DeVito, Daniella Picciotti, James Shore, Peter Clark, Sandra Storli, Pamela Carvell, Grant Short, and Allen Wong. The overall quality and technical content of this handbook has been greatly enhanced by the contributions of individual subject matter experts. We would also like to acknowledge Shawn Armstrong, CSQP, and Daniella Picciotti, CSQP for their very thorough and detailed review of the draft manuscript of the second edition.

Additionally, Kevin Posey, Steven Walfish, Kenneth Crow, Jose (Pepe) Rodriguez-Perez, Manuel E. Pena, Zubair Anwar Khurshid, John Freije, Doug Shifflett, and Feroz Aziz are to be recognized for their chapter contributions to the first edition as well as Srikanto H. Paul, Norman Pedersen, and Jeff Veyera for their detailed review of the first edition manuscript.

Mark Allen Durivage, ASQ Fellow and SRE Fellow
Editor, Lambertville, Michigan

Stephanie Parker
Editor, Raleigh, North Carolina

> *I would like to thank those who have inspired, taught, and trained me throughout my academic and professional career. And lastly, I would like to acknowledge the patience of my wife Dawn and my sons Jack and Sam, which allowed me time to organize, write, and edit this handbook.*
>
> – Mark Allen Durivage

> *Thank you to Michelle McGinn and Brenda Walton for teaching me how to write, to Bonnie Tripp, Forrest Hentz, and Charlie Rozanski for teaching me how to think critically, to Tara Bork for teaching me how to communicate ideas, and to Kay Yow for showing me how to inspire others. And, as always, thanks to my family for their unending support.*
>
> – Stephanie Parker

LIMIT OF LIABILITY/DISCLAIMER OF WARRANTY

Part I

Supplier Strategy

Chapter 1
Supply Chain Vision/Mission

> Assist in the development and communication of the supply chain vision/mission statement. (Apply)
>
> **Body of Knowledge I.A**

Today, the word *quality* appears on almost every product we come across—food, appliances, automobiles, aircraft, consumer goods, medical devices, pharmaceuticals, and software. No matter where you turn, you hear about quality. In fact, we demand a certain level of quality in everything we buy.

What is meant by the word *quality*? The American Society for Quality (ASQ) defines quality as "the totality of features and characteristics of a product or service that bear on its ability to satisfy given needs." The features and characteristics of a product or service are how we as consumers evaluate how good that product or service is. For example, in a restaurant, the amount we leave as a tip is a direct reflection on the quality of the food and the service. In a supply chain, the features and characteristics that define quality are translated into specifications and requirements.

We live in an age of globalization. Not only has the major industrial trend of outsourcing continued unabated, but outsourcing now has a truly global footprint. This has increased the importance of the supply chain in every industry, but especially in regulated industries like medical devices and pharmaceuticals. For example, a drug can now utilize a key starting material that is manufactured in a Chinese plant, received and converted into an active pharmaceutical ingredient in India, and ultimately formulated into a finished drug product in the United States to be distributed in the United States, Canada, and the European Union (EU). The regulations touching this one product alone, not to mention the complexities of the supply chain and the technical challenges of manufacturing, are staggering. The importance of a safe, reliable, cost-effective, high-quality, integral, and nimble supply chain cannot be overstated. Equally critical are the supply chain processes (e.g., strategic, procurement, quality assurance, technical, regulatory) that ensure these requirements are met. By extension, then, the supply chain professional—notably, the supplier quality professional (SQP)—is of paramount importance as a creator, driver, and full participant in these business-critical activities.

As supply chain professionals, it is our job to add value to our organizations by:

- Ensuring that the quality of purchased components, materials, parts, and finished goods meets requirements
- Complying with applicable regulations and voluntary standards (RoHS, REACH, Conflict Minerals, FDA, MDR, IVDR, GHTF, FMD, ICH, CMDR, and various ISO, ASTM, and other standards)
- Ensuring the integrity of the supply chain to maintain product purity (uncontaminated, not counterfeit)
- Minimizing interruptions and disruptions in the supply chain
- Adhering to manufacturing schedules and order fulfillment
- Ensuring cost-effective execution of processes (e.g., strategic, procurement, quality assurance, technical, regulatory)
- Effectively defining expectations, roles, and responsibilities through contracts and quality agreements
- Effectively managing the supply base over the lifecycle (selection, qualification, management, monitoring and measurement)

We can accomplish these objectives via a supply chain strategy.

SUPPLY CHAIN STRATEGY

> *This is the first time in the history of business that you can be great at what you do today and be out of business tomorrow.*
>
> —Ken Blanchard and Terry Waghorn, *Mission Possible*

As supply chain professionals, we espouse the ideal that our supply base is an extension of our internal manufacturing—indeed, an extension of our business. As an SQP, you are part of the supply chain team that is responsible for ensuring the smooth operation and high-value proposition of the supply base. So how do we make that ideal a reality? Certainly it will not happen organically or accidentally. How, then? Precisely by *managing it*. That is to say, we must have a supply chain strategy. Such a strategy allows us to plan how the supply base will unfold and add value to our organization. It also characterizes the supply base itself: What will it look like? With whom will we choose to partner? Engage with? How much development work will we invest in the supply base? How will the links in the various supply chains connect and secure themselves? Strategy will determine all of this and more. In fact, having a good supply chain strategy can be a competitive advantage, even a competitive imperative. It can be the key to long-term success.

Strategy means different things to different people. In its simplest form, it is a way of maximizing the actions you can control and minimizing those you cannot. This means that any business must have a set of guidelines that dictates where they want to be at a given point (a vision), and this must be clearly understood by everyone (deployment and communication). For example, the strategy could be that the business will achieve the highest quality possible, or that the product will be recognized by others as being the best in the field, or that the product will provide an increase in market share. It is the function of the supply chain professional, including the SQP, to draft and deploy a strategy that achieves the stated objectives of the organization, whatever they may be.

BENEFITS OF A SUPPLY CHAIN STRATEGY

Drafting and deploying an appropriate supply chain strategy yields many benefits, ranging from reductions in cost, lead time, and recalls to an overall increase in productivity and efficiency. When suppliers know what is expected of them (quality agreements) and how they are evaluated (measuring and monitoring), they can focus on improvements such as reductions in variation and cost. The cost reductions resulting from variation reduction will enable suppliers to enter into longer-term contracts while also accomplishing their financial goals. At the same time, suppliers are assured of a sustained partnership with the customer, a benefit to both parties.

Another important aspect of supply chain strategy is that, if properly communicated, it will provide organizational and cross-organizational *alignment*. This is beneficial for several reasons: effective deployment of the strategy, efficiency in operational execution of the strategy, and a clear, common direction, inside both the customer's organization as well as the suppliers'. In this way, the SQP can ensure that the effort towards the deployment of strategy not only goes smoothly, but also that all involved are working towards a common purpose and goal.

Another component of a good supply chain strategy involves making the supplier a part of the design team developing new products. In such cases the supplier is expected to lend expertise to the design process as part of a mutual commitment benefiting both customer and supplier: The customer gets the latest technology in the supplier's specialized area, and the supplier gets an opportunity to increase business and improve relations with the customer. The supplier participates in design reviews, specification development, prototyping products, measurement assurance, and life testing. The supplier is given unique opportunities to have some control over new products. Furthermore, in many cases suppliers are aware of new technologies in their field, as well as the capabilities and limitations of their current products. As they learn the expectations for the product they are helping to develop, they can recommend changes in the product that will enhance its performance. For example, say a certain plastic part is to go into a copier. The supplier is brought in and finds that the part is to be located near a heating element that could cause the plastic part to warp over time. The supplier recommends a different composition of resins with a higher heat resistance. This results in an improved product and significant savings in potential redesign and service costs.

As we can see, strategy is without a doubt one of the most critical aspects of any endeavor, including supply chain. Surely with the complexities of global commerce and supply chains that span the world, supply chain strategy has to be a primary concern and focus of the supply chain professional, and therefore the SQP. The SQP must partner with procurement and other sourcing professionals to develop the most effective, efficient, and value-added supply chain strategy possible. This involves understanding the needs and capabilities of all stakeholders—internal and external, customer and supplier oriented—as well as the general technological and market conditions relevant to the product lines under development, manufacturing, or service. Although the specifics of business and market conditions are beyond the scope of this certification, general strategic concepts are not, and these must be understood to create the most efficacious supply chain strategy possible. Two of the most effective tools in developing, communicating, and deploying a truly effective strategy of any kind are mission and vision.

MISSION AND VISION

Knowing where you're going is the first step in getting there.

—Ken Blanchard and Terry Waghorn, *Mission Possible*

Many incredible books have been written on the subject of mission and vision. One of the most useful and easily understood is *Full Steam Ahead!* by Ken Blanchard and Jesse Lyn Stoner (2011). Blanchard has been called "the dean of leadership," and for good reason. In *Full Steam Ahead!* he lends his talents to a parable-style narrative that explains not only what vision is but why it is critical to leaders and organizations, and how to create a vision that energizes. Another very useful work is *The Servant Leader* by James A. Autry. In it he writes, "The mission question is simply, 'What do we do?' By extension, it's 'what do we do to fulfill our purpose?'" (Autry 2001, 28). As we will see, this definition of mission dovetails nicely with Blanchard and Stoner's three key elements of a compelling vision. They begin the process of vision formulation with the concept of purpose as well.

Without the structure, definition, and direction that mission and vision provide, the supply chain will not function effectively and the organization's options will thus be limited, sometimes severely. The development, communication, and deployment of a robust strategy employing the tools of mission and vision ensures that an organization's supply chain will be consciously and deliberately developed and will therefore best serve the needs of the organization. The purpose defines the mission, and in turn the mission informs the vision. If crafted correctly, Blanchard and Stoner write, "The vision must benefit everyone it touches" (2011, 167).

Mission

If mission stems from purpose, and so too does vision, then our purpose is the foundation of our "why." Our purpose serves as our anchor and the reason we come to work every day. In every good organization, a functional mission and vision flows down from the organizational mission and vision. This provides alignment and enables efficient execution of objectives and goals. It also drives the appropriate key process indicators (KPRIs) and their associated metrics. Autry points out that the mission and resultant objectives can change over time, but the purpose should not (Autry 2001). Autry also agrees with Blanchard and Stoner (2011) that values are an important part of the mission and vision. Values are the bedrock to which the cultural anchors of an organization are fastened. Values provide stability, focus, and guidance, if not boundaries, over time. In the end, every professional needs to understand the purpose of their activities and their organization. To know and understand the purpose is to understand the mission. The purpose serves to ground people, anchor them, and focus their energies on the appropriate value-added activities. In this way, the supply chain professional understands each and every day what they are to accomplish and therefore what inputs will drive the desired outcome.

As a segue to the discussion of vision, let's look at a purpose that is germane to supply chain and create a good mission statement. For example, if the stated purpose was concentrated on ensuring a supply of high-quality and high-value components and materials, we might construct the following mission statement:

> To procure the highest-quality components from the most reliable suppliers at the best cost possible.

With this mission statement we can move forward to craft our vision statement. The vision statement is a fundamental leadership tool to ensure focus, cohesiveness, and cross-functional alignment. If crafted carefully and communicated and deployed effectively, it can be a powerful bonding agent between functions that must work toward a common goal. This is particularly true in the matrix organizational structures that are increasingly popular. Given the diversity of supply chains, cross-functional alignment can be an invaluable strategic asset. How, then, do we progress from understanding our purpose to developing a compelling vision that can serve to not only inform the development of the supply chain strategy but also unite the diverse functions involved in supply chain activities to effectively deploy that strategy?

Vision

According to Blanchard and Stoner, "Vision is knowing who you are, where you are going, and what will guide your journey" (2011, 79). Further, they assert there are three key elements of a compelling vision:

1. Significant purpose
2. A picture of the future
3. Clear values

It is easy to see how applying these concepts to a supply chain strategy would increase its robustness, flexibility, and adaptability to inevitable changes, even on a global scale, by defining, inspiring, and guiding supply chain professionals as they seek to engage suppliers and potential suppliers in the supplier lifecycle management process.

Using the example stated above as our mission statement, "To procure the highest-quality components from the most reliable suppliers at the best cost possible," the supply chain professional and SQP would focus their efforts on stable suppliers that have developed quality systems and are efficient and cost-conscious. Inherent in the second element is the notion that vision is by definition forward-facing. Since the future has not occurred yet, it is sometimes difficult for people to see it clearly. It is nonetheless critical in the process of constructing that future to have a clear idea of what it looks like. A carefully crafted picture, rich with detail, can serve as a conceptual model of the desired future state, and it can also inspire if done correctly. Appeals to people's passions, pain points, and better nature can be crafted into the future state that is painted in the picture. Using the same mission statement above, a picture could be painted in this way:

> Two years from now I see an uninterrupted supply chain providing on-time delivery of quality components—such high quality that incoming inspection will no longer be required. Downtime will be the exception rather than the rule. We will partner with our preferred suppliers to keep costs to a minimum by collaborating to maximize efficiencies, drive out waste, and streamline demand planning as a joint activity.

In this picture of the future state are clearly embedded values: waste reduction, efficiency, implicit trust of the supplier, and collaboration. These values will serve as cultural anchors in times of uncertainty, ambiguity, and change. These factors may distort the future-state picture as time goes by, but the values will remain timeless and immobile, providing secure ground from which to operate. Without this secure footing, time, conditions, and events might make it difficult to remain true to the vision and purpose of the supply chain organization. Clear values are the anchor to which we can all secure our understanding of our work, and they also give us all something to rally around.

Taking our purpose, picture of the future, and values, our vision statement might look something like this:

> Maximizing value by focusing on quality, cost, and reliability in our supply chain.

See Table 1.1 for a summary of the process of creating a vision statement.

Table 1.1 Summary table of steps in creating a vision.

Mission statement	Picture of the future	Clear values	Vision statement
To procure the highest-quality components from the most reliable suppliers at the best cost possible.	Two years from now I see an uninterrupted supply chain providing on-time delivery of quality components—such high quality that incoming inspection will no longer be required. Downtime will be the exception rather than the rule. We will partner with our preferred suppliers to keep costs to a minimum by collaborating to maximize efficiencies, drive out waste, and streamline demand planning as a joint activity.	Waste reduction, efficiency, implicit trust of the supplier, collaboration.	Maximizing value by focusing on quality, cost, and reliability in our supply chain.

To further illustrate the point, here are some examples of actual vision statements:

> To become a Center of Excellence in the supplier quality function by leading strategic partnerships with our suppliers and our internal customers through capable and committed people, focusing on improving product quality and compliant processes.
>
> Be the most effective management organization that supports global suppliers, our partners and our customers, and meets worldwide compliance through:
>
> - Supplier partnership
> - Consistent and standard processes
> - Reduction of costs
> - Continuous improvement
>
> Creating a reliable supply chain through unwavering commitment to quality and cost.

In summary, mission and vision are critical to providing structure, cultural anchors for stability, and commonality of purpose as well as a direction forward into the future. As we have seen, if crafted purposefully, the vision can also serve to inspire! This is critical if the supply chain strategy is to be deployed effectively.

Chapter 2
Supplier Lifecycle Management

World-class organizations understand that a planned and measured approach to managing suppliers is essential, and many models have been developed to support the need to manage suppliers. Supply chain management as a field has been around for decades, but the relatively new concept and analytical model of supplier lifecycle management (SLM) takes an integrative approach and explicitly plans for the acquisition of suppliers as well as natural (and unnatural) turnover in suppliers (see Figure 2.1). SLM supports and guides the business relationship from initial supplier discovery through qualification and onboarding to ongoing maintenance and possible termination or obsolescence. SLM is an integrated approach that considers business and quality needs of the organization. Its purpose is to

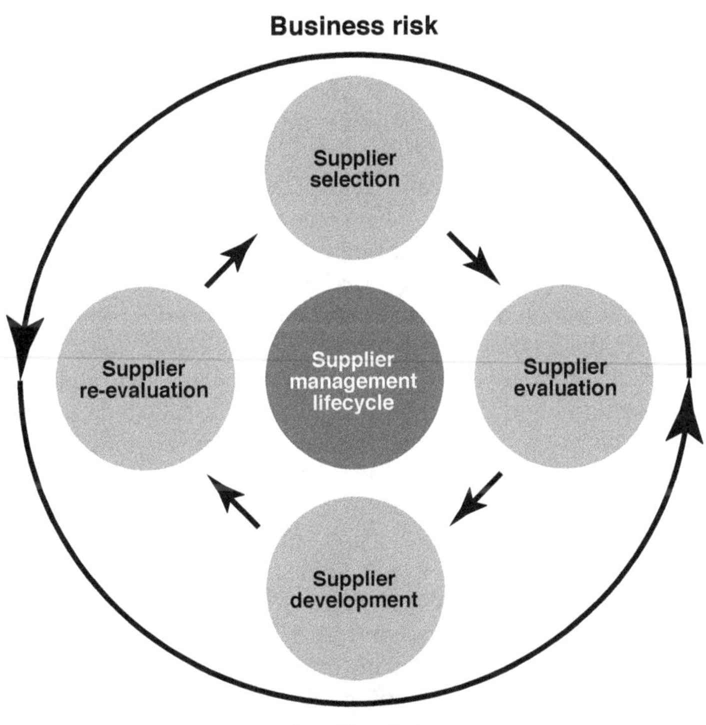

Figure 2.1 Supplier lifecycle management model.

recognize suppliers as a prime source of value to the organization and deliver that value by putting them at the heart of procurement strategy and management.

Many supply chain professionals are familiar with the supply chain management (SCM) model, and they may wonder: How does SLM differ from SCM? SCM is a set of system-focused approaches utilized to efficiently integrate suppliers, manufacturers, warehouses, and stores so that merchandise is produced and distributed in the right quantities, to the right locations, and at the right time in order to minimize systemwide costs while satisfying service level requirements. SLM is a systemic approach that is customer focused (service level) and intent on minimizing costs, but it considers all links in the supply chain that can affect either of the two main objectives. In this sense, there seem to be only minor differences between the concepts, perhaps the most significant being SLM's explicit planning for both the natural and unnatural turnover in suppliers. SCM focuses more explicitly on the entire supply chain, including lower-tier suppliers, across the entire enterprise, and focuses on managing and minimizing total systemwide costs.

Many supplier management professionals are also familiar with the supplier relationship management (SRM) model and may wonder how SLM differs from SRM. SRM is the systematic, enterprise-wide assessment of suppliers' assets and capabilities with respect to overall business strategy, the determination of which activities to engage in with different suppliers, and the planning and execution of all interactions with suppliers in a coordinated fashion across the relationship lifecycle to maximize the value realized through those interactions. The focus of SRM is developing two-way, mutually beneficial relationships with strategic supply partners to deliver greater levels of innovation and competitive advantage than could be achieved by operating independently or through a traditional transactional purchasing arrangement.

In many fundamental ways, SRM is analogous to customer relationship management. Just as companies have multiple interactions over time with their customers, so too do they interact with suppliers—negotiating contracts, purchasing, managing logistics and delivery, collaborating on product design, etc. The starting point for defining SRM is a recognition that these various interactions with suppliers are not discrete and independent. Instead, they are accurately and usefully thought of as comprising a relationship—one that can and should be managed in a coordinated fashion across functional and business unit touchpoints and throughout the relationship lifecycle.

In summary, SRM, as the name implies, is very much relationship focused, and less focused on regular transactions and their associated metrics.

Although it is not a complete supplier management paradigm, many supplier management professionals may have had exposure to supplier information management (SIM) solutions. SIM is a corporate or even enterprise-wide IT solution that manages, retrieves, and organizes supplier information and maximizes supplier engagement and data quality across a global supplier base while ensuring minimum risk and high efficiency. As supply chain strategies are implemented that reduce cost and lead times and increase service level, the timeliness and availability of relevant information is critical. One alternative to using an enterprise resource planning (ERP) solution for supplier information management is the implementation of a multimode, cloud-based SIM platform. A dedicated SIM platform has both advantages and disadvantages compared with an all-in-one ERP solution that also includes a SIM module. However, SIM only focuses on management of the information; it does not give strategic guidance on which information is important, unlike a full lifecycle model such as SLM.

The choice of an appropriate supplier management lifecycle and supplier management model is dependent on many things, including:

- Corporate strategy
  - May include strategic objectives to incorporate Six Sigma and lean methodologies, continuous improvement, or voice of the customer utilized separately from Six Sigma
- Company demographics
  - Size
  - Global presence
  - Company maturity
- Industry maturity and trends
  - Reliance on economies of scale for cost control?
  - Or push for one-piece flow for lean?
  - Demand stability and predictability
    - Demand for finished good versus aggregated component
    - Push-pull versus pull only or push only
    - Critical component/assembly lead times
  - Quality levels expected by customers
    - Tolerance for nonconformance
    - Taguchi loss function
  - Company location relative to critical resources
    - Skilled personnel
    - Universities and centers of innovation
    - Industry centers
    - Raw materials
  - Vertical integration/global insourcing versus contract manufacturing/outsourcing
  - Industry competition for goods as well as resources
  - Competition on price or differentiation of features
  - Cycle time to market
    - Short competitive cycle for "appearance" of new products?
    - Drive to deliver new real value that is less time driven?
    - Driven by technology cycle?

- — Disruption from new technologies/platforms
  - Virtual products replacing physical products
- Logistics considerations in the supply chain
  - — Modes of transport—land/sea/air
  - — Time in transport, economic value in transit
  - — Tax implications

SUPPLIER SELECTION

> Develop the process for supplier selection and qualification, including the identification of sub-tier suppliers using tools such as SIPOC, decision analysis, and total risk factor analysis. (Create)
>
> **Body of Knowledge I.B.1**

Regardless of the supplier management model chosen, supplier selection methods and criteria are a core component. In this section we will discuss how to develop the process for supplier selection and qualification, including the identification of sub-tier suppliers.

The ASQ Customer-Supplier Division outlines the supplier selection process using nine steps:

1. Identify all possible sources
2. Evaluate reputations
3. Prepare bidders list
4. Request bids
5. Evaluate pre-award data
6. Select the supplier
7. Update approved supplier list
8. Prepare contract
9. Accept contract

Identifying candidates may involve a multitude of evaluation criteria. Over time, providers of goods and services build a reputation. This reputation may be for price, durability, variety, or any number of different attractors for that supplier's chosen market. Customers may learn about a supplier's reputation in various ways, including:

- Experience (e.g., seeing the supplier's components in other assemblies they have worked on)
- Professional engineering societies (e.g., quality, mechanical, plastics)
- The supplier's marketing (e.g., print, television, internet)
- Recommendations from trusted friends or colleagues or customer feedback ratings
- Sales agents and distributors

All of this input allows customers to develop a list of possible suppliers, often called a bidders list. The customer now needs pre-award data to make its selection from the bidders list. A common way to obtain these data is to send out a request for bid to four to six possible suppliers. This is called the short list. The functional specifications are sent (minus any proprietary information), along with quantity and delivery date estimates. If a supplier wishes to perform the work, they send a bid back to the customer. Many types and pieces of information are included in the bid, so the customer can make an informed decision.

Pre-award data can also come from conversations with knowledgeable parties. Suppliers often have sales staff who can answer technical and pricing questions from existing and potential customers. Customers can also gather pre-award data by sending surveys (called self-assessments or questionnaires) to possible suppliers. Today, these questionnaires are often sent by e-mail or placed on a customer website for potential suppliers to complete online. Past performance history can also be used to make a supplier selection. If an item worked without problems before, it is likely to work again in the same environment. Sometimes the customer's customer provides a list of preapproved sources to use. This promotes efficiency in that the decision work has already been done. It also promotes consistency in that these preapproved suppliers are known performers. The practice of preapproving sub-suppliers is common in the automotive and aerospace industries.

One of the most common tools used for the supplier selection process is the Suppliers, Inputs, Process, Outputs, and Customers (SIPOC) diagram. A SIPOC diagram is a data collection tool used to assist in gathering information about suppliers, inputs, processes, outputs, and customers of a process. Figure 2.2 is an example SIPOC diagram applied to supplier selection process.

Suppliers	Inputs	Process	Outputs	Customer
Potential: **Supplier 1** **Supplier 2** **Supplier 3** **Supplier 4**	Cost Quality Location Reliability Payment options	Identify all possible sources Evaluate reputations Prepare bidders list Request bids Evaluate pre-award data Select the supplier Update approved supplier list Prepare contract Accept contract	Approved Supplier	Requester

Figure 2.2 Supplier selection process SIPOC.

Once all the data are collected on the short list of possible suppliers, tools such as decision analysis or multi-criteria decision analysis tools can be used to compare the suppliers against the critical criteria, which are often weighted to help further differentiate suppliers (see Table 2.1). For this example, a higher score is more desirable.

Risk

Supplier risk should significantly impact the supplier selection and qualification process and the selection criteria. For example, if a company ranks supplies as high, medium, and low risk, the supplier might use a risk-based scalable system as shown in Table 2.2 or Table 2.3 for critical and non-critical suppliers.

Regulations, standards, and contracts may indicate the need for additional qualification requirements, such as compliance with REACH, RoHS, Conflict Minerals regulations, and others. It is important for the organization to communicate these requirements as part of the selection and qualification process so there

Table 2.1 Weighted decision analysis table.

	Semiquantitative factors:					
	Cost	**Quality**	**Location**	**Reliability**	**Payment options**	
Weights:	**4**	**5**	**1**	**2**	**3**	**Total**
Supplier 1	4	0	0	2	9	15
Supplier 2	0	15	2	4	3	24
Supplier 3	8	10	1	6	0	25
Supplier 4	8	15	3	6	0	32

Table 2.2 Supplier qualification requirements based on supplier risk for high, medium, and low risk suppliers.

Supplier Qualification Requirements Based on Supplier Risk		
High Risk	**Medium Risk**	**Low Risk**
Questionnaire	Questionnaire	Questionnaire
On-Site Audit	ISO Certification	
ISO Certification	Non-Disclosure Agreement	
Nondisclosure Agreement		
Quality Agreement		

Table 2.3 Supplier qualification requirements based on supplier risk for critical and non-critical suppliers.

Supplier Qualification Requirements Based on Supplier Risk	
Critical	**Non-Critical**
Questionnaire	Questionnaire
On-Site Audit	ISO Certification
ISO Certification	Non-Disclosure Agreement
Nondisclosure Agreement	
Quality Agreement	

are no "surprises" and the supplier can determine whether they can meet the requirements.

There may also be times when the organization is required to manage sub-tier suppliers and customer directed suppliers. The management, including qualification and monitoring, of sub-tier supplier and customer-directed suppliers is usually driven by contractual obligations.

Supplier Qualification

Some options for the qualification of a supplier include self-assessment, detailed supplier questionnaire, due diligence, and on-site audit/inspection/preassessment.

Self-assessment and remote review may be an appropriate method for low-risk or off-the-shelf sourced products. Certificates of compliance or conformance to a quality management system (QMS) standard such as ISO 9001 may be sufficient to indicate that the company has controlled processes for quality conformance. For these products, the cost of additional qualification steps may not be warranted if information on the company and its products is readily available through public channels.

For higher-risk sourced materials, products, or services, additional evaluation and qualification is likely to be required. Also, if the material will be single sourced, business continuity concerns in the event of a supply disruption should be addressed.

On-site inspections allow the purchasing agent to determine the capability of the supplier to provide the material and to convey the seriousness of the possibility of doing business with the supplier. The site visit also serves as a means to impress on the supplier that the customer expects the supplier to provide exactly what the customer wants.

Avoid relying only on the presence of an ISO certification for critical supplier qualifications. Qualifying a high-risk or critical supplier based on an ISO certification alone can pose a risk to the organization due to the ever-present proliferation of certificate mills and fake certificates.

PERFORMANCE MONITORING

> Develop the supplier performance monitoring system, including expected levels of performance, process reviews, performance evaluations, improvement plans, and exit strategies. (Create)
>
> **Body of Knowledge I.B.2**

Once a supplier has been selected and brought on board, ongoing performance assessment is an important part of managing the supplier lifecycle. In this section we will discuss the development of a supplier performance monitoring system, including expected levels of performance, process reviews, performance evaluations, improvement plans, and exit strategies.

Depending on the risk level of the material, performance monitoring and supplier rating can take many forms, ranging from simple measures of delivered quality level and on-time delivery to more complicated systems that may also include measures for innovation, risk sharing, supply assurance/continuity, and active collaboration. In the end, it is important that we not only measure the few items that make the biggest impact but also set a baseline of measurement so each party gains from the relationship. Without a win–win approach to sourcing, we lose out on key factors such as trust, respect, understanding, and communication, which

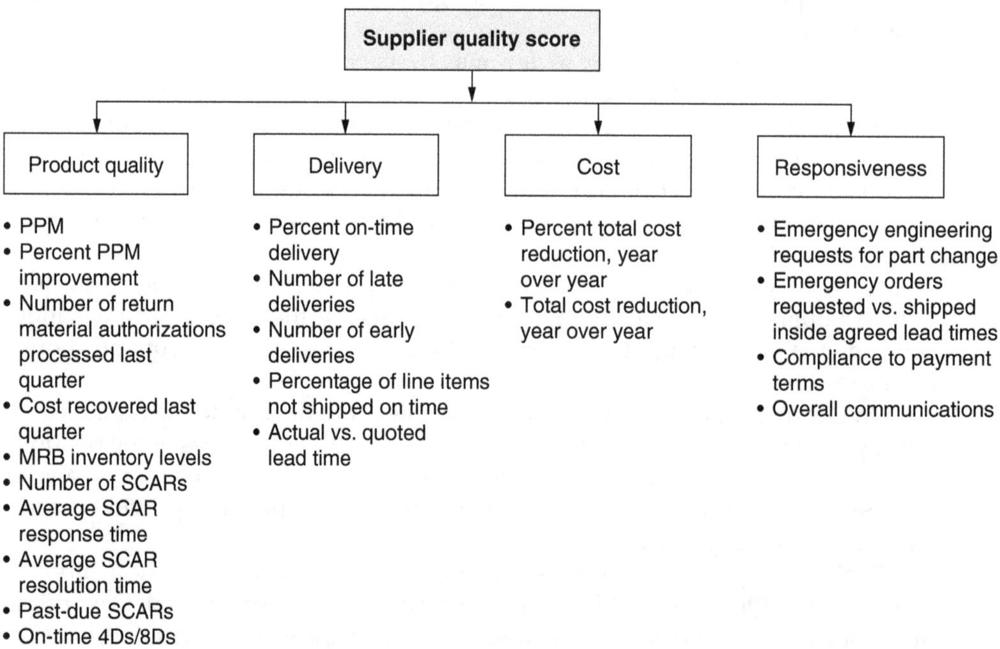

Figure 2.3 Supplier performance categories and metrics.

are necessary for a high-performing supplier relationship. At the most fundamental level, without measures of supplier performance, we would have little idea of when requirements are being met and when corrective action is required to address nonconformances or poor performance.

Expected performance levels for suppliers will generally align with both financial and quality corporate objectives. You may also want to consider some of the measures under the four categories of supplier performance shown in Figure 2.3.

A scorecard system (e.g., Kaplan and Norton's balanced scorecard) is a common method of combining data in such a way that performance can be understood at a

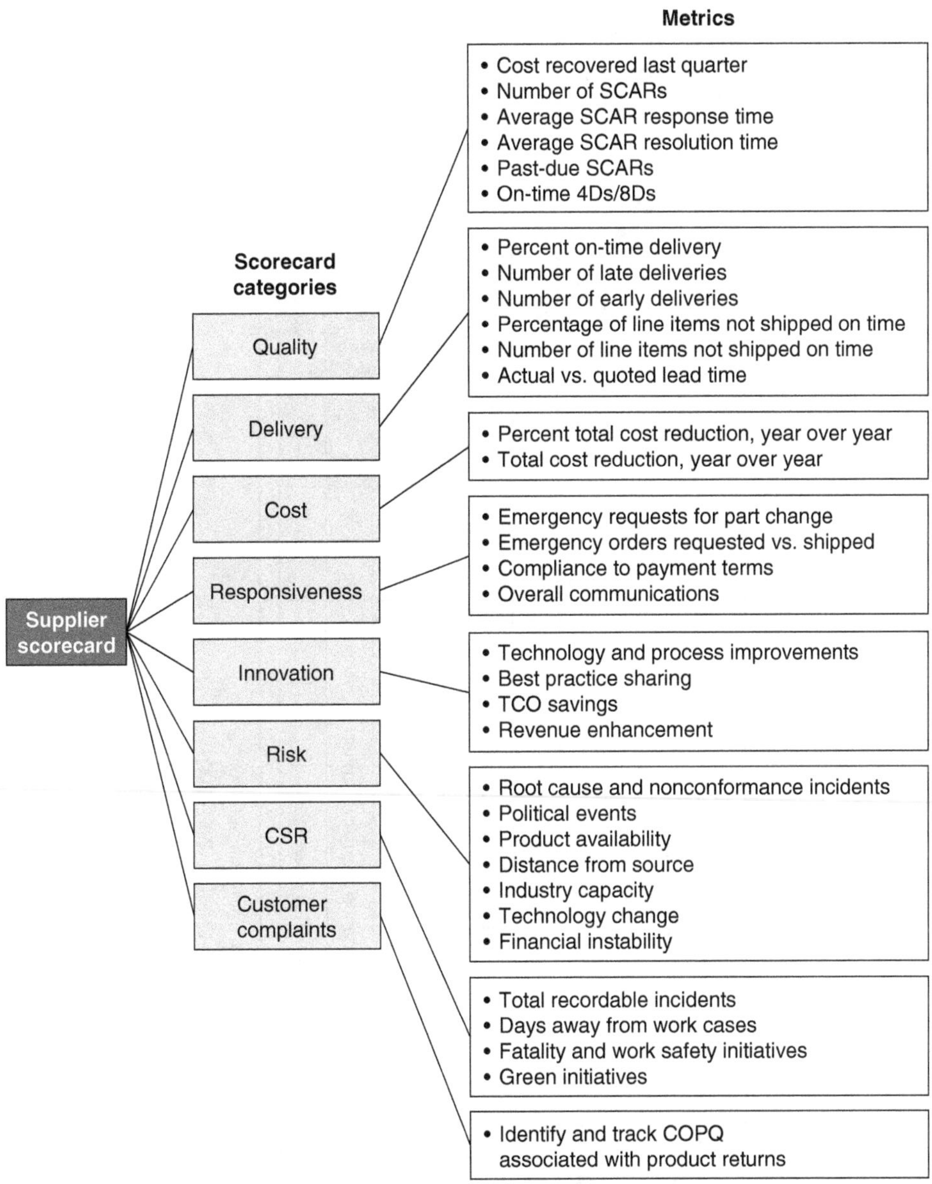

Figure 2.4 Supplier scorecard categories and metrics.

		KPI	Freq	Jan	Feb	Mar	Apr	May	Jun	Target
Financial	1	Supply chain cost (per unit volume)	m	$1.70	$1.69	$1.68	$1.65	$1.57	$1.55	$1.60
	2	Cash—debtor days	m	16.2	17.1	18.3	16.5	14.6	16.0	15.0
Customer	3	Distribution coverage	m	76%	74%	78%	81%	82%	85%	80%
	4	Customer service, OTIF	m	91%	92%	89%	93%	96%	94%	95%
	5	Stock availability at distributor	m	97.6%	94.3%	96.5%	98.1%	98.7%	98.4%	98.0%
Process	6	Total stock (days)—producer, distributor	m	18.6	18.9	16.7	17.5	15.5	14.3	12.0
	7	Sales forecasting accuracy	m	64%	59%	58%	65%	76%	81%	90%
Learning and growth	8	Performance appraisal status	q	25%	25%	25%	65%	65%	65%	90%
	9	Competency attainment	q	40%	40%	40%	50%	50%	50%	75%

Figure 2.5 Sample supplier balanced scorecard.

Source: Adapted from Kaplan and Norton, 1992–2008.

quick glance; it also allows comparison between suppliers (see Figures 2.4 and 2.5). Perhaps more importantly, scorecards that are updated on a regular basis provide a means of tracking and trending supplier performance over time, allowing for regular performance evaluations with the supplier.

Tracking and trending supplier performance may also provide advance warning of major problems before they occur, allowing for preventive or corrective actions. Continuous improvement plans can also use supplier scorecard status as a jumping-off point for the desired improvement. Scorecard warnings indicating negative trends can also be useful in determining when exit strategies should be considered with a troubled supplier, allowing time for the supplier management function to evaluate and bring online an alternate supplier with little or no impact to the rest of the supply chain.

While the eight supplier scorecard categories and metrics provided in Figure 2.4 can be used for all industries, the relative importance of the category and supporting metrics will vary by industry. For example, quality will probably play a more significant role in more regulated industries such as aerospace and medicinal products whereas cost would generally be a driving factor in general consumer goods.

The data used to compile and support supplier scorecards should be validated for accuracy. Inaccurate or incomplete data may lead to faulty decisions related to supplier performance potentially resulting in fracturing the relationship with the supplier. Additionally, the data should be easy to compile. Best practice is to leverage currently available data whenever possible so as not to impose additional burden on those responsible for entering, compiling, and reporting on supplier performance.

SUPPLIER CLASSIFICATION SYSTEM

> Define and develop a supplier classification system (e.g., non-approved, conditionally approved, approved, preferred, certified, partnership, and disqualified). (Create)
>
> **Body of Knowledge I.B.3**

In this section we will discuss ways to develop a supplier status and classification system. Supplier status is typically straightforward and follows the lifecycle steps: nonapproved or pending approval, approved, probation or warning, and phase-out, inactivated, or disqualified.

One method for establishing supplier classifications is to use the Kraljic Portfolio Segmentation Model, shown in Figure 2.6. This allows us to establish risk-based supplier selection, qualification, and performance monitoring processes to focus supplier lifecycle management activities on the riskiest and highest-impact products.

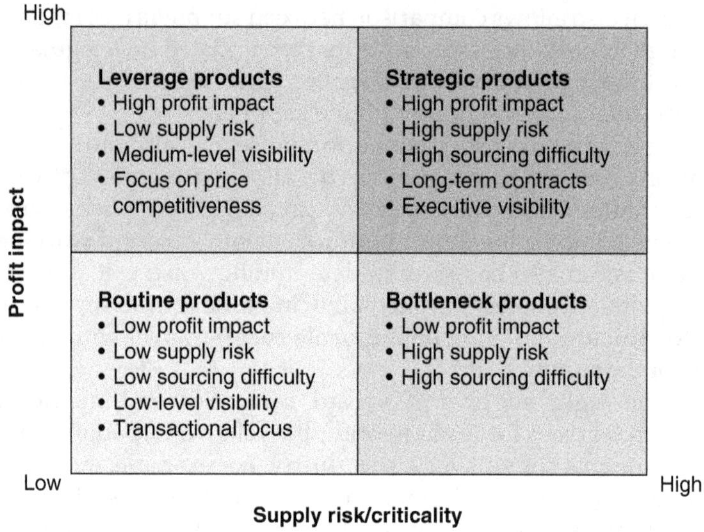

Figure 2.6 Kraljic portfolio segmentation model.

Source: Adapted from Peter Kraljic, "Purchasing Must Become Supply Management," *Harvard Business Review*, 1983 (https://hbr.org/1983/09/purchasing-must-become-supply-management).

Kraljic recommends the following purchasing approaches for each of the four product quadrants:

1. *Strategic products* (high profit impact, high supply risk). These products deserve the most attention from purchasing managers. Options include developing long-term supply relationships, regularly analyzing, and managing risks, planning for contingencies, and producing the item in-house rather than buying it, if appropriate.
2. *Leverage products* (high profit impact, low supply risk). Purchasing approaches to consider here include using your full purchasing power, substituting products or suppliers, and placing high-volume orders.
3. *Bottleneck products* (low profit impact, high supply risk). Useful approaches here include over-ordering when the item is available (lack of reliable availability is one of the most common reasons that supply is unreliable) and looking for ways to control vendors.
4. *Routine products* (low profit impact, low supply risk). Purchasing approaches for these items include using standardized products, monitoring and/or optimizing order volume, and optimizing inventory levels.

In addition to the Kraljic portfolio segmentation model for supplier classification, many organizations will classify suppliers as non-approved, conditionally approved, probationary, approved, preferred, certified, partnership, disqualified, and desourced. The supplier classification categories should be defined with the QMS documentation, consistently applied, and indicated on the Approved

Supplier List (ASL) or Approved Vendor List (AVL). Below are some commonly used terms which are used to define supplier status:

- Pre-qualified or conditionally approved – preliminary data indicates the minimum qualifications that can be met. Part or service qualification may be required
- Approved – Meets the minimum defined criteria
- Customer Directed – a supplier the customer has contractually required for use in the product/service
- Nonapproved – a potential supplier which fails to meet the minimum criteria
- Disqualified – existing supplier who is ineligible for continued business for not meeting expectations/requirements
- Desourced – existing supplier who is no longer awarded business for either for not meeting expectations/requirements or because there is no longer a business need for the goods or services provided
- Preferred or certified – existing supplier with excellent performance ratings over a defined period of time
- Partner – mutually beneficial relationship and interdependence
- Strategic alliance – strong collaborative relationship with information sharing and common goals

PARTNERSHIPS AND ALLIANCES

> Identify and analyze strategies for developing customer-supplier partnerships and alliances. (Analyze)
>
> **Body of Knowledge I.B.4**

Customer–supplier relationships can be adversarial in nature. Customers typically want high quality, short lead times, and the lowest price, whereas suppliers want adequate quality, long lead times, and the highest price. The Quality, Time, and Cost constraint diagram in Figure 2.7 depicts the relationship between quality, time, and associated costs. As can be seen in Figure 2.7, there is always a tradeoff between quality, time, and cost. For example, products with good quality and lower prices result in a longer delivery time; products with good quality and faster delivery result in higher prices, and faster delivery and lower prices result in poor quality.

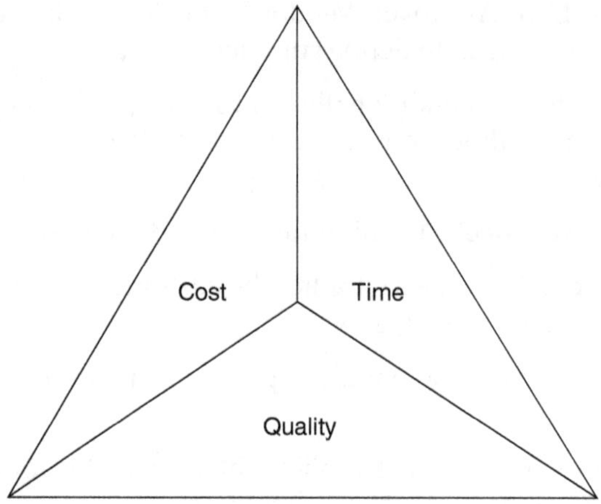

Figure 2.7 Quality, Time, and Cost constraint diagram.

When the customer supplier relationship is built mainly on cost, quality is generally the element that suffers and long-term relationships are not normally developed.

Successful progressive organizations work with suppliers to build mutually beneficial relationships and alliances to provide goods and services at a price that provides an acceptable profit, are motivated to build loyalty, and create a close-knit relationship to secure future business built on trust.

Alliances imply the customer and supplier are working together for a common goal. Partnerships between a customer and supplier imply a mutually beneficial relationship with both parties gaining financially and technologically. Strategic relationships involve technology, licenses, or supply, marketing, and distribution agreements.

Organizations may develop a customer–supplier alliance or partnership to:

- expand core competencies, capabilities, or capacity
- expand into other geographically markets
- offer new products or services
- combine resources to reduce costs
- leverage assets for potential mergers or acquisitions

We will now discuss how to identify and analyze strategies for developing customer–supplier partnerships and alliances. As shown in Figure 2.6, some sourced items will fall into the strategic products category. These products should be given more attention and are usually good candidates for developing long-term partnerships or alliances with the supplier. These partnerships may involve giving the supplier a preferred classification that offers them a significant advantage in winning additional business from the customer. The preferred strategy and choice of partnership or alliance may be influenced by the customer's relative purchasing strength compared with the supply market strength, as shown in Figure 2.8.

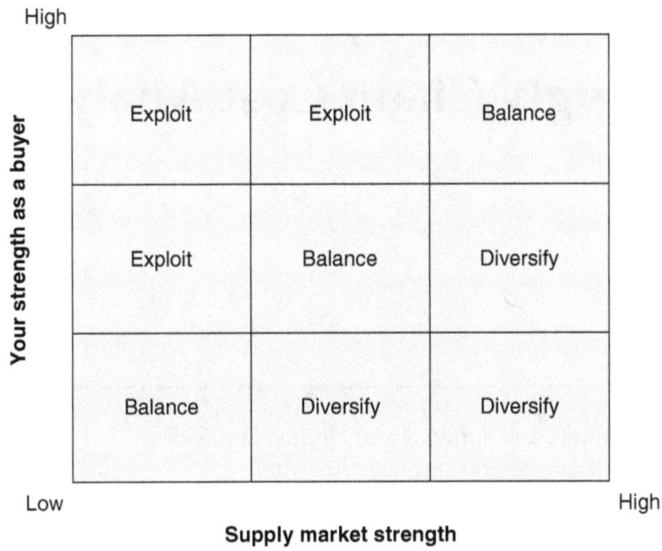

Figure 2.8 Kraljic purchasing portfolio matrix.

The other category of supplied products that may benefit from an alliance are the bottleneck products that have low profit impact but carry high supply risk. One way to help control vendor volatility is through an alliance, which could involve helping the supplier implement a stronger QMS, offering training and development in lean and Six Sigma techniques, or even providing financial support in the form of long-term supply contracts with regular payments.

Chapter 3
Supply Chain Cost Analysis

> Identify and apply relevant inputs to prioritize cost reduction opportunities. (Analyze)
>
> Body of Knowledge I.C.1

Sourcing strategy is critical to the SQP. The SQP must partner with procurement and other sourcing professionals and internal stakeholders to develop the most effective, efficient, and value-added supply base and sourcing strategy possible. This requires an understanding of the very real need for minimizing cost in the supply chain while maximizing value. Let's take a look at some of the concepts the SQP needs to be familiar with.

COST REDUCTION

Competition is global and very intense. Since a large part of the cost of goods sold is the cost of materials, the greatest opportunity to maintain a competitive edge is often found in the supply chain. As supply chain professionals, we are familiar with the reduction in spend targets given to procurement. Too often, these are based upon and focused upon PPV (purchase price variance). Ostensibly this is done because it is a significant cost component, but just as likely, it is because PPV is very accessible and easy to measure, and shows concrete cost reduction. However, it is important to understand the impact of various costs on the supply chain. This includes the more difficult to measure costs, such as COPQ (Cost of Poor Quality). The measures of cost and the COPQ categories discussed below represent an opportunity to maximize cost reduction or minimize excess costs. Keep in mind that some of these can be hard to measure. But the only way to reduce costs is to identify them and to focus on them.

Cost reduction should be approached as a project, utilizing proper project management principles and techniques. Further, cost reduction efforts need to be collaborative in nature, seeking a win–win outcome for customer and supplier. Internal collaboration is critical, too, ensuring that internal stakeholders (quality, operations, procurement, engineering/R&D, etc.) are engaged and committed and that we seek a win–win outcome for internal stakeholders. It is this win–win collaboration that is the key to minimizing cost while maximizing value in the supply chain. As we will see in this chapter, the opportunities for cost reduction in the supply chain are as real as the need for cost reduction in the supply chain.

Supply Chain Cost Concepts and Terminology

Spend

The total amount of purchased goods procured from a supplier constitutes the spend. This value is important and should be used to rate the criticality and importance of a supplier. It should indicate the amount of risk with that supplier as well. It is a key indicator of how much time, effort, and resources should be spent managing that supplier. It should also be a measure of the supplier's investment in the customer.

Logistics, Geography, and Geopolitical Considerations

Distance, transportation, and the current political climate are important considerations in any supply chain decisions and can have a profound effect on cost. Obviously transportation costs, import/export costs, and other logistics-related costs are important. Geography is also a consideration, especially in relation to travel costs, as well as the infrastructure of the countries involved. Related considerations include potential damage, number of reloads and transfers, supply chain security, material integrity (e.g., counterfeits, contamination), and theft. Natural disaster is also a consideration, although some natural disasters are more predictable than others. For example, if we are operating in an earthquake-prone area we can plan for that contingency, but it is difficult to prepare for an unexpected disaster such as the tsunami that hit Japan in 2011. Lastly, geopolitical instability is also a factor. Civil war and other conflicts can wreak havoc on infrastructure and production capacity and schedules, not to mention the availability of raw materials. All of these factors increase cost and risk. Much thought and planning must go into securing the supply chain, especially when considering this category.

Cost of Poor Quality Analysis

> *Quality is free. It's not a gift, but it's free. The "unquality" things are what cost money.*
>
> —Philip B. Crosby, *Quality Is Free: The Art of Making Quality Certain*

Traditionally there are four categories of COPQ: internal failure cost, external failure cost, appraisal, and prevention. Failure cost is any scrap, rework, waste (muda), returned material, rejected material, line downtime, and so on. Appraisal is the cost of deciding whether product meets requirements. Traditionally this involves inspection-related costs: inspectors, equipment, redundant activities (e.g., 200% inspection), etc. Prevention cost is really a misnomer, as prevention is an investment if done correctly, in the sense that these activities both prevent failure costs and minimize or eliminate the need for inspection (appraisal). Prevention includes activities such as validation, poka-yoke, quality by design (QbD), Design for Six Sigma (DFSS), appropriate use of failure modes and effects analysis

(FMEA), process characterization and optimization (e.g., use of design of experiments [DOE]), and process audits that focus on in-process control and prevention to reduce waste and rejections, such as the use of statistical process control (SPC) and automated inspections (e.g., vision systems, reversing conveyors, mechanical sorting, weighing).

It is important to remember that the more effort you put into prevention, the lower the cost will be for appraisal and for either type of failure. Keep in mind that external failure cost includes many intangibles, the most important of which is damage to the organization's reputation. The benefit of preventing this kind of damage is difficult to quantify, but certainly it will bear returns over the investment manyfold. If you were to Pareto chart the four categories of COPQ, ideally prevention would be the largest bar and failure and appraisal would be small. This would represent a mature system. If you are just starting out, you will first need to put effort into both appraisal and prevention to reduce external failure cost and then internal. Then you can gradually transition to greater investment in prevention, transferring the money spent on appraisal to designing in quality, poka-yoke, validation, and SPC and DOE.

It is also important to remember that, as with all waste, quality costs can be hidden, not unlike the costs in the "hidden factory" in a lean manufacturing analysis, as illustrated in D. C. Wood's *Principles of Quality Costs*. Wood uses the iceberg analogy—that most of the iceberg is hidden beneath the water's surface—to illustrate the concept of hidden quality costs. According to Wood, "90% of most quality costs are under water" (Wood, 2013, p. 7). Figure 3.1 contrasts the superficial and hidden costs, utilizing both data from Wood and data that are applicable to supplied material.

COPQ Analysis Applied to Suppliers

Applying the principles of quality costs to the supply base is just as important, and not that different from, internal COPQ analysis (COPQA). It is likely you will readily find cost reduction opportunities through application of COPQA principles to the supply chain.

It is important not to get bogged down in discussions of minutiae when compiling quality cost data, internally or externally. For the most part, COPQA should be used for comparative purposes and to assess the value that a supplier adds (or fails to add) from month to month, for both trend analysis and comparison

Superficial costs *(above the surface, easy to see and measure)*	**Hidden costs** *(below the surface, difficult to see and measure)*
Scrap Rework Warranty Incoming inspection rejections	Engineering time Time to process CAPAs/SCARs Management time Line and field downtime Increased inventory Decreased capacity Delivery problems Lost orders Lost reputation

Figure 3.1 Superficial and hidden costs.

with like suppliers. Therefore, 100% accuracy is not required. If the data are estimates, that's OK, as long as we always estimate in the same way. For example, the wages for internal rework of supplier lots are a common sticking point. Different operators might earn slightly different wages. That can be a nightmare from an accounting standpoint from one rework activity to another. Luckily we are not doing COPQA as an accounting activity. Simply assign an average wage, use it in *all* calculations, and go do some quality cost analysis. Even if the analysis is not accurate to the penny, it will be an apples-to-apples comparison and thus allow us to do both meaningful trend analysis and comparisons to like suppliers. After all, the point is not to invoice the supplier but to determine its overall value proposition. That being said, how do we do our COPQA? We start by choosing categories based on customer pain points.

Link COPQA to Scorecard/Monitoring Activities

Quality costs are important in judging a supplier's overall value proposition. One way to bring these data to light is to link them to the supplier scorecard. Categories could include nonconformance processing costs, excess inventory, reinspection costs, number of supplier corrective action requests (SCARs), line downtime, scrap costs, and any other category that represents a pain point to your organization. Socialization and voice of the customer (internal customers such as manufacturing and purchasing) will help to identify the pain points and consequent categories. Figure 3.2 shows a COPQA spreadsheet for a fictional Supplier A.

Figure 3.3 shows the same data for Supplier A graphically, this time as compared with overall spend (both in dollars and as percentages). This analysis is helpful when comparing the value propositions of like suppliers (e.g., comparing injection molders to each other).

Figure 3.4 shows these data incorporated into a monthly scorecard template. This is a balanced scorecard, containing categories for quality, manufacturing, and purchasing and based on a possible score of 100. Note the COPQA categories circled. These categories correspond to the spreadsheet and graph in Figures 3.2 and 3.3.

Roll-Up into Adjusted Purchase Price Variance and Total Cost of Ownership

It is almost certain your company knows the purchase price variance (PPV) for purchased goods. For emphasis and impact, the SQP can show an "adjusted PPV" by rolling in the COPQA data we just discussed. Focus on total cost of ownership (TCO) in your reporting to management.

Analysis and Proposals to Reduce Costs

As previously stated, cost reduction should be approached as a project, utilizing proper project management principles and techniques. Further, cost reduction efforts need to be collaborative in nature, seeking a win–win outcome for customer and supplier. Internal collaboration is critical, too, ensuring that internal stakeholders (quality, operations, procurement, engineering/R&D, etc.) are engaged and committed and that we seek a win–win outcome for internal stakeholders. It is this win–win collaboration that is the key to minimizing cost while maximizing value in the supply chain. We have examined opportunities to reduce costs using

COPQA spreadsheet for Supplier A						
Elements	**Jan**	**Feb**	**Mar**	**Apr**	**May**	**Jun**
Appraisal						
Supplier audits and/or visits (*travel expenses*)						
Normal incoming quality inspection (*purchased material*)	$ 1,998.85	$ 898.34	$ 1,547.25	$ 1,470.80	$ 1,636.59	$ 1,056.00
Tightened incoming quality inspection (*purchased material*)	$ 13.52					
Cost of samples (*if samples are not returned to the lot, i.e., destructive test*)		$ 4,373.94	$ 7,767.07	$ 7,715.96	$ 7,722.24	$ 4,801.80
Total appraisal costs	**$ 2,012.37**	**$ 5,272.28**	**$ 9,314.32**	**$ 9,186.76**	**$ 9,358.83**	**$ 5,857.80**
Internal failure						
Rejected lots/units (*includes scrap*)	$ 27.18	$ 28.87	$ 53.13	$ 52.36	$ 102.93	$ 66.63
Inspection (*labor and related activities cost*)		$ 1,082.27			$ 2,792.70	
Line downtime (*includes line clearance*)						
Expedited material (*due to supplier's delivery and/or quality issues*)			$ 7,205.00			$ 7,635.89
Total internal failure costs	**$ 27.18**	**$ 1,111.14**	**$ 7,258.13**	**$ 52.36**	**$ 2,895.63**	**$ 7,702.52**
External failures						
Complaints (*includes complaints and supplier QA labor*)	$ 1,132.30					
Field actions/recalls						
Total external failure costs	**$ 1,132.30**					
Total quality costs	**$ 3,171.85**	**$ 6,383.42**	**$16,572.45**	**$9,239.12**	**$12,254.46**	**$13,560.32**

Figure 3.2 COPQA example.

COPQA. Supply chain rationalization is another tool, although it yields far more value than mere cost reduction. We also will look at make/buy decisions. These decisions can save a great deal of money if made wisely.

The key is to look at the data from your various analyses and show return on investment (ROI). Every cost reduction is an investment—it takes resources to plan and execute as well as to maintain safety stock. Never forget excess inventory, which ties up cash and reduces cash flow. Proposals must appeal to the business sensibilities of your partners. Each proposal should clearly show how much will

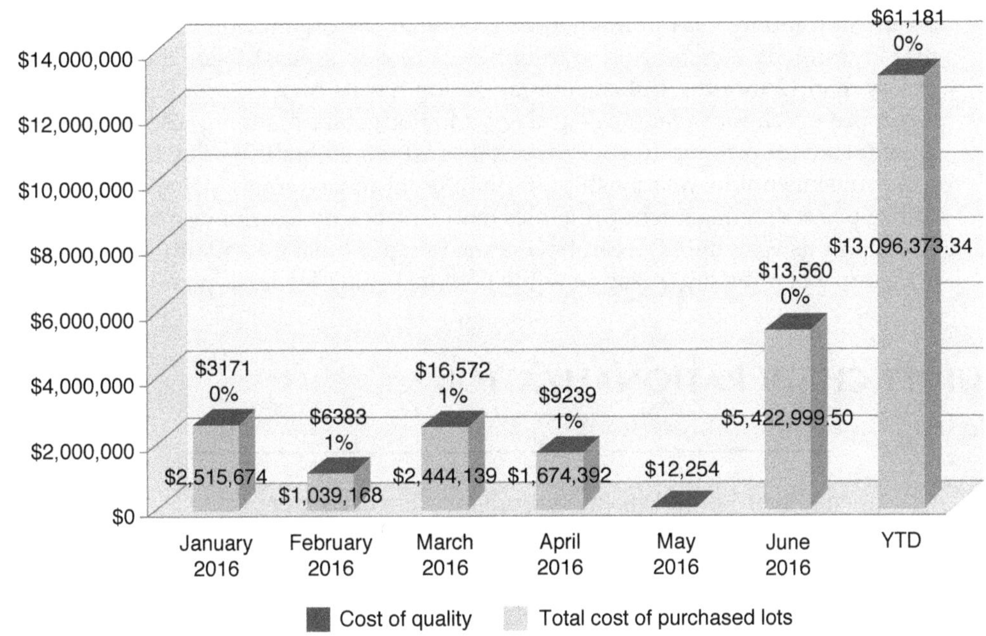

Figure 3.3 Quality costs vs. overall spend for a supplier.

		Jan	
Incoming lots			
Rejected lots			
Reinspected lots			
Metric	**Points**	**Occur**	**Score**
Supplier corrective action request (SCAR)	15		0
Repeated SCAR (last 12 months)	10		0
SCAR responsiveness	10		0
Reworked/reinspected lots	5		0
Lot conformance	5		0
Quality and operations	**45**		**45**
Appraisal	5		0
Internal failures	5		0
External failures	10		0
Costs	**20**		**20**
On-time deliveries	10		0
Expedited shipments	10		0
Purchase orders with documentation issues	10		0
Responsiveness	5		0
Purchasing	**35**		**35**
		Jan	
Overall performance	100		100%

Figure 3.4 Monthly scorecard template incorporating COQA categories.

be spent in money and human effort and what the organization will recover for that investment. A good rule of thumb is to recoup at least twice the investment in less than 12 months, but each organization has its own criteria for ROI. Yearly advanced operations planning, strategy, and budgetary planning sessions can be a great forum for presenting your proposals. Just remember these key things: Leverage your stakeholder relationships (both internal and external), strive to add value, utilize project management principles, and collaborate to achieve a win–win outcome. In this way the SQP can have a meaningful business, quality, and compliance impact on the organization while demonstrating leadership and vision.

SUPPLY CHAIN RATIONALIZATION

> Interpret and analyze the optimization of a supply base to improve spending and leverage investments into supplier quality or risk reduction. (Analyze)
>
> **Body of Knowledge I.C.2**

As supply chains grow over time, suppliers for certain core manufacturing or service processes will proliferate. It is important to look at the supply base from time to time to understand what the right size is for your particular supply chain needs and make corrections if necessary. As one might deduce, this is an important aspect of supply chain strategy, as supply chains and product life cycles mature. This involves not only looking at each category of suppliers but also harmonizing the supply chain with your strategy. How many suppliers do you need for process X? (What is the right number of suppliers? Is dual sourcing or multiple sourcing appropriate?) Who among them will help you achieve your strategy and vision? (Who are the right suppliers? Who meets your requirements best? Whom might you wish to partner with?) How can you maximize your value proposition by increasing leverage (competitive price, good service), increasing influence, and balancing that with contingency planning to protect the integrity of the supply chain (dual versus single sourcing in the event of an emergency)? This is the essence, and the value, of supply chain rationalization.

Optimization

The objectives of supply base optimization are cost reduction, risk reduction, and increased quality. The utilization of data and a data-driven decision process are the most critical components of any rationalization effort. Obviously quality costs are a major data source in this decision process. Consolidation, partnerships, and collaboration will help to optimize the supply chain.

Consolidation of Suppliers and SKUs

Consolidation of suppliers is one way to reduce costs and increase leverage and importance, investment and ROI, and incentive and cooperation. Consolidation and standardization of stock keeping units (SKUs) will save money and requires a partnership with R&D/design and procurement. Consolidation of suppliers is a matter of finding the right number of suppliers in each spend category or outsourced process (e.g., plastics or injection molders). Then work with partners internally to find the best suppliers in those categories and transition to them. This will take buy-in and planning from all stakeholders. Working together, you can determine the right suppliers and the correct number of suppliers to ensure supply integrity, decrease costs, and increase quality and leverage with the suppliers in one fell swoop. This can lead to partnerships, drive truly value-added investment, and reduce the overall risk in the supply chain.

The same process used for make/buy decisions (discussed later in this chapter) can be used to determine the best suppliers. Obviously COPQA data will be useful in this decision as well. Raw material rationalization and centralized buying are also great optimization opportunities. You can reduce costs and simplify the overall supply chain by leveraging economies of scale made possible by increasing spend on core materials by consolidating them and by securing the best price. This effort could go hand in hand with consolidating SKUs. Simplification always brings cost reduction opportunities. Any of these initiatives will help to optimize the supply chain.

Partnerships and Collaboration

The importance of partnerships and collaboration, both internal and external, cannot be overstated. But it is worthwhile to consider the words of Dr. W. Edwards Deming on this topic. We would be remiss if we failed to point out that customer–supplier collaboration is not a new concept. In fact, in his classic book, *Out of the Crisis*, Deming taught us that collaboration with suppliers bears significantly more fruit than basing the procurement relationship on price alone. His famous 14 points include a plea to "end the practice of awarding business on the basis of price tag alone. Instead, minimize total cost" (Deming, 1982, p. 31). *Out of the Crisis* provides an extensive explanation of this concept, focusing on the differences between the procurement strategies of Western and Japanese firms in the late twentieth century. Deming viewed sourcing strategy as a strategic advantage as well as a benefit to any firm. Some of his major arguments include:

> We can no longer leave quality, service and price to the forces of competition and price—not in today's requirements for uniformity and reliability. (Deming, 1982, pp. 31–32)
>
> What one company buys from another is not just material: it buys something far more important, namely engineering and capability. (Deming, 1982, p. 40)

MAKE/BUY DECISIONS

> Provide input on make/buy decisions using internal and external capability analysis. Apply tools such as SWOT analysis and use historical performance to analyze requirements. (Analyze)
>
> Body of Knowledge I.C.3

To outsource or not to outsource? Make/buy decisions are perhaps the most crucial any organization can make. To make the best decision, the following tools should be applied equally and without bias to both the external and internal organizations.

Process Analysis

Core Competencies

The core processes and services offered by the supplier need to be analyzed in order to understand the value the supplier offers, which should be compared with the internal value offered. A subject matter expert (SME) or technical expert (TE) should be involved in this effort to ensure a thorough and accurate analysis. A complete process audit should be conducted focusing on the supplier's technical competence in the particular process (manufacturing or service) being considered. Technical competence includes current best practices, knowledge, training, education, and experience of technical staff, as well as proper deployment and execution of sound scientific and engineering practices. This includes personnel and practices from the functions of manufacturing, procurement, engineering/R&D, quality, and maintenance. Additionally, applicable regulations and voluntary standards must be considered as a facet of technical and core competency. If environmental systems are important, then competency with ISO 14000 is essential. In a regulated environment such as medical or pharmaceutical, competency with current good manufacturing practices (cGMP) regulations is essential. So if the supplier is a medical molder, then that supplier must be an expert in cGMPs and molding. Consequently, so does your company.

Control Systems

The same principles apply to control systems. Control points, decision points, and methods for preventing defects, mix-ups, errors in traceability and labeling, and other critical errors must be incorporated into the competency audit. If there is a control plan, utilize it as an anchor in the process audit, along with any other inspection planning documentation. Ask where decisions (e.g., accept/reject) are made, if they occur in the right stages in the process, and if they are effective. Look for poka-yokes and automated inspection. Don't overlook process controls such as critical processing parameters being controlled and validated statistically, as well as safeguards against process errors such as using cameras to prevent a "mold crash."

Internal and External Capability Analysis

Of course, one of the most effective indices of competency is capability. Statistical data on control (statistical process control or SPC) and capability (short-term process capability [C_{pk}] and long-term process performance [P_{pk}]) need to be analyzed. Compare the internal and external processes and their respective capabilities.

Historical Performance Analysis

Key data sources for analyzing historical performance include SCARs, corrective and preventive action (CAPA), internal nonconformances, historical capability, and results from regulatory, ISO, and customer inspections and audits. Planned or scheduled maintenance, age of equipment, historical downtime, first time yield, and rolled throughput yield are also great predictors of success. In the end, it is most likely that future performance will reflect historical performance.

SWOT Analysis

It may be helpful to use strengths-weaknesses-opportunities-threats (SWOT) analysis as another decision input. For the project, or alternately for each organization, list the associated strengths, weaknesses, opportunities, and threats. In this way a comparison can be made. The SWOT analysis usually takes the form of a box divided into four quadrants; see Figure 3.5 for an example.

Pareto Analysis

Pareto analysis may also be useful. Pareto analysis is a ranking of categories of interest used to identify the "vital few" categories in order to focus on them, thereby maximizing effectiveness and results (see Figure 3.6). An example might be to rank historical data and compare two organizations.

Strengths	**Weaknesses**
• Processing, training competence • Quick changeover expertise	• Regulatory risk • Infrastructure risks
Opportunities	**Threats**
• Simplify the supply chain: rationalize and reduce costs • Eliminate logistical issues by insourcing	• Low-cost competition • Long lead times

Figure 3.5 SWOT analysis example.

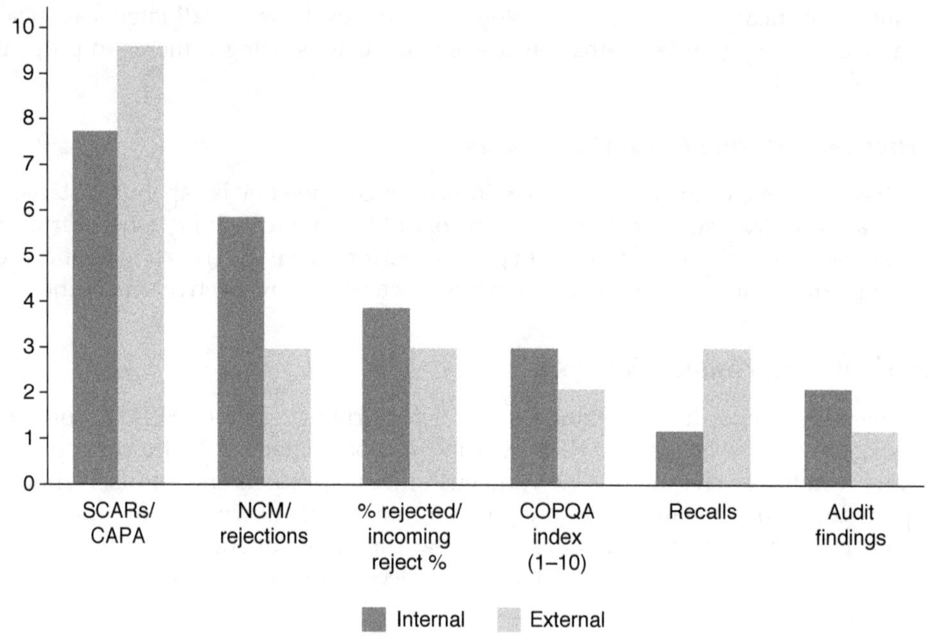

Figure 3.6 Pareto analysis of historical data.

Chapter 4
Supplier Agreements or Contracts

TERMS AND CONDITIONS

> Review and provide input for developing terms and conditions that govern supplier relationships to ensure quality considerations are addressed. (Apply)
>
> **Body of Knowledge I.D.1**

Supplier Agreements establish the terms on which a supplier will supply products or services to a customer. Defining the terms and conditions clearly to both parties improves the likelihood of success that an agreement will benefit both parties.

Most organizations have a standard set of terms and conditions that govern supplier agreements. These vary depending on the size of the organization, the annual spend amount, the amount of leverage the organization has, and regulatory requirements.

The Statement of Work: Key Points

In developing requirements, the statement of work (SOW) identifies the part of the contract where all the operational aspects of a service or product are described. This document contains—expressed in terms your organization understands—the definitions developed in brainstorming sessions of what a good service is. Drafting a good SOW is critical to success.

In some cases, the SOW is a separate exhibit; in other cases, it is embedded in the body of the contract. Whichever form is used, always bear in mind that a contract is a single entity, and due attention must be given to the balance and harmony of all its parts. Blind cut-and-paste exercises, where a SOW is inserted into some boilerplate clauses, are the shortest path to disaster. Nevertheless, for our discussion, it helps to think of the SOW as a separate document that must have certain traits in order to be successful.

It is normal for a supplier to offer its own version of the SOW. Though this document should be considered as background information, it is not advisable to sign it without close scrutiny.

A detailed study on how to draft SOWs is beyond the scope of this chapter; however, there are some key points the SOW must cover. Not all of them are relevant in every situation, but it is always a good idea to go through this checklist:

- *Service definition and scope.* There must be a clear definition of the service you are contracting. All relevant contractor obligations must be listed. Emphasis must be placed on the end results to be accomplished, giving the contractor as much leeway as practical to decide how to organize its work.

- *Deliverables.* Define clearly what service deliverables your company expects to receive. The contractor must be made responsible for these deliverables, and its compensation (payments) must be tied as much as possible to their timely production. Also, acceptance criteria must be set forth for these end products.

- *Company-provided inputs.* In some cases, your company will provide goods or services necessary to perform the work. Examples might be office space, design parameters, computers, travel expenses, measurement fixtures and/or analysis, and electricity. These must be clearly defined.

- *Supplier-provided inputs.* Typically, the supplier's responsibility is to provide certain materials or intellectual services. For example, in outsourced plant maintenance services, the deliverables might include provision of spare parts and lubricants. For products, the deliverables may include tooling, spare parts, and quality services like part inspection and certificates of conformance. Once again, be careful to list these.

- *Performance evaluation—bonus/penalty clauses.* Performance review mechanisms must be defined and applicable penalties or bonuses stated. A correct definition of performance indicators is vital to close the feedback loops essential for a successful outcome.

- *Price and payment.* The amount of money to be paid for the services or product and the form and date of payment must be specified.

- *Warranties.* Any warranties must be clearly spelled out. For example, in computer programming, the contractor might agree to correct any software bugs that appear within the first 12 months of software operation.

- *Contract changes.* The only sure thing in business nowadays is that everything changes. Consequently, contracts must allow for flexibility.

- *Confidentiality.* Clear rules must be set forth regarding what use the contractor can make of information your company supplies. You must indicate that all blueprints, data sheets, and related information cannot be shared with third parties without your organization's written consent. Nondisclosure agreements are designed to protect both parties' intellectual property and trade secrets.

- *Key personnel.* In specialized, high-value services (for example, consulting), service quality is heavily dependent on the individuals assigned to the project. The contracting organization must, in these cases, specify the persons who will perform the work and set down its right to request personnel changes if performance is not acceptable.
- *Software and patents.* If the work involves using software, provisions must be drafted to cover its use. Similar concepts apply to any patented processes, procedures, tools, and other intellectual property the contractor will use for its work.
- *Termination clauses.* In case of substandard performance and/or repeated failure to comply with the agreed performance level, provisions must be in place for contract termination. Minimum performance levels must be defined such that, should the contractor not be up to par, corrective action can follow.

Potential suppliers need to take care in reviewing the purchaser's terms and conditions prior to signing agreements. Some purchasers include an evergreen clause, which means the contract will automatically be renewed unless one party disagrees, sometimes within a limited renewal time frame.

When there is potential for significant warranty or recall costs, agreement on any limitation of supplier liability for the remedies in the event of a purchased product or service failure should be agreed upon up front. Different industries handle warranty differently. In the passenger car market, the automakers provide warranty for their product, including purchased product for end users. In the heavy truck and aerospace industries, parts suppliers offer warranty to their customers' customers.

This limits warranty liability to the prime contractors or original equipment manufacturers (OEMs) but results in higher piece prices for purchased products. Regulatory requirements may also specify how product or service guarantees or remedies will be handled for a specified industry or commodity.

When it comes to purchased product, it has been said that he who has the gold wins. Organizations with a large annual purchasing spend can dictate more terms and conditions than others. This is generally an advantage for the large organization, but large organizations can fall into the trap of creating too many customer-specific requirements, which actually drives up cost. When a supplier has multiple customers in the same or a similar industry, additional resources are needed to meet customer-specific requirements. The cost of these extra resources has to be recovered from one or more customers.

Some purchasers now have environmental requirements for suppliers. The World Trade Organization (WTO) has cautioned that these can impede trade and even be used as an excuse for protectionism. It adds that WTO member governments consider the protection of the environment and health to be legitimate policy objectives, and thus they recommend taking an approach that helps exporters meet standards and requirements.

To avoid proliferation of customer-specific requirements, organizations should participate in and make use of any standards that are applicable and available. There are standards at the international, national, industry, and product levels. Standards development organizations such as the International Organization for

Standardization (ISO), the American National Standards Institute (ANSI), the American Society of Mechanical Engineers (ASME), and the Society of Automotive Engineers (SAE) offer many voluntary standards that organizations can use in specifications for their purchased products. Industry trade associations (e.g., the Automotive Industry Action Group [AIAG]) also produce harmonized customer requirements that can be used to avoid non-value-added costs in the supply chain.

SUPPLIER AGREEMENTS

> Identify elements of supplier agreements (e.g., business and legal approach/requirements). (Understand)
>
> Body of Knowledge I.D.2

Principal supply contracts are vital for manufacturing organizations seeking to assure a consistent quality and flow of incoming materials. The agreements provide a mutual understanding of the expectations for the supplier and customer.

Agreement terms should be understood by all involved in the supply chain to minimize risk while reaching the desired outcome. Too often the agreement negotiation focuses on price alone to the exclusion of an understanding of quality, production, safety, and delivery risks. Taking a supply team approach by incorporating the viewpoints of manufacturing, quality personnel, designers, supply chain logistics, purchasing, and legal allows for risks to be reviewed and reduced, assuring a smooth supply flow for long-term results for all parties.

Customers that wish to protect their intellectual property may require a nondisclosure agreement (NDA) to be signed by a potential supplier. An NDA—also known as a confidentiality agreement (CA), confidential disclosure agreement (CDA), proprietary information agreement (PIA), or secrecy agreement (SA)—is a legal contract between at least two parties that outlines confidential material, knowledge, or information that the parties want to share with one another for certain purposes while restricting its access by third parties. It is a contract through which the parties agree not to disclose information covered by the agreement. An NDA creates a confidential relationship between the parties to protect any type of confidential and proprietary information or trade secrets. As such, an NDA protects nonpublic business information. NDAs can be "mutual," meaning both parties are restricted in their use of the materials provided, or they can restrict the use of material by a single party ("unilateral").

Other liability agreements that may come into play before business parties start to share requirements or visit each other's facilities include indemnity agreements and hold harmless agreements. Indemnity is compensation for damages or loss. Indemnity in the legal sense may also refer to an exemption from liability for damages. The concept of indemnity is based on a contractual agreement made between two parties, in which one party agrees to pay for potential losses or damages caused by the other party. The agreement may include requiring specific types and minimum amounts of insurance the supplier must carry before they can do business. In addition, a properly drafted agreement will indicate that the supplier must furnish evidence that they are complying with the indemnification and insurance requirements. This is typically accomplished by requiring a certificate of insurance from an insurance company or agent. A hold harmless agreement is a contract between two parties designed to release one or both parties from legal claims. Most often, one party agrees not to sue the other party for any expenses, damages, or losses arising from a transaction or activity between the two parties.

Requirements are identified in different manners. The requirements can be distributed to the supplier on a separate supplier requirements document, which is then referenced in the purchase order, contract, or agreement. In some cases, the requirements are stated directly on the purchasing document, which suppliers should review to ensure they can satisfy the stated requirements.

Initiating the Supplier Selection Process

An organization can use a number of documents to initiate the supplier selection process. They may include:

- *Request for information (RFI).* Used to collect information from potential suppliers regarding products, services, and options
- *Request for proposal (RFP).* Formal request for suppliers to submit a quote to supply a particular product or service

These documents may include information regarding:

- Product or service characteristics
- Product or service performance requirements

Sourcing, contracts, and other purchasing decisions are at the tactical level of supply chain management, as are production decisions, inventory decisions, and transportation strategy. When an organization has many potential suppliers and significant leverage, however, sourcing can also be a very strategic decision. A contract is a legally enforceable promise or set of promises made by one party to another. It is a legally binding agreement concerning a bargain that is essentially commercial in nature and involves the sale or hire of commodities such as goods, services, or land.

Team Supplier Selection

Purchased product often represents a significant cost to an organization. When this is true, the sourcing selection should be the result of work of a cross-functional team. The criteria for sourcing should include quality and service as well as price. This requires the purchasing function to consult with and comprehend feedback from other disciplines such as quality, manufacturing, operations, risk, engineering, and after-market service in the sourcing decision.

Define Requirements

The supplier is not obligated by contract to satisfy the expectations of the organization if they are not properly identified and so is not bound by contract to any requirements that are not documented. If there are any ambiguous, deficient, or inaccurate supplier flow-down requirements, the supplier must contact the organization (the supplier's customer) to gain further detail for clarification prior to accepting the order, because ambiguous requirements can put both the organization and the supplier at risk.

When an order is placed with a supplier for any product or service, the organization should give the supplier a clearly written specification with a sufficiently detailed description of needs and requirements as well as terms and conditions (the output of the design control program).

It is very important to ensure that the supplier has received, reviewed, and acknowledged or confirmed the requirements. By doing so, the organization will have a greater degree of confidence that the supplier is capable of fulfilling the order requirements. If the purchasing requirements are not properly managed, this will have a cascading negative effect on delivery, cycle time, quality, and cost throughout the supply chain.

Purchasing information describes many aspects of the product and services to be delivered by the supplier, including, but not limited to, the following:

Purchasing focus deliverables:

- Description of the product
- Quantity
- Delivery due dates
- Agreed-on price, rate, or fee
- Payment terms such as invoicing, timing, and method
- Shipping requirements including imports, exports, and customs

Quality focus deliverables:

- Approval of product, procedures, processes, and equipment
- Qualification of personnel
- Management systems such as those for quality, safety, environmental, and risk management
- Relevant drawings and process data (technical data), and revision level
- Requirements for test specimens
- Record retention
- Sub-tier supplier flow-down requirements
- Notification of changes in management or management systems
- Right of access
- Notification of nonconformances

- Shipping, labeling, and packaging requirements
- Product shelf life and age control requirements
- First article inspection (FAI) requirements and number of samples to be inspected
- Source inspection
- Certificate of conformance, analysis, or compliance requirements
- Environmental requirements/regulations, such as Restriction of Hazardous Substances (RoHS) or Conflict Minerals Regulation
- International Traffic in Arms Regulations (ITAR) and Export Administration Regulations requirements
- Identification and traceability requirements for possible recall and investigation

Competition

Competition in economics is a term that encompasses the notion of individuals and firms striving for a greater share of a market to sell or buy goods and services. A supplier that monopolizes the market for its product or service has no incentive to reduce cost or improve quality or service. Competition makes all competitors better because they have to improve to survive. Organizations should develop a number of potential suppliers for all critical purchased products. The efforts will pay off in both the short term and the long term. Even when an organization elects to award all its volume for a given purchase to a single supplier, having a number of bidders will ensure that the successful supplier understands the requirements and is providing a competitive price.

A practice is anticompetitive if it is deemed to unfairly inhibit free and effective competition in the marketplace. Examples include cartels, restrictive trade agreements, predatory pricing, and abuse of a dominant position.

Understanding Cost

Cost control is important to organizations. Cost plus margin equals price. Two main elements of cost control are cost estimating and actual cost tracking. Organizations must account for costs and have a way of recovering them in their revenue. Some industries use competitive benchmarking and tear-down of competing products to better understand the design and estimate costs. This can involve tracking and use of data on costs such as raw materials and labor by region. At a minimum, organizations should estimate potential costs, then compare actual costs with the estimates and make adjustments to pricing or costs as necessary to ensure organizational commercial viability going forward.

Most sectors once operated with a pricing model based on cost plus margin. In sectors in which competition is significant, the pricing model has to be market based—what customer organizations are willing to pay regardless of cost. This puts incentive on the producer to reduce costs to maintain or increase margin.

QUALITY AGREEMENTS

> Analyze the elements of quality agreements/ requirements (e.g., other levels of approval/review). (Analyze)
>
> **Body of Knowledge I.D.3**

Specifications and Design Responsibility and Approach

Prior to the Industrial Age, customer organizations and suppliers were close enough geographically to do business personally, face-to-face. The Industrial Revolution created a proliferation of manufacturing organizations and purchased product requirements, which created a need for a new intermediary tool, the specification, to replace the direct communication between customer and supplier. In its essence, a specification describes and prescribes requirements for a purchased product or service that can be used in a contract with a supplier.

In many sectors today, design responsibility for a product or service lies with the supplier rather than with the customer organization. The party in the contract with responsibility and authority for establishing and maintaining the design records (i.e., the specification) is responsible for the design. Where this is the customer organization, it is best to not overengineer the design or specification up front. Even when suppliers are not responsible for the design, they often have valuable input to offer for design consideration. For example, they often can provide input that would improve the design or the cost of manufacture, such as:

- The manufacturability of the design (Design for Manufacturability, DFM)
- Establishing nominal specifications and tolerances on various performance-related specifications such as dimensions, percentage of active ingredients, or features
- Alternate materials or manufacturing processes

A supplier might have certain equipment and capabilities that influence the optimum design approach or requirements. When this is the case, the customer should send out an RFP rather than a request for quote (RFQ) to solicit ideas from potential suppliers to incorporate into the final design for later quoting. On the other hand, the customer should avoid specifications that only one supplier can meet.

When a supplier is providing a new product, the customer will typically require a first article inspection report (FAIR) to validate that all the dimensions on the part are within specifications. An initial report may begin with one inspection or multiple inspections in the case of a multiple cavity tool—for example, an eight- cavity injection mold. Besides the physical measurement inspection, several samples will typically be visually inspected to verify surface conditions and textures. Best practice is to provide measurements and samples to the customer and have the customer sign off on the acceptance of the product.

The next step of the quality agreement is the qualification of the process to ensure that reliability and quality targets are met. The qualification process can require multiple, substantial production runs to verify that a process is capable. The quality agreement can include costs for the measurements and reports as well as a price for the number samples created since some of the samples can be saleable parts. It is not uncommon to have a separate quote and contract for the qualification to ensure that the proper number of samples are produced and measured. Different industries have established qualification methods to suit their needs. For instance, the automotive industry, through AIAG, uses the Production Part Approval Process (PPAP), and the Food and Drug Administration uses Installation Qualification, Operational Qualification, and Performance Qualification methods. See Chapter 10 for more details on Part, Process, and Service Qualifications.

One of the major challenges of achieving qualification success is navigating the conflict between business goals—and the deadlines attached to them—with everything needed to build a complete technical file.

To resolve this issue: 1) challenge any assumptions being made early in the product realization process; 2) lay out the full set of requirements for a given product with input from every impacted department; and 3) acknowledge that few decisions can ever be responsibly made in a silo, especially early on.

The effectiveness of qualification drives down to the approach of quality by design—doing things in ways that have proven to be effective time and again.

FINALIZATION CONTROLS

> Describe controls used to finalize terms and conditions that govern supplier relationships (e.g., agreements, contracts, and purchase orders). (Understand)
>
> **Body of Knowledge I.D.4**

Purchasing Function

The purchasing (also called procurement) function in an organization has two primary and essential tasks: to select and contract with suppliers, and to establish terms of purchased goods and services needed by the organization. Some organizations also assign responsibility for supplier development and improvement to the purchasing function.

Many organizations labor under the false belief that they have no options other than to buy from the first supplier they can find or, in some cases, their traditional supplier, and to pay the supplier's price. Some industries still have the luxury of passing costs on to one or more end customers or users. These are industries in which customers have poor leverage, a concept that is discussed later in this chapter. Other industries do not have this luxury and have to work diligently on cost reduction and efficiency to compete.

Where the organization has an option, it is best to purchase needed products directly from the supplier that has the most "value add" in the product. For example, when buying a metal casting that must be subsequently machined, if the machining costs more than the casting, the organization should purchase from the machining company to maximize leverage. In some cases, it makes sense to purchase from distributors, but organizations will not have as much leverage on price with them as they will with original suppliers. There may be other benefits of or reasons for using distributors, however. In the case of an electrical connector, the manufacturer will produce large quantities of connectors to achieve economy of scale. It may not be interested in selling small quantities of connectors to end users, so use of a distributor would be the best, if not the only, option.

Legal Aspects of Purchasing

Since contracts are enforceable in a court of law, they are subject to regulatory requirements. These requirements can vary by jurisdiction. For purchases on which you can't afford to lose, you should seek legal advice from a competent source when developing terms and conditions to govern the transaction, including any remedies that may be needed.

The US Uniform Commercial Code (UCC) is a set of statutes governing the conduct of business, sales, warranties, negotiable instruments, loans secured by personal property, and other commercial matters. It has been adopted with minor variations by all states except Louisiana. The UCC is a model code, so it does not have legal effect in a jurisdiction unless UCC provisions are enacted by individual legislatures as statutes.

The UCC consists of rules for different transactional areas under articles based on types of transactions. For supply chain management, these four areas are especially relevant:

1. Warranties
2. Transportation terms and risk of loss
3. Seller's rights
4. Buyer's rights

Award Contracts

The UCC establishes each party's rights and obligations according to common business practices based on the principles of fairness and reasonableness. Fairness and reasonableness also serve when a particular term is not defined in the contract. The law is not applicable outside the United States, but most other countries have similar commercial codes. English contract law is a body of law regulating contracts in England and Wales. It shares a heritage with countries across the British Commonwealth such as Australia, Canada, and India.

EU law (commonly referred to as Union Law and historically called European Community law) is a body of treaties and legislation such as regulations and directives that have direct or indirect effect on the laws of EU member states.

The United Nations Commission on International Trade Law developed its Contracts for International Sale of Goods (CISG) treaty in 1980. As of May 2016, it had been ratified by 85 countries. It is similar to the UCC.

Types of Agreements

There are several tools an organization can use to make an authorized purchase. These typically depend on the spending level and include:

- Spot buys (onetime purchases for special or emergency situations)
- Contracts (formal documented arrangements with a supplier due to risk)
- Expense reports (minor office and travel expenses charged against a budget)
- Acquisition cards (a preauthorized spending tool)
- Evergreen contracts (formal arrangements with suppliers that automatically renew)
- Blanket contracts with releases (formal arrangements with suppliers that require specific follow-up information before execution of the contract, such as releasing quantities over time)

Negotiation and Buyer Leverage

Chester Karrass is credited with saying, "In business as in life, you don't get what you deserve, you get what you negotiate." Negotiation is a learned skill. While some are naturally better at it than others, everyone can benefit from some training in the art of negotiation. The results of your negotiations with suppliers will depend on your negotiation skills and your leverage. Because leverage is usually on the customer's side prior to awarding business to a supplier, this is likely the best opportunity to reduce the cost of purchased products. Negotiation techniques are discussed in more detail in Chapter 19.

Buyer leverage is the amount of bargaining power buyers have when purchasing goods and services. The amount of buyer leverage relative to the bargaining power and leverage of the seller depends on the information the seller and buyer have about the product, the relative scarcity or abundance of the product, the availability of product substitutes, and many other factors. In some cases, there may be grants, discounts, or government subsidies to promote trade for certain products, services, or industry sectors. The relative leverage of buyers and sellers determines the price and terms of transactions and the nature of business relationships.

Many organizations work to create or improve leverage for their purchases. There are some things you can do to get a better deal for your organization. These include:

- *Part number rationalization.* Part number proliferation fragments your total volume requirements across part numbers and drives up cost by causing extra manufacturing setups, start-ups, changeovers, inventory, warehousing, scheduling, tracking, and servicing. When it comes to part numbers, fewer is better.
- *Bundling awards.* Organizations that make repetitive purchases or use several purchased products that could come from one type of supplier can create leverage by requesting bids on more than one item at a time or by sourcing one item number on a long-term contract rather than as a spot buy.

Performance Rules

Supplier performance results should be used by the customer to prioritize supplier development, establish an escalation or exit strategy, and influence future sourcing. Some customers effectively use contract incentives such as higher pricing for delivery by a specified date to drive better performance or, conversely, pricing penalties for missing a required timing deadline.

Supplier performance, including pricing, delivery, and service, is key to customer satisfaction. Customers and suppliers both need to work on it whenever possible. Open and timely communication and collaborative planning and problem solving are necessary to achieve the best performance over the life of a contract.

Purchase Order Process

To secure good quality of purchased materials, adequate and complete documentation should be issued to the supplier. A signed purchase order is a contract. To start the purchase order process, an individual or group (operations, engineering, district office) that requires the materials or components should prepare purchase requisitions and forward them to the procurement or purchasing manager. The purchase requisitions should reference the following:

- Material specifications with revision level
- Identification requirements
- Documentation requirements
- Need to send an inspector to witness source inspection

The purchase requisitions and purchase order or contract must incorporate these important points:

- All technical documentation defining the products, such as standards, specifications, and drawings, should be clearly identified, on the correct revision level, and enclosed when required.
- Adequate quality records, such as testing or inspection certificates, statistical process data, quality system certificate, and warrants, should be explicitly required in the purchase order.

- Requirements for notification of changes including source and/ or composition of materials; manufacturing location; production, processing, or testing; certifications; or licensure.
- When purchasing toxic, hazardous, or otherwise restricted substances, the procurement or purchasing manager should include in the purchase order a request that the supplier provide a warrant or certificate that the substance and its packaging comply with governmental and safety regulations. Safety Data Sheets for all components should be supplied and reviewed by the firm's safety and environmental personnel to assure limited risk.

The procurement or purchasing manager should review and approve all purchasing documents prior to release, issue the inquiry to the suppliers in accordance with the purchase requisition, and select and issue the purchase order to the most appropriate supplier. Routine or repeat purchases can be approved for a specified quantity or period of time (blanket purchase orders). Sufficient data on requirements for product characteristics or service specifications should be clearly defined or attached.

Risk Management

Risk management and organization continuity planning should significantly drive purchasing strategy. When it comes to purchased goods and services, an organization can do a few things to mitigate risk. The following strategies have advantages and disadvantages:

- *Multiple sourcing.* Multiple or dual sourcing reduces supply risks, but it is expensive. When purchased product and services have to be qualified or validated, this requires redundant activities that drive up cost. It also divides the organization's total volume requirements over more than one supplier, which can result in forfeiture of any volume discounts. In special cases where the customer controls the design of a product, a duplicate set of tools, molds, or production equipment should be kept in a separate location to protect against potential disasters such as earthquake, fire, or flood. Having tooling and molds manufactured in an emerging market may be more cost-effective but may not provide comparable quality. One of the benefits of multiple sourcing is the ability to determine if changes in the process are operations related compared with supplier related if the multiple supplier's components are traceable in the process.
- *Build inventory.* Inventory is not necessarily a bad thing. Excess inventory is a bad thing, whether it is raw material, work in process, or finished goods. Excess inventory is one of the types of waste recognized by lean manufacturing, but carrying some inventory provides safety stock in case of an unforeseen problem. Dividing inventory among multiple distribution warehouse sites, including requiring a supplier to maintain some ready stock, may reduce risk. When customers elect to

re-source a current production part or material, building inventory to use while qualifying the new supplier is usually a necessity.

Like multiple sourcing and dual tooling, building inventory can be costly. An organization should determine which of the options is most effective in mitigating risk at the optimal cost.

- *Qualify alternate part numbers or material.* This approach is used in industries such as commercial electronics in which several commodities can be readily and effectively substituted for a given application. This can also add cost by requiring multiple validations or qualifications. Many customers require suppliers to receive customer approval prior to shipping a new part or material to their locations. This requires the supplier to complete the customer-specific requirements for this approval, which also adds time and cost. In exchange, both the customer and supplier have a ready alternative if needed.
- *Supplier management requirements.* Contracts can include requirements to implement and maintain a management system certification. Assessment/certification of a supplier's management system gives the customer more confidence that the supplier will meet ongoing customer requirements but does not necessarily mitigate the risk of a sporadic outage or shortage.

Some sectors have developed sector-specific quality standards, such as IATF 16949 (automotive), AS9100 (aerospace), and ISO 13485 (medical devices), with third-party certification requirements for suppliers. Use of standards as a baseline for supplier management can minimize cost to organizations but usually falls short of including all the customer-specific requirements they would like.

Contracts can also include requirements for the supplier to maintain a documented and effective disaster recovery plan or to demonstrate it is prepared for a fire, cyberattack, or natural disaster.

Contract/Order Review

The objective of contract review by the organization is to ensure that the contract requirements are adequately defined and the supplier has the ability to meet the defined requirements. This review ensures three things:

1. Order requirements, including delivery schedule and the requirements for delivery and post-delivery activities such as handling, storage, installation, operation, and maintenance, are adequately defined and documented
2. Contract or order requirements differing from those previously expressed are resolved
3. The supplier has the ability to meet the defined requirements

Many low-risk and common products or services are purchased based on supplier catalog descriptions. Even for such products or services, however, the descriptions should be clearly referenced when ordering.

If a supplier received the order verbally and does not have the order requirements documented, there is a possibility that the supplier's performance will be unsatisfactory and result in a dispute or legal action.

Confirm Requirements

The supplier should confirm the order requirements before acceptance by documenting the contract terms and specifications in some appropriate manner with the supplier's signature and submitting it for confirmation.

Amendments

When the organization changes product or service requirements, suppliers need to know about the change. These changes are normally amended to the original contract as a change order.

For customized products such as chairs, every order must be reviewed and verified due to lot-to-lot or batch-to-batch variations in color, size, mechanisms, fabric, and ergonomics.

In summary, a good supply agreement assures all parties have a thorough understanding of what is expected of each party for the mutual benefit of both.

Chapter 5

Deployment of Strategy and Expectations

> Communicate strategy internally and communicate expectations to suppliers externally. (Apply)
>
> **Body of Knowledge I.E**

Supplier strategy captures the set of guidelines that dictate definitions and expectations. Such expectations may include specifications, deliverables, process or product yields, and other measurable outputs. Though at times the intent may appear one-sided, favoring the business (i.e., the customer), the ideal supplier management relationship is based on mutual benefit. The customer provides the requirement of meeting consistently high quality, and in the process of achieving this, the supplier learns to increase productivity in its operation, which results in reduced internal costs. The customer is then able to control the incoming product quality (Bossert, 2004, p. 9). How can this ideal state be achieved?

The key to successful deployment of supplier strategy is good communication, internal and external. It is best to think of the relationship as more of a synergistic one, wherein the actions of the business (i.e., the internal side) are linked to the supplier (i.e., the external side). Figure 5.1 shows how the two are linked and the various points that an organization should consider when seeking a successful approach to the supplier management relationship.

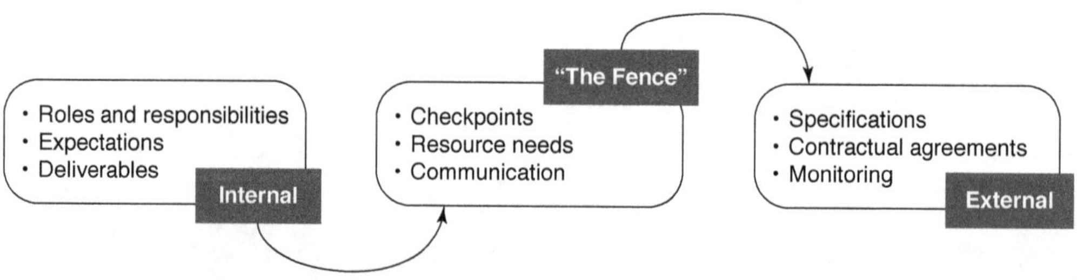

Figure 5.1 The link between the internal and external partners in developing a successful supplier management relationship.

INTERNAL COMMUNICATION

The need for a supplier typically arises in the product development phase. At this point the business should begin to evaluate what its exact need is, whether it be a material or a service.

Identifying the roles and responsibilities of the business is the next step in internal alignment. In this discussion, the business should determine what boundaries to set in the supplier relationship. In cases where the business is simply looking for a solution and has no set requirements, the provision of materials or services originates from the supplier. It is difficult to evaluate quality and acceptability in these circumstances, as the business may not have the necessary technical knowledge and competence. On the opposite end of the spectrum, a business may produce specifications (e.g., design, working parameters) with an expected output in mind. They are able to provide acceptability levels and easily identify opportunities during the production process. In both cases, the business needs to identify and agree upon what the expected output should be. Otherwise, the end result will look like what is shown on Figure 5.2 and the business will have one unhappy customer. Start with the voice of the customer to design the concept and ensure that material or service requirements have been defined and approved by the customer. Once aligned, its various roles and associated responsibilities from the beginning of the supply chain to the end should be discussed internally and conducted according to expectations.

Once roles and responsibilities are established, expectations can be laid out as to how execution will occur. The business should list attributes leading to what it wants, assess potential areas of risk or failure, perform any necessary financial analysis (e.g., budget review), and determine when the business has achieved its goals. Once the responsibilities and expectations are aligned, a set of deliverables can be documented and passed on to the supplier, along with any other set product specifications.

Knowing what the supplier needs to achieve and having organizational alignment is the ideal scenario. The deliverables should include the material or

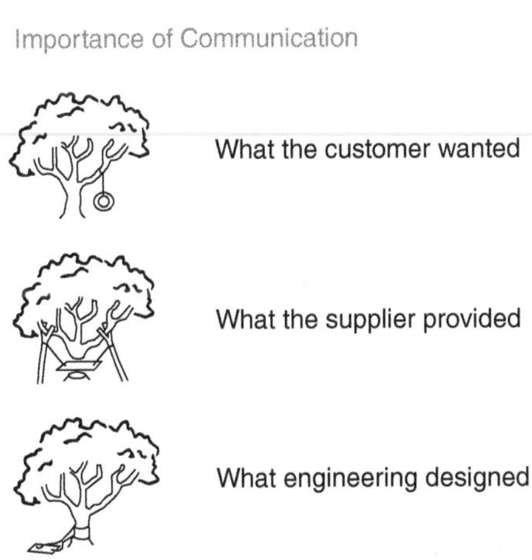

Figure 5.2 Interpretation of expectations from different stakeholders.

service requirements at a given time and at a given price. When obtaining consulting services, for example, the business must define what exactly the consultant will be doing and for how long, what the end goals are, and what the rate and maximum amount are.

THE FENCE: BRIDGING THE INTERNAL AND EXTERNAL

Once aligned, the business is ready to execute the next phase of supplier deployment. It is at the "fence" where the project manager (or whoever is leading the supplier development) must prepare the information for their external partner (i.e., supplier).

As a verification that the business is ready to deliver its deployment strategy, various criteria may be recorded, evaluated and confirmed to be a "go" in the list of checkpoints. Figure 5.3 shows a table that captures items that are part of a business's "go-live."

The team should also determine whether additional resources may be needed at either end (i.e., internally or externally). If the roles and responsibilities in the supplier engagement add work that none of the internal team members can take on, the business must think of a way to address the issue. One potential solution is to pass this work on to the supplier. However, the business needs to recognize that this will be an added cost to the project.

When suppliers provide a bid or a standard price for what it would take to deliver a material or service, they generally consider how the work will impact their existing resources. They ask themselves the following questions:

- Would we need inventory space?
- Would we need to hire more people?
- Can we source the material that our customers want at the anticipated price point?
- Can we meet the customer's deadline with our existing setup?
- Would additional or new equipment be needed?
- Do we have the technology needed to deliver what the customer wants?

Enter √	Must-haves for deployment
	Business points of contact: quality, operations, procurement (roles and responsibilities)
	Definitions and terms
	Material/service specifications
	Pricing
	Contractual agreements (drafted and include all requirements and deliverables for supplier review and acceptance)
	Terms of completion (should also specify how incomplete or unacceptable work will be managed)

Figure 5.3 Go-live checkpoints.

For this reason, pricing may vary from supplier to supplier. As a best practice, three supplier bids are taken as part of the purchasing process, as shown in Figure 5.4. In addition to pricing, the supplier quality professional should consider what else the supplier may bring to the upcoming business relationship. Some things to take note of are the maturity of the supplier's quality management system, production controls, and customer service.

Nothing beyond what is disclosed in the statement of work (SOW) is accounted for in the process. This is why the business has to be transparent about its needs to ensure a successful deployment. This includes the request of meeting specific quality requirements, if needed. Product certification, for example, may appear to be a simple task, but the additional time needed to collect the data, analyze them, and document for certification at every order is, again, an added cost to the supplier which may inherently be passed down to the business.

Communication is key to success in any partnership, including one between the business and supplier. The information provided must be delivered in whole and in written agreement. Negotiations may take place between the business and supplier in order for both parties to meet at an agreeable point; this is expected. However, this will not happen until the next phase, when the business moves forward with its selected external partner and delivers the information.

EXTERNAL COMMUNICATION

When a business comes to a supplier with a need, the supplier sees an opportunity—an opportunity for additional income as well as an opportunity to showcase its expertise. A mature supplier organization, however, knows that its success can be achieved only through preparation, planning, and organization.

As SMEs, suppliers know how to provide the end product. It is the details that they are most concerned with, since they need to ensure that customer requirements are met. For this reason, the first information the supplier should receive from the customer are specifications. Specifications should include the acceptance criteria, tolerances (if any), material limitations (e.g., substance maximums/minimums, microbial organisms), test standards, quality requirements, and any other

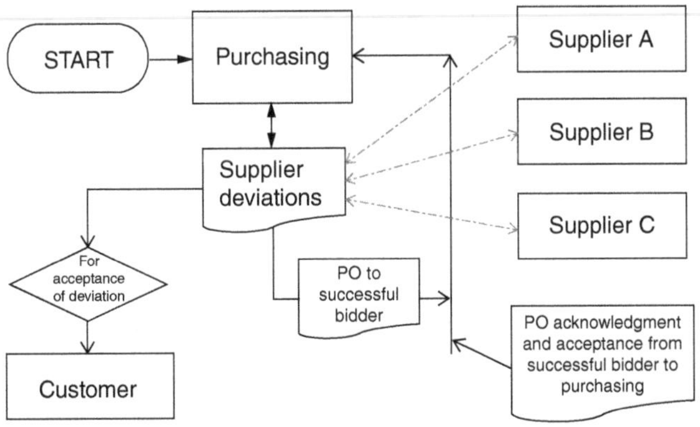

Figure 5.4 External communication flow.

applicable information. This allows the supplier to know whether it can meet the customer requirements.

When a supplier has determined that it can meet the requirements, external communication should next include contractual agreements. The agreements convey to the supplier the terms of providing the material or service of choice. The terms, as mentioned earlier, may be subject to negotiation. Some terms may be non-negotiable. An example of this is ensuring confidentiality between the parties involved in which case a supplier's legal representative may also be asked to review the document.

Once everything is agreed upon, the success of the supplier deployment can be measured through monitoring. It is best practice to consider the type of monitoring and the metrics to be monitored. The supplier truly becomes a partner when it is aware of all expectations, especially those that are measured and to which the supplier will be held accountable. In some cases, the monitoring phase becomes an incentive for suppliers to improve their processes and hopefully gain more business. The feedback begins a two-way conversation when requirements are not met. The reality is that expectations in writing may not always match up with actual production data. The practice of making changes without proper documentation can lead to issues that may damage the relationship.

Successful supplier deployment strategies must originate from the business internally before expectations are delivered externally, as communication is only as good as the information being provided. Ideal situations may not always occur, but by using a proper approach, it is possible to meet a material or service need without causing frustration on the customer or supplier side.

Part II

Risk Management

Chapter 6
Strategy

SYSTEM

> Develop a risk-based approach to manage the supply base, including business continuity, contingency planning, and supply chain resilience. (Create)
>
> **Body of Knowledge II.A.1**

The supply chain is the series of processes and/or organizations that are involved in producing and delivering a product or service to a customer (see Figure 6.1) (Westcott, 2014). The globalization of the manufacturing and supply of products that people consume poses unique and demanding challenges throughout the

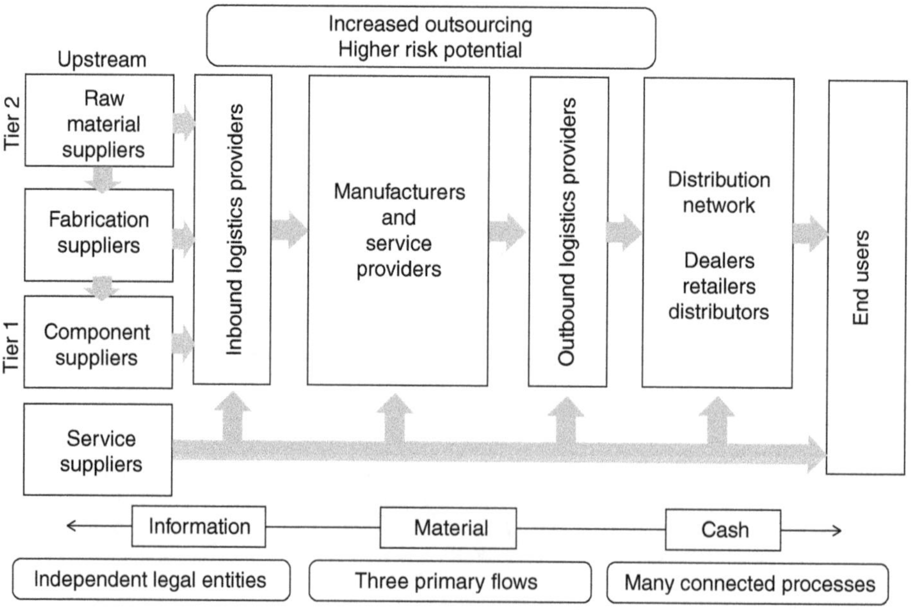

Figure 6.1 Supply chain overview.

entire supply chain. In the complex and rapidly changing environment driven by globalization, manufacturers cannot rely on traditional approaches such as inspection and sampling at the points of entry. Rapid globalization requires that manufacturers implement new approaches such as including business continuity, contingency planning, and supply chain resilience, and conduct activities to effectively monitor the supply chain. Supply chain safety and security relies on risk-based prevention with a verification-focused approach to hold all segments of industry accountable for ensuring that their products and services meet safety and quality requirements.

Business Continuity Planning

When businesses are disrupted by supply chain and other issues, the resulting impact can affect revenue, profits, and customer good will. Business continuity planning is essential for businesses to not permanently suffer from manmade and natural disasters. A business continuity plan allows a business to function after an unplanned event has occurred (see Figure 6.2).

Contingency Planning

A contingency plan documents the actionable items that can be enacted once an unplanned event has occurred. Contingency plans may consider the following manmade and natural disasters:

- Facility fire
- Bomb threats
- Hazardous materials accident
- Tornados/thunderstorms
- Blizzards
- Tsunamis
- Earthquakes

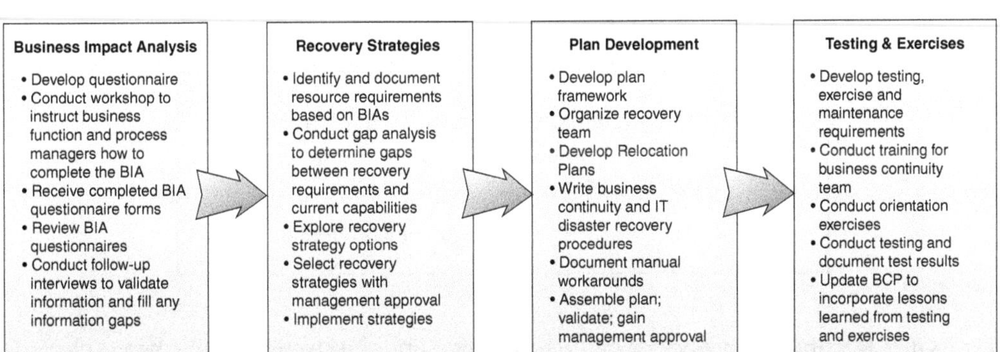

Figure 6.2 Business continuity planning process diagram.
Source: https://www.ready.gov/business-continuity-plan.

- Hurricanes
- Floods
- Power failure
- Water failure
- Gas failure
- Violence in the workplace
- IT data server failure
- Cybersecurity
- Equipment failure
- Resources
- Medical emergency
- Transportation issues
- Supplier issues
- Ownership issues
- Succession plan
- Terror attack
- Civil unrest
- Phone system failure

A comprehensive contingency plan identifies the event (manmade and natural disasters), develops a response, and documents the preparation for the event. Additionally, contingency plans may determine the severity and likelihood of the event to help prioritize responses and preparations.

PRODUCT/SERVICE RISK MITIGATION

> Develop and implement a risk mitigation plan for predicting, minimizing, monitoring, and/or controlling risks. (Create)
>
> **Body of Knowledge II.A.2**

Once the product or service has been defined and the suppliers have been selected, it is important that the manufacturer create a risk mitigation plan to minimize, monitor, and/or control risks. In this section, we will cover the essential requirements in the development of risk management plans.

Risk Management Standards

ISO 31000 Risk Management - Principles and Guidelines and ISO 14971 Medical Devices - Application of Risk Management to Medical Devices are two commonly used standards used for risk management activities whether the risks have positive or negative consequences.

ISO 31000 provides principles and guidelines on risk management which can be used by any organization or industry. ISO 31000 can be applied to strategies, decisions, operations, processes, functions, projects, products, services, and assets.

The implementation and maintenance of risk management plans need to consider the varying needs of the organization including objectives, context, structure, operations, processes, functions, projects, products, services, assets, and practices.

ISO 31000 is intended be utilized to harmonize risk management processes in existing and future ISO standards and provides a common approach in support of ISO standards managing specific risks.

ISO 14971 specifies a procedure by which a manufacturer can identify the hazards associated with medical devices and their accessories, including in vitro diagnostic medical devices.

ISO 14971 can help estimate and evaluate the risks, control these risks and monitor the effectiveness of the control throughout the life cycle of a medical device. ISO 14971 does not apply to clinical decisions relating to the use of a medical device and does not specify acceptable risk levels.

Risk Management Plans

ISO Guide 73:2009 defines a risk management plan (RMP) as a "scheme within the risk management framework specifying the approach, the management components and resources to be applied to the management of risk." Furthermore, it says that "management components typically include procedures, practices, assignment of responsibilities, sequence and timing of activities" and also that "the risk management plan can be applied to a particular product, process and project, and part or whole of the organization." Risk management activities must be planned. Therefore, for the particular product being considered, the manufacturer must establish and document an RMP in accordance with the risk management process. The RMP must be part of the risk management file. This plan must include at least the following:

- The scope of the planned risk management activities, identifying and describing the product and the lifecycle phases for which each element of the plan is applicable
- Assignment of responsibilities and authorities
- Requirements for review of risk management activities
- Criteria for risk acceptability based on the manufacturer's policy for determining acceptable risk, including criteria for accepting risks when the probability of harm occurring cannot be estimated
- Verification activities
- Activities related to collection and review of relevant production and post-production information

If the plan changes during the lifecycle of the product, a record of the changes must be maintained in the risk management file. Compliance is checked through inspection of the risk management file. The RMP can be a separate document, or it can be integrated into other documentation, such as QMS documentation. The content and level of detail for the plan should be commensurate with the level of risk associated with the product. The following sections explain the minimum requirements for an RMP. Manufacturers can include other items such as time schedule, risk analysis tools, or a rationale for the choice of specific risk acceptability criteria.

Scope of the Plan

The scope identifies and describes the product and the lifecycle phases for which each element of the plan is applicable. All elements of the risk management process should be mapped to the manufacturer's defined product lifecycle. Some of the elements of the risk management process will occur during the phases of the manufacturer's established product realization process, such as design and development control. The remaining elements will occur during the other lifecycle phases through to product decommissioning. The RMP provides this mapping for a specific product either explicitly or through reference to other documents. Although all risk management activities need to be planned, a manufacturer can have several plans covering different parts of the lifecycle.

Assignment of Responsibilities and Authorities

The plan should identify the personnel with responsibility for the execution of specific risk management activities, such as reviewers (including independent ones not directly involved in the process under evaluation), SMEs, and individuals with approval authority.

Requirements for Review of Risk Management Activities

The plan should detail how and when risk management reviews will occur for the specific product. The requirements for the review of risk management activities can be part of other quality system review requirements (e.g., design and development review activities).

Criteria for Risk Acceptability

Criteria for risk acceptability are derived from the manufacturer's policy for determining acceptable risk. The criteria can be utilized for similar categories of the product. Criteria for risk acceptability can be part of the manufacturer's established QMS, which can be referenced in the plan.

Verification Activities

The plan should specify how two distinct verification activities will be carried out. The first verification is required to demonstrate that the risk control measure has been implemented. The second verification is required to ensure that the implemented measure is effective (i.e., it actually reduces the risk). Verifying the effectiveness of risk control measures can require the collection of different types of data (manufacturing data, customer data, etc.). The plan can detail the verification activities explicitly or through reference to the plan for other verification activities.

Methods of Obtaining Relevant Post-Production Information

The method of obtaining post-production information can be part of established QMS procedures (e.g., monitoring and measuring). Manufacturers must establish procedures to collect information from various sources, such as service personnel, training personnel, incident reports, and customer feedback. While a reference to the QMS procedures will suffice in most cases, any product-specific requirements should be explicitly stated in the RMP. The plan should include documentation of decisions, based on a risk analysis, about what sort of post-market surveillance is appropriate for the product, such as whether reactive surveillance is adequate or whether proactive studies are needed.

PREVENTION STRATEGIES

> Identify and evaluate strategies and techniques such as supply chain mapping, avoidance, detection, and mitigation used to prevent the introduction of counterfeit parts, materials, and services. (Evaluate)
>
> Body of Knowledge II.A.3

Quality risk management supports a scientific approach to decision making. It provides documented and reproducible methods to accomplish steps of the quality risk management process based on current knowledge about assessing the severity, probability, and detectability of the risk. Traditionally, industry and regulators have assessed and managed risk in a variety of informal ways based on, for example, compilation of observations, trends, and other information. Such approaches continue to provide useful information that might support topics such as handling of complaints, quality defects, deviations, and allocation of resources.

Quality risk management methods and the supporting statistical tools can be used in combination (e.g., probabilistic risk assessment). Combined use provides flexibility that can facilitate the application of quality risk management principles. These techniques can be complementary, and it may be necessary to use more than one of them. The basic principle is that the chain of events is analyzed step by step. The degree of rigor and formality of quality risk management should reflect available knowledge and be commensurate with the complexity and/or criticality of the issue under evaluation. In this section, we will cover the selection and types of risk assessment techniques, the application of risk assessment during the lifecycle phases, and the supply chain risk considerations.

Selection and Types of Risk Assessment Techniques

Risk assessment may be undertaken in varying degrees of depth and detail using one or several methods ranging from simple to complex. The form of assessment and its output should be consistent with the risk criteria developed as part of establishing the context. In general terms, a suitable technique should have the following characteristics:

- It should be justifiable and commensurate with the situation under consideration
- It should provide results in a form that enhances understanding of the nature of the risk and how it can be treated
- It should be capable of being used in a manner that is traceable, repeatable, and verifiable

Once the decision has been made to perform a risk assessment and the objectives and scope have been defined, the techniques should be selected based on applicable factors, such as:

- Objectives of the study
- Type and range of risks being analyzed
- The potential magnitude of the consequences. The decision on the depth to which risk assessment is carried out should reflect the initial perception of consequences (although this may have to be modified once a preliminary evaluation has been completed).
- Degree of expertise and resources needed. A simple method done well may provide better results than a more sophisticated procedure done poorly, as long as it meets the objectives and scope of the assessment. The effort put into the assessment should be consistent with the potential level of risk being analyzed.
- Regulatory or contractual requirements

Various factors influence the selection of an approach to risk assessment, such as the availability of resources, the nature and degree of uncertainty in the data and information available, and the complexity of the application.

Availability of Resources

Several factors related to available resources and capabilities may affect the choice of risk assessment techniques, including:

- Skills, experience, capacity, and capability of the risk assessment team
- Constraints on time and other resources within the organization
- Budget if external resources are required

The Nature and Degree of Uncertainty

The nature and degree of uncertainty requires an understanding of the quality, quantity, and integrity of information available concerning the risk under consideration. This includes the extent to which sufficient information is available regarding the risk, its sources and causes, and its consequences. Uncertainty can stem from poor data quality or a lack of essential and reliable data. Uncertainty can also be inherent in the external and internal context of the organization.

Available data do not always provide a reliable basis for predicting the future. For unique types of risks or for new product types, historical data may not be available or there may be different interpretations of available data by different stakeholders (industry and regulators, for example). Those undertaking risk assessment need to understand the type and nature of the uncertainty and appreciate the implications for the reliability of the risk assessment results. These should always be communicated to decision makers.

Complexity

Risks can be complex in themselves; for example, in some complex systems, the risks must be assessed across the system rather than treating each component separately and ignoring interactions. In other cases, treating a single risk can have implications elsewhere and can impact other activities. Consequential impacts and risk dependencies need to be understood to ensure that managing one risk does not create an intolerable situation elsewhere. Understanding the complexity of a single risk or of an organization's portfolio of risks is crucial for the selection of the appropriate risk assessment techniques.

Application of Risk Assessment During Lifecycle Phases

Many activities, projects, and products can be considered to have a lifecycle, starting from initial concept and definition through realization to final completion, which might include removal of the product from the market. Risk assessment can be applied at all stages of the lifecycle and is usually applied many times with different levels of detail to assist in the decisions that need to be made at each phase.

Lifecycle phases have different needs and require different techniques. For example, during the concept and definition phase, when an opportunity is identified, risk assessment may be used to decide whether to proceed. Where several options are available, risk assessment can be used to evaluate alternative concepts to help decide which provides the best balance of risks. During the design and development phase, risk assessment contributes to ensuring that system risks are tolerable and to identifying risks impacting subsequent lifecycle phases.

SUPPLIER RISK IDENTIFICATION AND CATEGORIZATION

> Identify supplier risks and develop categorization (e.g., organizational, business, security, and product) using tools and models, such as the Kraljic portfolio segmentation model. (Create)
>
> **Body of Knowledge II.A.4**

Globalization presents serious challenges to the ability of the product safety system to protect consumers. Among these challenges are the growth in the number of products manufactured abroad, the increasingly complex path that products travel from source materials to consumers, and the greater chance for the intentional substitution of ingredients for profit (economic adulteration). The Kraljic Portfolio Segmentation Model, shown in Figure 6.3, allows for the establishment of risk-based supplier selection, qualification, and performance monitoring processes to focus supplier lifecycle management activities on the riskiest and highest-impact products. The Kraljic Portfolio Segmentation Model can help organizations manage the global supply chain by ensuring products are safe and protecting consumers by minimizing the intentional substitution of ingredients for profit (economic adulteration).

The ultimate goal of the product safety system is to protect people from contamination, diversion, counterfeiting, and other risks that could harm them. *Supply chain safety and security* refers to minimizing risks that arise anywhere along the supply chain continuum, from sourcing a product's raw material, ingredients, and components through the product's manufacture, importation, sale, and distribution. Addressing supply chain safety and security presents some of the greatest challenges. To address these challenges and protect public health, four primary strategies must be adopted:

1. Focus on prevention
2. Enhance regulatory information
3. Improve scientific and analytic capabilities
4. Expand risk-based inspection and enforcement

Supply chain safety and security is the shared responsibility of not only the producer but the country of origin, the importer, the importing country, and the

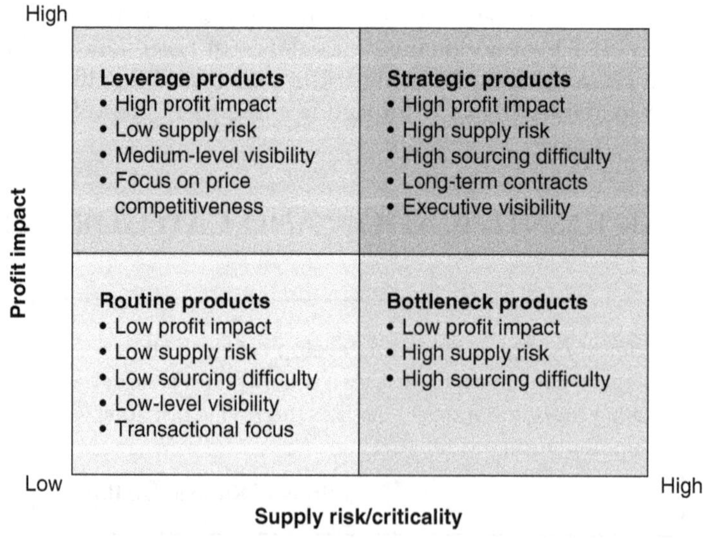

Figure 6.3 Kraljic portfolio segmentation model.

Source: Adapted from Peter Kraljic, "Purchasing Must Become Supply Management," *Harvard Business Review*, 1983 (https://hbr.org/1983/09/purchasing-must-become-supply-management).

final company in the supply chain. Companies that sell contaminated products because of loose supply chain oversight need to face serious penalties and cannot excuse themselves by blaming their suppliers.

Companies must have a strong supply chain management system in order to withstand the increased scrutiny by regulators due to increased globalization and outsourcing. Companies should start reviewing their supply chain management program now, identify gaps, and address them before a regulator's inspection. Additionally, those companies that mostly use contract manufacturers to produce and distribute their products must ensure that they have tight control over those contractors.

Companies must ensure that their supplier management programs are compliant, maintain oversight, and facilitate strong supplier relationships. They should also ensure that their suppliers are evaluated and that regulations and guidance are followed according to the product or service they provide. Companies should focus on several areas in their supplier programs, such as:

- Knowledge about their supply chain, including participants and vulnerabilities
- Risk management of the supply chain
- Qualification of prospective suppliers
- Audits and testing programs based on identified risks
- Quality and performance of suppliers
- Building relationships with suppliers (e.g., partnerships and alliances)

More and more, regulators are seeing companies that do not have control over their supply chain. These companies need to ensure compliance and begin treating their suppliers as an extension of their own company.

Chapter 7
Analysis and Mitigation

There are many tools that can be used to identify, assess, and prioritize risks to supplier quality. In this chapter, we will cover decision analysis, Failure Modes and Effects Analysis (FMEA), fault tree analysis (FTA), preliminary hazard analysis (PHA), risk ranking and filtering, and process audits. These tools provide a solid foundation for analyzing and mitigating risk.

ANALYSIS

> Identify, assess, and prioritize risks to supplier quality using tools such as decision analysis (DA), failure mode and effects analysis (FMEA), fault tree analysis (FTA), and process auditing. (Evaluate)
>
> **Body of Knowledge II.B.1**

Decision Analysis

Decision analysis refers to a systematic approach to evaluating multiple options to arrive at a best solution. In the context of supplier quality and supply chain management, multiple options may be considered for a product solution, and/or multiple suppliers may be considered to provide a selected product or service. A decision matrix may be used to evaluate different options against multiple, weighted criteria to determine and select the best option. A benefit of using a decision matrix is that it allows a cross-functional team of stakeholders to participate in the analysis and selection. A decision matrix also helps the team evaluate all potential solutions against all of the stated criteria simultaneously to ensure that all needs and requirements are considered.

To complete a decision matrix, the problem must be defined; what is the team trying to identify or solve? The team then defines the criteria for evaluating any potential solutions, weights each of the criteria in terms of relative importance, and rates each potential solution against each of the stated criteria, ultimately

Solution	Ease of Use	Integration between Modules	Future Expandability	Integration with ERP system	Data Mgmt/Extraction	Dashboard/Reports - customizable	Web Accessibility	App Available	License Structure	Search Functionality	Structure Customization	Implementation Support	Tech Support (ongoing availability)	Total	
Weight	10	8	6	2	9	5	8	4	8	9	7	10	9	Score	Rank
Option 1	1	0	1	1	1	1	9	1	1	5	5	9	5	331	4
Option 2	9	9	5	1	5	9	9	9	1	9	5	9	9	687	2
Option 3	5	5	1	5	5	5	5	9	5	9	1	5	5	475	3
Option 4	9	9	9	1	5	9	9	0	9	9	9	9	9	767	1
Total	240	184	96	16	144	120	256	76	128	288	140	320	252		

Figure 7.1 Decision matrix.

arriving at a score for each option that can be used to rank the potential solutions against each other.

As an example, consider Figure 7.1, which illustrates a decision matrix used by a company that is evaluating a software solution to be implemented company wide. The criteria and weights are determined by a cross-functional team of users, including IT, Production, Finance, Purchasing, and Quality, and are listed across the top of the matrix in the columns. The potential solutions are listed down the side in the rows. The ratings for each option against each criterion are entered into the cells, and the sum of each option's rating multiplied by the associated weight is calculated to give the total score for that option. All options are then ranked on the basis of the total scores. In this example, Option 4 is the best choice, followed by Option 2.

Different rating scales may be used by different companies or for different projects, as long as each person participating in an analysis is using the same scale and the scale is understood by each team member. For this example, ratings used were 9 (fully addressed), 5 (mostly addressed), 1 (partially addressed), and 0 (not addressed). Teams may choose to use other scales such as 0-10, 0-5, 0/1/3/5, etc.

Failure Modes and Effects Analysis

Failure Modes and Effects Analysis (FMEA) provides for an evaluation of potential failure modes for processes and their likely effect on outcomes, product performance, or both. Once failure modes are established, risk reduction can be used to eliminate, contain, reduce, or control the potential failures. FMEA relies on product and process understanding. FMEA methodically breaks down the analysis of complex processes into manageable steps. It is a powerful tool for summarizing the important modes of failure, factors causing these failures, and the likely effects of these failures. FMEA can be used to prioritize risks and monitor the effectiveness of risk control activities.

FMEA is an analysis technique that facilitates the identification of potential problems in a design or process by examining the effects of lower-level failures. Actions are recommended or mitigation provisions made to reduce the likelihood of the problem occurring and to mitigate the risk if it does occur. There are many different types of FMEAs, such as design FMEA, process FMEA, and system FMEA. They are inductive techniques using the question, "What happens if . . . ?" They analyze elements of the process (e.g., components) one at a time, thus they generally look at a single-fault condition. This is done in a bottom-up fashion, for example, following the procedure to the next highest functional system level.

FMEA can be applied to equipment and facilities and might be used to analyze a manufacturing operation and its effect on a product or process. It identifies elements or operations within the system that render it vulnerable. FMEA results can be used as a basis for design or further analysis or to guide resource deployment. While FMEA can be utilized for failures and risks associated with manufacturing processes, it is not limited to this application. The output of an FMEA is a relative risk score for each failure mode, which is used to rank the modes on a relative risk basis. FMEA can provide input to other analysis techniques such as FTA at either a qualitative or quantitative level. Table 7.1 lists some strengths and limitations of the FMEA tool.

Table 7.1 Strengths and limitations of FMEA.

Strengths
• Provides a basis for identifying potential root causes and developing effective preventive actions. • Identifies reliability/safety-critical components. • Is widely applicable to human, equipment, and system failure modes and to hardware, software, and procedures. • Provides a foundation for other maintainability, safety, testability, and logistics analyses. • Identifies component failure modes, their causes, and their effects on the system and presents them in an easily readable format. • Avoids the need for costly equipment modifications in service by identifying problems early in the design process. • Identifies single-point failure modes and requirements for redundancy or safety systems. Provides input to development of monitoring programs by highlighting key features to be monitored.
Limitations
• It is difficult to deal with redundant (backup) systems, the incorporation of repair or preventive maintenance actions, and the restriction on single-fault conditions. • Can only be used to identify single failure modes, not combinations of failure modes. • FMEA/FMECA studies can be time-consuming and costly. • FMEA/FMECA studies can be difficult and tedious for complex, multilayered systems.

The FMEA Process

FMEAs require information about the elements of the system in detail sufficient for meaningful analysis of the ways in which each element could fail. Information may include:

- A flowchart of the process or system being analyzed and its components
- An understanding of the function of each step of the process or component of the system
- Details of environmental and other operational parameters that may affect operation
- An understanding of the results of particular failures
- Historical data on failures, including failure rate data (if available), which will allow determination of the probability of failure

There are several ways meaningful analysis may be done. The most common tool used to classify each of the identified failure modes according to its criticality is the risk priority number (RPN), which is a semi-quantitative measure of criticality obtained by multiplying numbers from rating scales (usually from 1 to 5) for consequence (severity) of failure, likelihood of failure, and ability to detect the problem. A failure is given a higher priority if it is difficult to detect. Once potential failure modes and mechanisms are identified, preventive actions can be defined and implemented for the more significant ones.

The FMEA process is depicted in Figure 7.2. Details regarding potential scales for severity, occurrence, and detectability ratings are included in Chapter 10.

The primary output of FMEA is a list of potential failure modes—the failure mechanisms and effects for each component or step of a system or process (which may include information on the likelihood of failure). Information is also given on the causes of failure and the consequences to the system as a whole as well as the detectability of the failure mode. See Figure 7.3 on p. 71 for an example of FMEA output for a tablet packaging process.

For risks identified in suppliers' processes that are determined to be unacceptably high, the customer must work with the supplier to determine the actions and controls that will be implemented to reduce the risk level. The supplier is responsible for managing its own processes by completing the implementation, while the customer is responsible for verifying that the implemented actions and controls have mitigated the risks to an acceptable level.

Fault Tree Analysis

Fault tree analysis (FTA) is a type of analysis in which a failure is analyzed using Boolean logic (AND/OR) to combine a series of lower-level events (causal factors) until their root causes are reached. This analysis method was originally developed to quantitatively determine the probability of a safety hazard in the field of safety engineering. FTA provides a method of breaking down chains of failures. A key addition permits the identification of combinations or interactions of events that cause other failure events. There are two types of interactions:

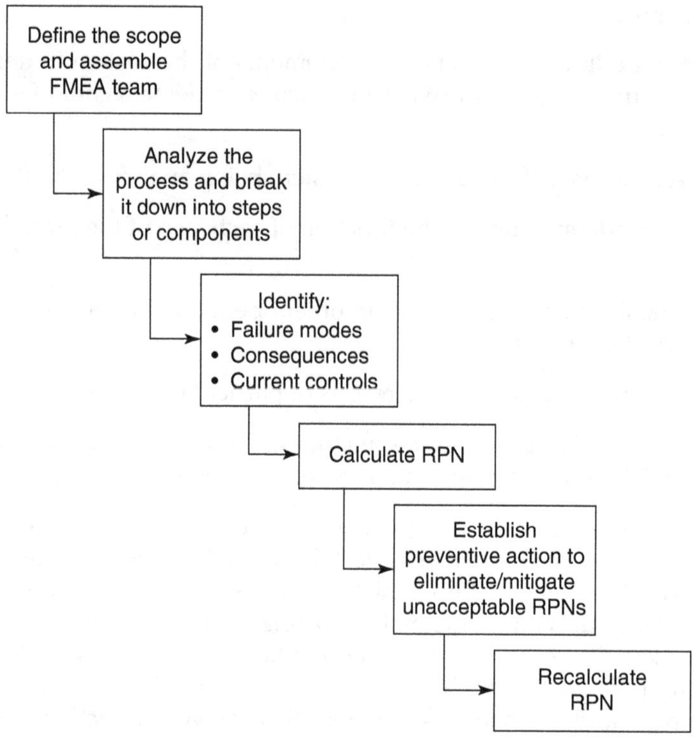

Figure 7.2 FMEA process flow.

1. *Several* items must fail together in order for another item to fail ("AND" combination)
2. Only *one* of a number of possible events must happen in order for another item to fail ("OR" combination)

Figure 7.4 shows an example of the FTA tool applied to the situation of the mix-up of two different products. By applying Boolean logic, we can find the root causes that may lead to the failure.

The AND and OR symbols are called gates. They prevent the failure event above them from occurring unless their specific conditions are met. When several factors must happen simultaneously (an AND relationship), we can avoid the failure simply by controlling one of them (the easiest or the cheapest). When any of several causal factors can create the failure (an OR relationship), then we must control all of them.

The tree is constructed working backward from a known event or failure and asking why it happened. The answer will represent the factor that directly caused the failure. Continuing with the "why" questioning will allow us to reach fundamental events or root causes. FTA is a very good tool for helping us understand how an event occurred. It is best used when working with complex issues with several interrelated causes of failure.

FTA is a deductive, top-down approach to failure modes analysis aimed at analyzing the possible causes of an undesired event or failure. In contrast, FMEA

Process step/input	Potential failure mode	Potential failure effects	SEV	Potential causes	OCC	Current controls	DET	RPN	Actions recommended	Resp.	Actions taken	SEV	OCC	DET	RPN
What is the process step/ input under investigation?	In what ways does the key input go wrong?	What is the impact on the key output variables (customer requirements) or internal requirements?	How severe is the effect to the customer?	What causes the key input to go wrong?	How often does cause or FM occur?	What are the existing controls and procedures (inspection and test) that prevent either the cause or the failure mode? *Should include an SOP number.*	How well can you detect cause or FM?		What are the actions for reducing the occurrence of the cause, or improving detection?	Who's responsible for the recommended action?	What are the completed actions taken with the recalculated RPN? *Be sure to include completion month/year.*				
Filling/feeder	Excessive vibration	Unfilled cavities	10	Lack of adequate preventive maintenance	8	Automatic inspection	7	560	Include the machine in the PM program	M. Peña	Machine was included in monthly PM program	10	2	3	60
Filling/feeder	Excessive vibration	Cavities with more than one tablet	10	Lack of adequate preventive maintenance	8	Automatic inspection	7	560	Include the machine in the PM program	M. Peña	Machine was included in monthly PM program	10	2	3	60

Figure 7.3 Partial example of FMEA for tablet packaging.

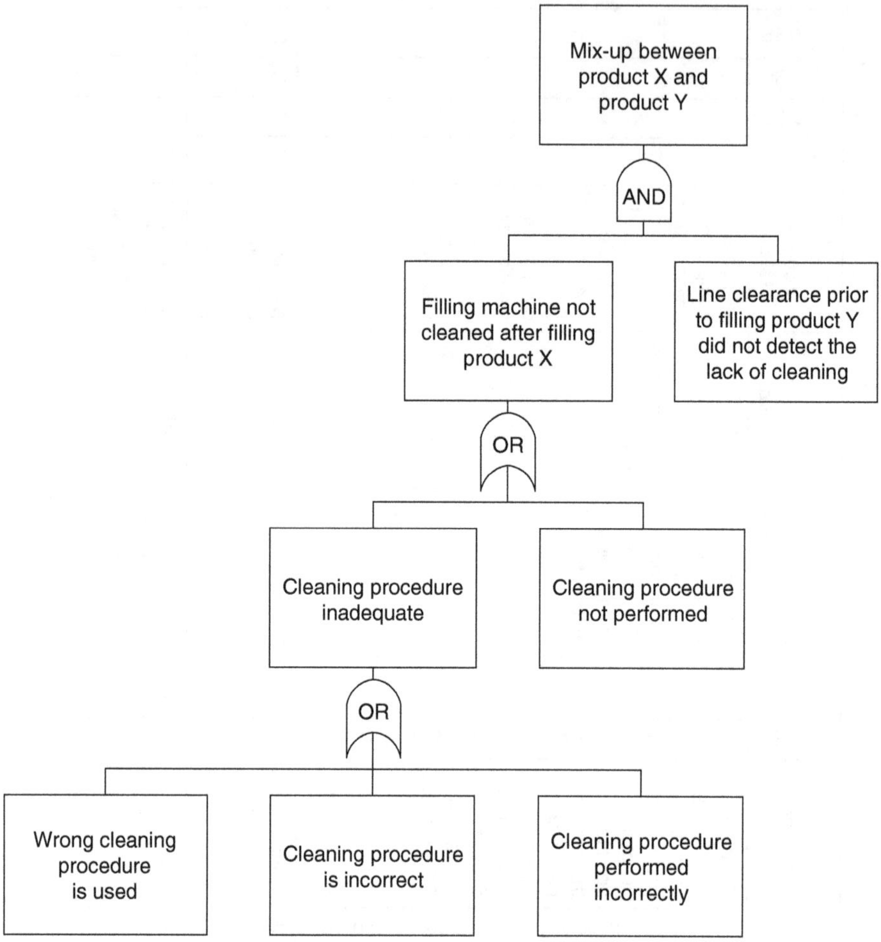

Figure 7.4 Fault tree analysis example.

is an inductive, bottom-up analysis method aimed at analyzing the effects of single-component or function failures on equipment or systems. We can define FTA as a reactive investigation tool (i.e., the failure already happened) whereas ideally FMEA will be used proactively during the design phase of a process to anticipate failure modes and generate preventive actions.

FTA is a technique for identifying and analyzing factors that can contribute to a specified undesired event (called the top event). In a deductive manner, starting with the top event, the possible causes or fault modes of the next lowest functional system level causing the undesired consequence are identified. The factors identified in the tree can be events that are associated with component hardware failures, human errors, or any other pertinent events that lead to the undesired event. FTA can be used to establish the pathway to the root cause of the failure. It can be used to investigate complaints or deviations in order to fully understand their root cause and to ensure that intended improvements will fully resolve the issue and not create other problems. FTA is an effective tool for evaluating how multiple factors affect a given issue. The output of an FTA includes

Table 7.2 Strengths and limitations of FTA.

Strengths
• FTA is a disciplined approach that is highly systematic but at the same time sufficiently flexible to allow analysis of a variety of factors, including human interactions. The top-down approach focuses attention on those effects of failure that are directly related to the top event. • FTA is especially useful for analyzing systems with many interfaces and interactions. • The pictorial representation leads to an easy understanding of the system behavior and the factors included, but as the trees are often large, processing of fault trees may require computer systems. • Logic analysis of fault trees is useful in identifying simple failure pathways in a very complex system where particular combinations of events that lead to the top event could be overlooked.
Limitations
• Uncertainties in the probabilities of base events are included in calculations of the probability of the top event. This can result in high levels of uncertainty where base event failure probabilities are not known accurately; however, a high degree of confidence is possible in a well-understood system. • In some situations, causal events are not bound together, and it can be difficult to ascertain whether all important pathways to the top event are included. • The fault tree is a static model; time interdependencies are not addressed. • Fault trees can only deal with binary states (failed/not failed). • While human error modes can be included in a qualitative fault tree, in general, failures of degree or quality, which often characterize human error, cannot easily be included. • A fault tree does not enable domino effects or conditional failures to be included easily.

a visual representation of failure modes. It is useful both for risk assessment and in developing monitoring programs. Table 7.2 lists some strengths and limitations of the FTA tool.

Preliminary Hazard Analysis

PHA is a simple, inductive method of analysis whose objective is to identify the hazards that can cause harm for a given activity, facility, or system. It is usually carried out early in the development of a product or process when there is little information on design details or operating procedures. It can be a precursor to further studies or provide information for specification of the design of a system.

The tool consists of:

- Identification of the likelihood that the risk event will happen
- Qualitative evaluation of the extent of possible injury or damage to health that could result
- A relative ranking of the hazard using a combination of severity and likelihood of occurrence
- Identification of possible remedial measures

Table 7.3 Strengths and limitations of PHA.

Strengths
• Can be used when there is limited information. • Allows risks to be considered very early in the system lifecycle.
Limitations
• Provides only preliminary information. • Is not comprehensive; does not provide detailed information on risks and how they can best be prevented.

The results obtained can be presented in different ways, such as tables and trees. Table 7.3 lists some strengths and limitations of the PHA tool.

Risk Ranking and Filtering

Risk ranking and filtering is a tool for comparing and prioritizing (i.e., ranking) risks. Risk ranking of complex systems typically involves evaluating multiple diverse quantitative and qualitative factors for each risk. The tool breaks down a basic risk question into as many components as needed to capture factors involved in the risk. These factors are combined into a single relative risk score that can then be used to rank risks. Filters in the form of weighting factors or cutoffs for risk scores can be used to tailor the risk ranking to management or policy objectives.

Risk ranking (Figure 7.5) and filtering (Figure 7.6) can be used to prioritize manufacturing sites for inspection/audit by regulators or industry. Risk ranking methods are particularly helpful when the portfolio of risks and the underlying consequences to be managed are diverse and difficult to compare using a single tool. Risk ranking is useful for management in evaluating both quantitatively assessed and qualitatively assessed risks within the same organizational framework.

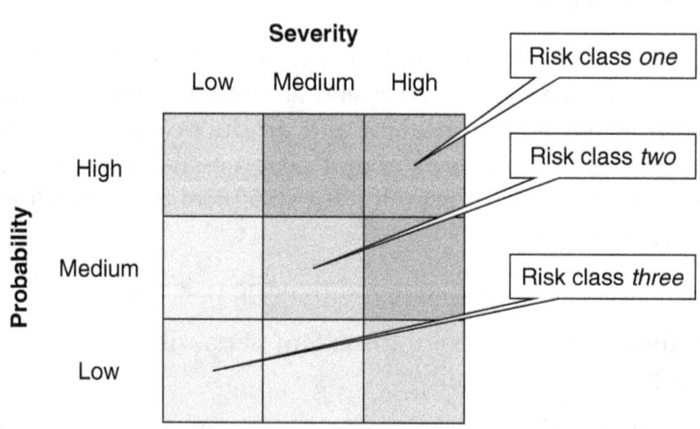

Figure 7.5 Risk ranking.

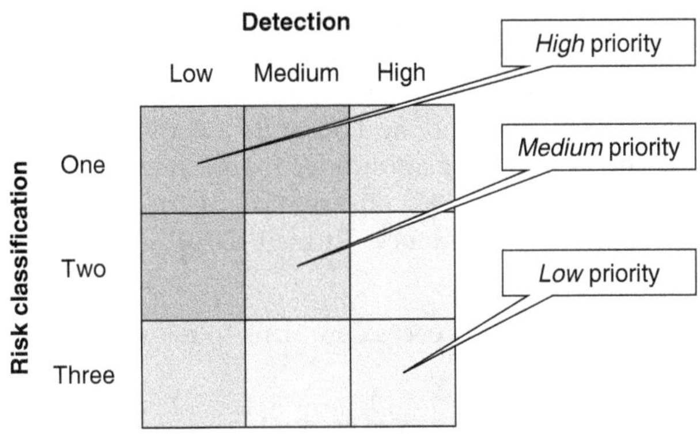

Figure 7.6 Risk filtering.

Table 7.4 Strengths and limitations of risk ranking and filtering.

Strengths
• Can be used for highly complex issues. • Very flexible. • Scalable to include multiple risk factors.
Limitations
• Sometimes difficult to establish risk factors and adequate evaluation criteria. • Difficult to correlate results with absolute risk.

Table 7.4 lists some strengths and limitations of the risk ranking and filtering tool.

Process Audits

While many supplier audits focus on a supplier's overall quality system, process audits may also be used to analyze specific supplier processes to identify supplier quality risks and potential improvement opportunities.

Process audits may be managed in a similar manner to systems audits, and Chapter 14 contains more details regarding audit stages, mechanics, and reporting. The main difference between a process audit and a systems audit lies in the audit scope. According to J.P. Russell (2010, p. 14), a process audit is "an audit of a process against agreed-upon requirements". Process audits focus on any specific manufacturing, software, and service process, where a process is defined as a series of steps that transform inputs into a desired output. By conducting a process audit, the auditor understands the steps involved in the process, the inputs and outputs of that process, and the controls that are in place.

A process auditor can best define the scope of the audit by creating a SIPOC diagram for the process in question (see Chapter 2). The SIPOC diagram will define where the process begins and ends, as well as all of the tangible and intangible inputs and outputs of the process along with the suppliers of the inputs and the customers of the outputs. The auditor can also use this diagram to begin to identify particular steps, controls, and other items of interest to explore during the audit phase as well as potential candidates for audit interviews.

The core of a process audit is observation of the process in action to evaluate the process and collect evidence. Russell (2010) describes four objectives for this phase:

- Verification that the process conforms to relevant internal and external standards,
- Confirmation that objectives are met,
- Validation that controls are adequate and effective in mitigating risk,
- Identification of potential improvement opportunities.

Auditors may choose to observe the process by tracing the process through all of its steps, paying particular attention to the placement, design, and use of process controls. The auditor can use information and data gathered from these observations to determine if the process and its controls are effective and/or if there are any potential areas where improvements should be developed and applied in order to mitigate risks from weaknesses in the process or from subpar performance.

MITIGATION CONTROL

> Develop and deploy controls such as inspection and test plans. Prioritize mitigation activities and sustain a risk mitigation plan appropriate to the risk of the product/service. (Create)
>
> **Body of Knowledge II.B.2**

Once all the risks have been identified and analyzed, they must be controlled. Mitigation control activities must be prioritized based on the level of risk. In this section, we will cover risk control (risk reduction and risk acceptance) and risk documentation and communication.

Risk Control

Risk control includes decision making to reduce and/or accept risks. The purpose of risk control is to reduce the risk to an acceptable level. The amount of effort dedicated to risk control should be proportional to the significance of the risk. Decision makers might use different processes, including cost–benefit analysis, to determine the optimal level of risk control. Table 7.5 includes some examples of risk control activities. Risk control might focus on the following questions:

Table 7.5 Risk control examples.

Product/process	Hazard	Inherent safe features and protective measures	Information for safety
Single-use medical device (sterile syringe)	Contamination (biological and cross-contamination)	Sterilization pouch, single use	Warning on labeling against reuse
Over-the-counter product	Criminal contamination	Tamper-evident packaging	Warning on labeling regarding tamper-evident features

- Is the risk above an acceptable level?
- What can be done to reduce or eliminate risks?
- What is the appropriate balance between benefits, risks, and resources?
- Are new risks created as a result of the identified risks being controlled?

Risk Reduction

Risk reduction focuses on processes for the mitigation or avoidance of quality risk when it exceeds a specified (acceptable) level. Risk reduction might include actions taken to mitigate the severity and probability of harm. Processes that improve the detectability of hazards and quality risks might also be used as part of a risk control strategy. The implementation of risk reduction measures can introduce new risks into the system or increase the significance of other existing risks. Hence, it might be appropriate to revisit the risk assessment to identify and evaluate any possible change in risk after implementing a risk reduction process. Risk treatment involves selecting one or more options for modifying risks, then implementing those options. Once implemented, treatments provide or modify the controls. Risk treatment involves a cyclical process of:

- Assessing a risk treatment
- Deciding whether residual risk levels are tolerable
- If risk levels are not tolerable, generating a new risk treatment
- Assessing the effectiveness of that treatment

Selecting the most appropriate risk treatment option involves balancing the costs and efforts of implementation against the benefits derived, with regard to legal or regulatory requirements. A number of treatment options can be considered and applied either individually or in combination. The organization can usually benefit from the adoption of a combination of treatment options. When selecting risk treatment options, the organization should consider the values and perceptions of stakeholders and the most appropriate ways to communicate with them. In cases where risk treatment options can have an impact on risk in other units of the organization or on stakeholders, these parties should be involved in the decision.

Though risk treatment options are equally effective, some are more acceptable to stakeholders than others. The treatment plan should clearly identify the priority order in which individual risk treatments should be implemented. Risk treatment itself can introduce risks. One significant risk is the failure or ineffectiveness of the risk treatment measures. Monitoring needs to be an integral part of the risk treatment plan to ensure that the measures remain effective. Risk treatment can also introduce secondary risks that need to be assessed, treated, monitored, and reviewed. Treatment plans should be integrated with the management processes of the organization and discussed with appropriate stakeholders.

Risk Acceptance

Risk acceptance is a decision to accept risk. Risk acceptance can be a formal decision to accept the risk or it can be a passive decision in which risks are not specified. For some types of harm, even the best quality risk management practices might not entirely eliminate risk. In these circumstances, it might be agreed that an appropriate quality risk management strategy has been applied and that quality risk has been reduced to a specified (acceptable) level. This level will depend on many parameters and should be decided on a case-by-case basis.

Risk Documentation and Communication

Risk communication is the sharing of information about risk and risk management between the decision makers and others. Stakeholders can communicate at any stage of the risk management process. The output of the quality risk management process should be appropriately communicated and documented. Communications might include those among interested stakeholders—for example, between regulators and industry; between industry and the customer; or within a company, industry, or regulatory authority. The included information might relate to the existence, nature, form, probability, severity, acceptability, control, treatment, detectability, or other aspects of risks to quality. Communication need not be carried out for each and every risk acceptance.

As with any important activity related to any product, risk management activities must be traceable. Decisions concerning the creation of records should take into account:

- The organization's need for continuous learning
- Benefits of reusing information for management purposes
- Costs and efforts involved in creating and maintaining records
- Legal, regulatory (e.g., retention period), and operational needs for records

The risk assessment process should be documented together with the results of the assessment. Risks should be expressed in understandable terms, and the units in which the level of risk is expressed should be clear. The extent of the report will depend on the objectives and scope of the assessment. The content of the report must be described within a standard procedure, and, except for very simple assessments, the documentation should include:

- Objectives and scope
- Description of relevant parts of the system and their functions
- The situation, system, or circumstances being assessed
- Risk criteria applied and their justification
- Assessment methodology
- Risk identification results
- Data, assumptions, and their sources and validation
- Risk analysis results and their evaluation
- Critical assumptions (including uncertainty analysis) and other factors that need to be monitored
- Discussion of results, and conclusions and recommendations

If the risk assessment supports a continuing risk management process, it should be performed and documented in such a way that it can be maintained throughout the lifecycle of the system, organization, equipment, or activity. The assessment should be updated as significant new information becomes available and the context changes, in accordance with the needs of the management process.

Inspection and Test Plans

Once process or product risks have been identified and prioritized, controls are implemented to mitigate risks as appropriate. Many of these controls will involve the use of inspection and testing activities. An inspection and test plan is a document that describes all of the tests and inspections that are required in a process to ensure that products or services conform to requirements, as in Figure 7.7. An inspection and test plan describes when, what, and how to check, following the sequence of steps in the process, as well as what specific requirements or objectives to look for and the criteria for acceptance. The inspection and test plan also identifies the specific people, by roles, that are responsible for signing off at each control point and where the results of each inspection or test are to be recorded.

While the contents of an inspection and test plan are similar to a control plan, and many companies may use the terms interchangeably, the key difference is

Inspection and Test				
Fabricate Cover				
Process Step	Method	Acceptance Criteria	Record	Responsible to complete
Cut material	Measure CTQ dimensions of piece with caliper	Meets dimensions specified in drawings	FRM-10-0042	Laser cutting operator
Bend material	Measure CTQ dimensions of piece with caliper and digital protractor	Meets dimensions specified in drawings	FRM-10-0043	Press brake operator

Figure 7.7 Inspection and test plan.

that a control plan is focused on identifying the process characteristics that must be controlled, while the inspection and test plan is a type of work instruction that places these controls in sequence. One control point on an inspection and test plan may address multiple process characteristics called out on a control plan. Also note that an inspection and test plan is not the same as a checklist. A checklist is a form that is used to record details and results of an actual inspection or test, and a completed checklist is maintained as a record. The inspection and test plan refers to the specific, separate checklist that is to be used for each control point.

MITIGATION EFFECTIVENESS

> Verify the effectiveness of the control plan and improve, if necessary, using continuous improvement methods such as plan-do-check-act (PDCA), lean, and product auditing tools. (Create)
>
> **Body of Knowledge II.B.3**

Once all the mitigation controls have been implemented, the effectiveness of those controls must be monitored and improved (if necessary) using continuous improvement methods. The effectiveness of the mitigation controls must be evaluated and communicated to stakeholders.

Risk Monitoring

Risk management should be an ongoing part of the quality management process. A mechanism to review or monitor events should be implemented. The output of the risk management process should be reviewed to take into account new knowledge and experience. Once a quality risk management process has been initiated, that process should continue to be utilized for events that might impact the original quality risk management decision, whether these events are planned (e.g., results of product review, inspections, audits, change control) or unplanned (e.g., root cause from failure investigations, recall). The frequency of any review should be based on the level of risk. Risk review might include reconsideration of previous risk acceptance decisions.

As part of the risk management process, risks and controls should be monitored and reviewed on a regular basis to verify the following:

- Assumptions about risks remain valid
- Assumptions on which the risk assessment is based, including the external and internal context, remain valid
- Expected results are being achieved
- Results of risk assessment are in line with actual experience

- Risk assessment techniques are being properly applied
- Risk treatments are effective

Accountability for monitoring and performing reviews should be established as part of a formal procedure. The risk assessment process will highlight context and other factors that might be expected to vary over time and that could change or invalidate the risk assessment. These factors should be specifically identified for ongoing monitoring and review so that the risk assessment can be updated when necessary. Data to be monitored in order to refine the risk assessment should also be identified and collected. In addition, the effectiveness of controls should be monitored and documented in order to provide data for use in risk analysis. Accountabilities for creating and reviewing the evidence and documentation should be defined. Progress in implementing risk treatment plans provides a performance measure. The results can be incorporated into the organization's overall performance management, measurement, and external and internal reporting activities.

Lean and Risk Monitoring

While lean principles may be incorporated into risk mitigation approaches, these principles may also be useful in monitoring the effectiveness of any implemented actions. For instance, visual factory concepts could be used such that it is easy to see and recognize when something has gone wrong so that the situation can be addressed and corrected. Similarly, poke-yoke techniques may be implemented to control or eliminate a cause of risk, so any difficulty in performing the poke-yoked process is in itself an immediate, obvious indication that the process is not performing and that the situation needs to be reviewed and addressed before continuing.

Verification of Effectiveness

Once implemented, controls to mitigate risk must be reviewed to ensure that they are effective and that they have not had a negative effect on another part of the process. Verification can be considered to be the Check part of the plan-do-check-act (PDCA) cycle. Where previously, a team identified and evaluated a risk, created a plan to control the risk (Plan), then implemented the plan (Do), now the team must evaluate the effectiveness of the implemented actions and controls (Check) and then adjust the team's approach based on the result (Act).

Verification Audits

The most common methods for evaluating effectiveness of a control are process audits and product audits. Process audits for verification of controls would be conducted as previously described in this chapter. Product audits are conducted on actual output from the process being controlled. Product audits may involve verification that the product conforms to stated requirements as well as validation that the product performs as intended. These types of audits should also be completed when there has been a change to the form, fit, or function of the product (Russell, 2010).

The stages of a product audit are similar to those of a process audit. Due consideration should be given to the sampling plan and method; how does the auditor plan to select samples for the verification and/or validation activities (Russell, 2010)? If auditing a supplier, the customer must specify the sampling plan and method and ensure that the supplier provides appropriate samples for auditing.

The auditor must have access to the requirements or criteria for the product, which may include specifications, drawings, testing requirements and criteria, and any other requirements stated in contracts or regulations (Russell, 2010). If auditing product from a supplier, the auditor should obtain these requirements and criteria from the supplier to verify that the supplier is using the same set of requirements that the customer provided. Product is evaluated against the requirements, and the results are communicated through the audit report.

Part III

Supplier Selection and Part Qualification

Chapter 8
Product/Service Requirements Definition

INTERNAL DESIGN REVIEWS

> Identify and apply common elements of the design review process, including roles and responsibilities of the participants. (Apply)
>
> Body of Knowledge III.A.1

Organizations offer products and services as the basis of doing business. The features associated with the offered products and services help the organization to sustain itself and compete in the market.

The design phase of products and services is usually conducted by a cross-functional team of experts from marketing, supply chain, quality, regulatory, engineering, production, manufacturing, finance, and related departments. Customers and suppliers can and should be included in the design phase. Concurrent activities by all these stakeholders make the design process robust and rigorous.

In overall product lifecycle, the design process is paramount. Product development is essentially a series of interdependent and frequently overlapping activities that transform an idea into a prototype and eventually a marketable product or service. The design stage is also an appropriate point for controlling the cost of making a product. Cost control during the design phase may result in either competitive or noncompetitive product to realize the full profit potential.

There are several types of new product and service design, including new products or services or adaptations or expanded features of a previous design. Technology advancement, process optimization, and new market requirements are the main drivers in new product design.

Design and Development Cycle

ISO 9001:2015 lays out a framework for determining the requirements for products and services. The framework requires that an organization define its products and services, comply with applicable statutory and regulatory requirements, and ensure it can meet the claims it makes for the products and services it offers. The organization is required to establish, implement, and maintain a design and development process that ensures the subsequent provision of products and services.

Design and Development Planning

In determining the stages and controls for design and development, the organization is required to consider the nature, duration, and complexity of the design and development activities; the required process stages, including applicable design and development reviews; the required design and development verification and validation activities; the responsibilities and authorities involved in the design and development process; the internal and external resource needs for the design and development of products and services; the need to control interfaces between persons involved in the design and development process; the need for involvement of customers and users in the design and development process; the requirements for subsequent provision of products and services; the level of control expected for the design and development process by customers and other relevant interested parties; and the documented information needed to demonstrate that design and development requirements have been met.

Design and Development Inputs

The organization is required to determine the requirements essential for the specific types of products and services to be designed and developed. The organization shall consider functional and performance requirements, information derived from previous similar design and development activities, statutory and regulatory requirements, standards or codes of practice that the organization has committed to implementing, and potential consequences of failure due to the nature of the products and services. Inputs must be adequate for design and development purposes, complete, and unambiguous, and conflicting design and development inputs must be resolved.

Design and Development Controls

The organization is required to apply controls to the design and development process to ensure that the results to be achieved are defined, reviews are conducted to evaluate the ability of the results of design and development to meet requirements, verification activities are conducted to ensure that the design and development outputs meet the input requirements, validation activities are conducted to ensure that the resulting products and services meet the requirements for the specified application or intended use, and any necessary actions are taken on problems determined during reviews or verification and validation activities.

Design and Development Outputs

The organization must ensure that design and development outputs meet the input requirements; are adequate for the subsequent processes for the provision of products and services; include or reference monitoring and measuring requirements and acceptance criteria, as appropriate; and specify the characteristics of the products and services that are essential for their intended purpose and their safe and proper provision.

Design and Development Changes

The organization must identify, review, and control changes made during or after the design and development of products and services to the extent necessary to ensure that there is no adverse impact on conformity to requirements. The organization must retain documented information on design and development changes, the results of reviews, the authorization of changes, and the actions taken to prevent adverse impacts.

Design Reviews

Design reviews are systematic reviews of design and development projects. Formal design reviews are conducted at various stages, sometimes referred to as phases. Figure 8.1 provides an overview of the various design and development phases.

The Project Leader creates the design and development plan and schedules regular project meetings, including project design and phase reviews, to facilitate the progress of the project activities in accordance with the project plan and to implement the requirements during product development.

The Project Team completes assigned design and development deliverables and ensures project requirements are adequately fulfilled.

The Design Review Team is a cross-functional team that reviews the project at various design stages or phases to assess if the design and development will meet requirements and to identify and initiate any changes that may be necessary. The Design Review Team is most effective when comprised of individuals representing the various departments within the organization. Additionally, the Design Review Team may be supplemented by the customer (in the case of a contracted design) and subject matter experts.

Phase 0 – Concept/Feasibility

Research may be performed without a formal project plan. Proof of concept and feasibility (research and development) addresses the need for a new product,

Figure 8.1 Design Review Phases.

service, or process and can be based on customer requests, market conditions, new product, service, or process ideas, new equipment, or other situations when the detailed requirements are not already established or defined by customers or by other interested parties.

Phase 1 – Project Planning

The Project Leader, with the assistance of the Project Team, creates the design and development plan at the start of the project. The plan includes documentation of the goals and objectives of the project; identification of tasks, resources, individual responsibilities, and organizational responsibilities; a schedule to demonstrate conformance to time constraints; and identification of major reviews, decision points, and approvals as a project evolves. Phase 1 – Project Planning identifies the plan for the development of products and services.

User needs are documented and analyzed for the product or service and are translated into design inputs. Various studies may be performed based on the complexity of the project. Design inputs evaluate the defined user needs and translate them into quantifiable design inputs that can be used for Phase 2 – Design and Development and Phase 3 – Verification and Validation.

Phase 2 – Design and Development

The main design and development effort is usually completed in Phase 2 of the product development cycle. Near the end of Phase 2 – Design and Development, the product development team reviews all proposed changes that affect safety, risk, and efficacy of the product/service.

Phase 3 – Verification and Validation

Design verification and testing is conducted according to internally approved protocols to demonstrate that design input requirements have been met. Design verification and validation testing normally occur after a design freeze—the point where no new design and development activities occur.

Design validation considers usability issues with the design considering user needs and intended uses. Simulated use testing should be commensurate with the risk involved from actual use of the product. Design verification and validation documents may be in the form of reports and other documents as appropriate.

Phase 4 – Design Transfer

Design transfer is the collective set of activities conducted to transfer the product to production. The transfer activities ensure that the functional specifications are properly transferred into production specifications. Transfer activities help confirm that the manufacturing process is repeatable and produces units that are safe and effective for its intended use. The design of the production process can be largely concurrent with the design of the product.

Phase 5 – Post Market Surveillance

The post market surveillance phase consists of the continued monitoring and maintenance of the risk management files. It is expected that the risk files are reviewed annually or when significant events or trends indicate that the risk profile/assumptions may have changed or that new, previously unidentified risks have emerged.

IDENTIFYING REQUIREMENTS

> Identify and apply internal requirements (e.g., interrelated functional business units) for product or service in collaboration with stakeholders, including the requirements for supply chain, sub-tier suppliers, and manufacturability evaluation. (Evaluate)
>
> **Body of Knowledge III.A.2**

The marketing function is primarily responsible for defining new products and services. This involves market research to determine the actual functional requirements. These requirements are balanced against technological and organizational provisions, innovation, creativity, and cost. Modern methods of defining new products and services include customer involvement during the

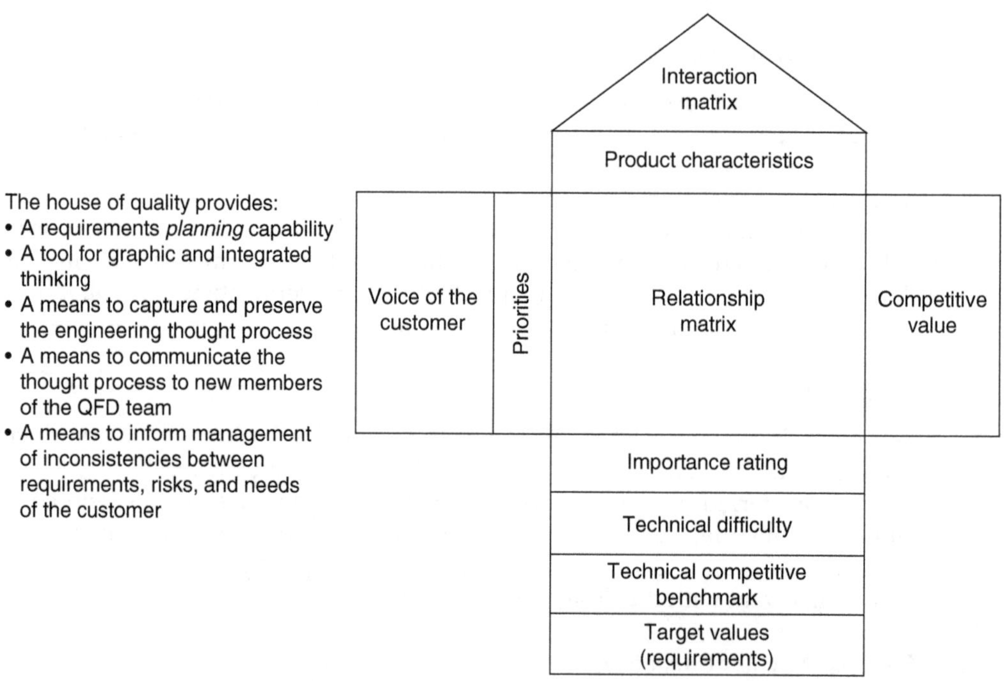

Figure 8.2 House of quality template and benefits.

Source: Adapted from ASQ, http://asq.org/img/laq/house-of-quality-figure-lg.gif (used with permission).

concept phase of the product. Quality function deployment (QFD) and Advanced Product Quality Planning (APQP) are examples of such methods. QFD, otherwise known as house of quality, is a focused methodology for carefully listening to the voice of the customer and then effectively responding to those needs and expectations (see Figure 8.2). APQP is a quality process used for developing new products that uses up-front quality planning and evaluates the output to determine whether customers are satisfied (see Figure 8.3).

Another important step is to determine which technology and supplier should be selected for an earlier and effective involvement. Organizations seek commitment from suppliers to review the design before it is finalized. Managing supplier involvement in the early stages of product development is a challenging and complex task. It requires a comprehensive understanding of supplier capabilities, design expertise, technology risk management, and expected performance against the business plans and projected revenues.

Early Supplier Involvement

Organizations may decide to work with selected outstanding suppliers to develop technology, product, and parts strategically.

Organizations commonly have processes in place such as registration and prequalification of suppliers and the use of RFIs to get to know available suppliers. Information is gathered to determine the supplier's ability, capability, and

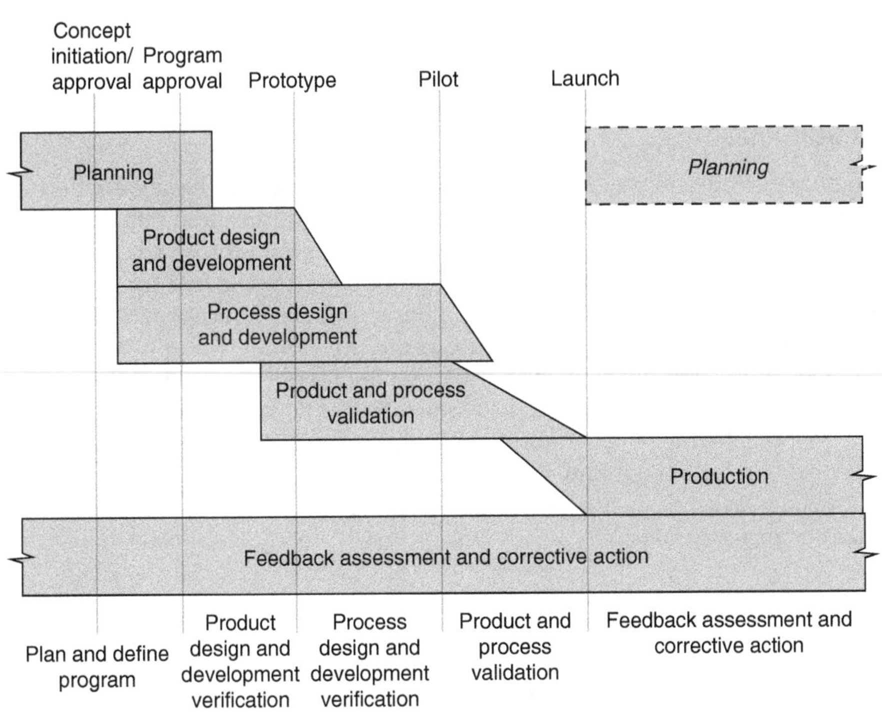

Figure 8.3 APQP process.

Source: Adapted from AIAG, *Advanced Product Quality Planning Manual*, 2nd ed. (2008).

potential to meet the functional, technological, quality, and cost requirements for the new products. Early supplier involvement (ESI) reduces the start-up difficulties of matching the design, process, and delivery compatibilities.

Development Phase

As proof of concept, prototypes are built to demonstrate the design of new products. Prototyping can be used as an iterative process to demonstrate a product as well as to develop the documentation, engineering design and drawings, and list of materials. Subsequently, design reviews are conducted to check for compliance with specification and standards. Computer-aided design and manufacturing (CAD/CAM) simulations provide myriad choices for reviewing designs and prototypes in a short time, saving resources and providing the opportunity to decide among multiple alternatives.

Qualification testing further substantiates the product performance. Failure testing can be used to determine tolerance, robustness, and reliability aspects of the product.

In order to optimize the design and associated costs, value analysis and engineering techniques are used. What-if scenarios are studied to ascertain whether there is room for further optimization before the product enters the large-scale production phase. Design review, supplemented by value analysis and value engineering, ensures that all the parts function as expected and work effectively and efficiently with other parts. Quality and cost are considered while conducting these analyses.

Checklists are developed and standardized to conduct value analysis and value engineering.

Production Phase

Manufacturing and production plans are developed by the production department in close coordination with the marketing and sales and supply chain functions. Based on the available capacity, a bill of material is provided to procurement to acquire the raw material. Obviously, other departments and functions also provide inputs to the production plan spanning the sales forecast, workforce requirements, financial resources, warehouse space, etc.

Due to ESI, supply chain and procurement identify and select the appropriate suppliers and obtain supply plans along with the contingencies.

Manufacturing and production acquire requisite knowledge from similar previous undertakings and their involvement in the design reviews. A specific knowledge base is developed through training, machine and plant operation, and maintenance manuals.

Part and Process Control

Managing the production process effectively and efficiently is a major challenge. If a flaw in the design is revealed, this could be detrimental and highly costly to the organization. In order to avoid such risks, a rigorous design review is conducted that involves production. Supplier part failure is usually treated separately from the design review process. Nevertheless, in the production process, there is always a focus on process improvement plans.

Product Lifecycle

Contemporary organizations are impacted significantly by very strong competition. Rapid technological developments, process improvements, and new market requirements compel organizations to introduce new products such as new models, new versions, and upgrades. Organizations are thus engaged in offering innovative, alternative, and enhanced products. Customer awareness and higher expectations also drive organizations to enhance features and optimize products.

Organizations thus are required to manage the product lifecycle for their product portfolios. Organizations do this using cross-functional teams composed of their functional experts complemented by supplier support to ensure the supply matches the product lifecycle road maps.

As shown in Figure 8.4, the product lifecycle consists of development, introduction, growth, maturity, saturation, decline, and cancellation phases. Organizations adopt a different set of strategies in each phase for marketing, distribution, and supply chain functions.

ESI takes place during the design/development phase. Contracts are furnished as the introduction phase begins and the production cycle starts. Raw material requirements vis-à-vis supplies ramp up during the growth phase and become consistent during the maturity phase. In the maturity phase, orders are commonly consistent since demand becomes predictable and organizations may pursue strategies to procure materials for production and deliver to the market to

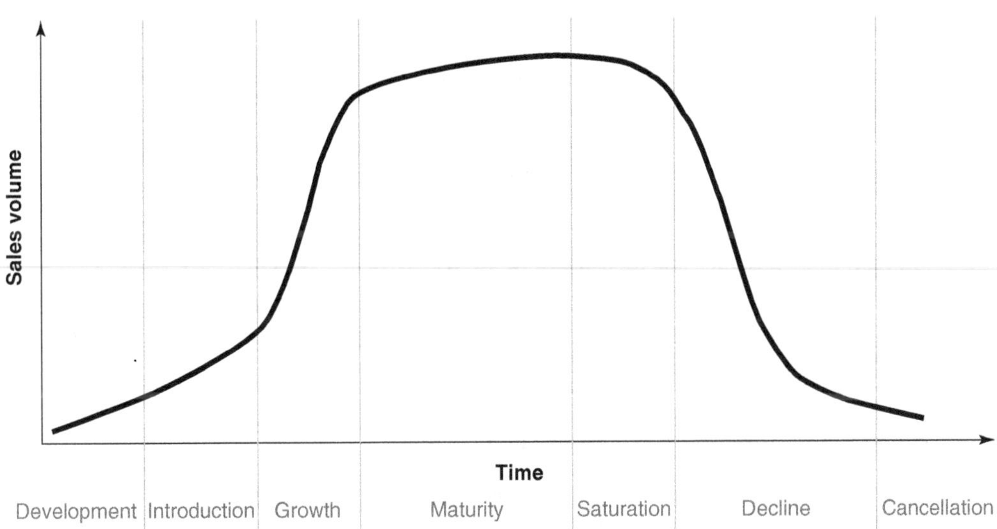

Figure 8.4 Product lifecycle.

Source: Adapted from Aron Chibba, "Measuring Supply Chain Performance Measures" (licentiate thesis, Luleå University of Technology, 2007).

achieve an optimum efficiency point. Organizations strive to lengthen this phase and adopt cross-selling and upselling activities to earn maximum revenues. This involves enhancing features, updating versions, and developing upgrades and alternative products.

During the growth phase, suppliers are developed and certified. This helps in smoothing and optimizing the procurement cycle for replenishment during the maturity phase. During the maturity phase, organizations maintain the approved supplier and qualified product lists for their production requirements.

When an organization is unable to create demand and product sales start declining, it retires the product and revokes its support.

Industry and technology lifecycles also impact the product lifecycle and may lengthen or shorten the life of the product. Supplier support is managed accordingly.

Finalizing Requirements

As the new product development process progresses from the design phase to the final product requirements, a statement of needs, desires, and objectives is developed. Needs are based on marketing perceptions, knowledge of what customers want, or direct input from customers. These needs are assessed for compatibility with the organization's objectives and resources. An organization's objectives include profit potential and sales volume while resources include personnel, machines, and management. Product objectives, including performance, price, quality, and market availability, are established and become the criteria that guide subsequent design, planning, and decision making.

ESI and information on cost, performance, market availability, quality, and reliability are used in finalizing the requirements for the new product. It is also important to establish a realistic target cost along with a selling price and target profit.

Finalizing product requirements is essentially conducted as a collaborative process whenever the cost of doing so is justified. Through collaboration between various departmental representatives and key suppliers, the final requirements and specifications can reconcile the goals that often conflict with each other. Performance goals such as quality and delivery should be balanced against cost.

Individual department goals as well as supplier goals should be considered. This balancing process is best done in an atmosphere of collaboration and mutual desire to develop and finalize requirements where win–win opportunities are maximized. The objective in collaboratively developing the requirements is to simultaneously meet quality, delivery, and cost goals.

Organizational Requirement Planning

Organizations have processes for planning their material and item requirements. These are known as material requirements planning (MRP) and manufacturing resource planning (MRP II). The materials management department carefully analyzes a forecast and rationalizes it according to material availability in warehouses. Historical trends are observed before finalizing the requirements. Cross-functional teams are engaged to finalize the requirements, which are then sent to the procurement, purchasing, or contracting department for purchasing. For purchasing of strategic items that have high risk and criticality, requirements are prepared in a comprehensive document known as an RFP.

Depending on the strategy, an organization decides whether to produce or provide the product or service or to outsource it. This strategic decision is supplemented by information such as total cost of ownership (TCO), taking into account the organization's area of core expertise and business or operations model.

Procurement Purchase Function and Scope

Procurement or purchasing control should be applied for all materials, components, parts, and services to be purchased, procured, or outsourced and then incorporated into the organization's final product. Such items are directly related to the organization's product or service. This control is applied not only to ensure good final product or service quality but also to ensure the quality of the organization's own procurement function.

Procurement control consists of the activities or processes for defining product or service requirements, evaluating or assessing suppliers, maintaining approved supplier lists, controlling order placement, accepting ordered items, and change management. It may be applied to equipment, parts, and materials to be used for the organization's own purposes as well as for manufacture of customer items or delivery of a service to a customer.

The purchasing of equipment to be used for a production or service operation process should be controlled through a predetermined method such as an equipment and materials control procedure. The purchase of office supplies should be controlled separately. Purchasing should be based on total value rather than just initial price. A supplier with the lowest quoted price may not present the lowest cost to the customer if defect costs and loss of customer goodwill are considered.

The requirement and needs process may include the following steps:

- The customer organization defines requirements
- The customer organization requests quotations from potential suppliers
- The suppliers review the requirements and submit offers or solutions to the customer organization
- Customer organization personnel review the offers and start negotiations
- A supplier is selected
- A contract, purchase order, or other agreement is issued to the supplier
- The supplier accepts terms and conditions
- The supplier and customer organization monitor compliance with terms and conditions
- The supplier reports any changes and their effects on the customer organization
- The customer organization renegotiates as needed to meet objectives

The customer organization should document the process for establishing a relationship with a supplier. The documentation should include required records. Establishing an agreement forms the basis for making changes through amendments or adjusting to market changes. Requirements for traceability and proper authorizations should be defined.

The organization must identify all the requirements related to the product or service being purchased. Some requirements may be critical to the form, fit, or function of a product. Requirements may be characteristic or performance based or both. Requirements may relate to the supplier's belief that process thinking leads to a product or service that meets specified requirements. There may also be statutory and regulatory requirements or additional requirements that are not stated but are known and considered necessary by the supplier. Table 8.1 provides examples of typical requirements.

Requirements and needs will vary based on the nature of the product or service. Requirements must be measurable, and the unit of measurement should be specified (e.g., response time in minutes or weight in kilograms).

For contracts requiring supplier engineering or design activities, determination of the requirements should be initiated through fact finding, question-and-answer sessions, or information exchanges with stakeholders or other interested parties. Data gathering practices include interviews, questionnaires, user observation, workshops, brainstorming, use cases, role playing, and prototyping.

If a supplier receives an order verbally and does not have the order requirements documented, the supplier may get into trouble at a later date by not meeting expectations. The supplier should confirm the order requirements before acceptance by documenting the contract terms and specifications in some appropriate manner with a signature and submitting it for confirmation. The descriptions should be clearly referenced when ordering.

Purchasing information specifies many aspects of the products and services to be delivered by the supplier, including:

- Description of the product
- Quantity
- Delivery due dates
- Agreed-on cost or price
- Approval of product, procedures, processes, and equipment
- Qualification of personnel
- Management systems such as those for quality, safety, environmental, and risk management
- Relevant drawings and process data (technical data) and revision level
- Requirements for test specimens
- Record retention
- Supplier flow-down requirements
- Notification of changes in management or management systems

Table 8.1 Product/service requirement examples.

Characteristic/ performance	• Dimensions • Weight • Height • Tolerances • Activity • Strength • Biological characteristics • Hardness	• Viscosity • Solubility • Revolutions per minute (RPM) • Speed • Size • Fatigue • Response time
Storage/delivery/ distribution	• Conditions • Turnover • Refrigerated truck • Mode of delivery • Delivery time • Maintenance in storage	• Shelf life • Shipping requirements • Traceability • Installation • Service • Permits
Documentation	• Instructions • Disposal • Identification • Bar code	• Drawings • Technical data • Records retention • Origin
System/process	• Process capability • Management system standards	• Training • Design • Qualifications of personnel
Packaging/labeling	• Placement • Size	• Contents
Safety/environment	• Hazardous vapors • Water or air pollution • Disposal	• Warnings • Protective equipment • Emissions
Testing/inspection	• First article • Source inspection • Sampling plan	• Test methods and equipment • Certificate of compliance/ conformance

- Right of access
- Notification of nonconformances
- Shipping, labeling, and packaging requirements
- Product shelf life and age control requirements
- FAI
- Source inspection

- Certificate of conformance, analysis, or compliance requirements
- Invoice and payment terms
- Shipping requirements, including imports, exports, and customs
- ITAR and Export Administration Regulations requirements
- Identification and traceability requirements for possible recall and investigation

These requirements are identified in different ways. Some can be distributed to the supplier on a separate supplier requirements document, which is then referenced in the purchase order, contract, or agreement. Other requirements are stated directly on the purchasing document, which suppliers should review to confirm they can satisfy the stated requirements.

The supplier is not obligated by contract to satisfy the expectations of the customer if they are not properly identified and is not bound by contract to any requirements that are not documented. A supplier may contact the customer for more details or clarification on the purchase order or agreement prior to fulfilling the order. Ambiguous, deficient, or inaccurate supplier flow-down requirements can put both the customer and the supplier at risk.

When an order is placed with a supplier for any product or service, the customer should give the supplier a clearly written specification with a sufficiently detailed description of needs and requirements as well as terms and conditions (the output of the design control program).

It is very important to ensure that the supplier has received, reviewed, and acknowledged or confirmed the requirements. By receiving confirmation, the customer will have a greater degree of confidence that the supplier is capable of fulfilling the order. If the purchasing requirements are not properly managed, this will have a cascading negative effect on delivery, cycle time, quality, and cost throughout the supply chain.

When the production of a new product or process is outsourced, requirements should be validated in accordance with procedures.

Chapter 9
Supplier Selection Planning

The planning process for supplier selection is integral to the assessment of existing and potential suppliers for a new procurement opportunity. The complexity of the new procurement opportunity is a key determinant of the level of planning to be involved and the supplier processes to be evaluated when selecting the best supplier. Typically, the supplier selection process is used for customized processes, products, and services; it is not typically used for standard or catalog type products. Before the supplier selection process begins, the process and the functions involved should be defined to consistency and an objective outcome. In this chapter, various elements of supplier selection are presented to provide a guide for the planning of these activities.

SUPPLIER COMPARISON

> Evaluate existing suppliers' and distributors' capabilities, capacities, past quality, delivery, price, lead times, and responsiveness against identified requirements. (Evaluate)
>
> **Body of Knowledge III.B.1**

Determining current suppliers' (includes manufacturers, distributors, and service suppliers) qualifications to provide goods or services for a new procurement opportunity leads customer organizations to evaluate suppliers against a set of objective criteria encompassing both current and expected performance. Current suppliers need to be evaluated if the new product or service is of a different nature than what they currently provide. The following are areas to consider during supplier selection planning and evaluation of existing suppliers.

Capabilities

Supplier capabilities include facilities, personnel, equipment, QMS, industry specific certifications, and product/process/service qualifications. Manufacturing-based suppliers and distributors have different capability matrices than service-based suppliers, but the assessments are similar. Manufacturing-based suppliers and distributors have process-based capability assessments while service-based suppliers have competency-based assessments.

Criteria for assessing capabilities include determining whether the existing equipment is appropriate, whether test or equipment validations are commensurate, and whether the supplier has the flexibility to provide sufficient staffing to support the new opportunity. Process capability (e.g., Cpk, or Cp) or process performance (e.g., Ppk and Pp) indices can provide an indication of the ability of a stable process to generate output within specification limits for a particular characteristic or feature that is key or critical to quality. Process capability studies are considered an effective technique for identifying processes capable of producing quality parts.

It is not always required that a customer audit a supplier's capability, especially when third-party certifications (ISO 9001, ISO 13485, ISO 15378, ISO 17025, AS9100, IATF 16949 etc.) or process certifications (American Welding Society [AWS], National Aerospace and Defense Contractors Accreditation Program [NADCAP], MedAccred etc.) are available from the supplier. Third-party certifications are not intended to be the sole determinant of capabilities; rather, they are used to support the supplier's capability and quality history. It should be noted that even if the supplier is certified, there is no guarantee of compliance to the requirements of the new opportunity.

Capacities

Evaluation of an existing supplier's ability to deliver a new offering (process, product or service) should consider current capacity.

> Can the existing equipment or personnel handle an increase in demand or output? Can the supplier deliver in the time frame required?

Determining the impact of increasing demand on the existing supplier is critical so as to avoid disrupting the current supply chain.

> Will the supplier need to add shifts, hire new personnel, or invest in new equipment?

Key to the capacity evaluation process is the risk of adding new products or services to the current products and services procured from the existing supplier.

Quality

An existing supplier's Quality Management System (QMS) and current quality performance are factors for consideration and comparison in the supplier selection process. Third-party certifications (e.g., ISO 9001, ISO 13485, AS9100, IATF 16949, etc.) are an indication of the supplier's commitment to quality and continuous improvement as well as its integration of industry specific quality system requirements into its QMS. If the current supplier has a certified QMS, the customer organization should consider the requirements of the certification standard (e.g., ISO 9001, ISO 13485, AS9100, IATF 16949, etc.) as it relates to the scope of the new process, product, or service opportunity. Does the new opportunity have industry specific requirements that may not be embedded in the supplier's QMS? For example, an ISO 9001 general quality management system may not address industry specific requirements noted in ISO 13485 (medical devices) or IATF 16949 (automotive). Without a formal certification, the supplier may not be considered for the new product or service; this is also known as a barrier to entry into the supplier selection process.

Historical supplier data, including quality and delivery performance, are critical to supplier performance oversight. Evaluation of existing suppliers includes a thorough review of product and/or process performance including, but not limited to, first-pass acceptance, production and/or installation performance, field performance, waivers/deviations/concessions, recalls, warranty work, and returns. Not all performance criteria are equal; each customer organization must evaluate the impact of the different criteria on the product or service to be procured.

Delivery

Customer organizations and consumers are increasing their focus on delivery—more than just on-time delivery, the focus is on the right thing being delivered at the right time at the right price. Evaluations of supplier delivery include method or type of shipping, location of services to be provided, and any special packaging needed. Additional consideration should be given to existing suppliers that are integrated with the shipping requirements of the supply chain.

Price

When purchasing commodity products, a low-cost provider is often the best solution. However, supplier selection based on price alone can lead to poor quality, delivery, and service. When evaluating an existing supplier, attention should be paid to the total cost of procurement, taking into consideration volume discounts, which can reduce the cost of other items procured from the supplier. The total cost of procurement includes the price of the item, shipping, handling, restocking fees, inventory handling or storage costs, minimum order quantities, and other nontraditional costs associated with procurement. A common supplier selection method is to create a "best value" matrix that compares the advantages of working with a high-performing supplier with the cost of working with a poor-performing supplier.

In a best value system, a supplier is selected through a process of researching suppliers before the procurement is made. Typically, values are assigned to factors such as price, past performance, and schedule. These values are tabulated for each potential supplier, and one will come out on top. This system can be beneficial because it requires less decision making, prepares the customer organization for the future, and minimizes risk. The best value selection system has been implemented by various governments as a means to ensure an objective procurement and selection process. One of the most important aspects of best value procurement is looking at past performance. For example, if a customer is looking to significantly increase volume or shorten delivery time, it is important to see if potential suppliers have performed on such a scale. The best value selection process prevents customer organizations from selecting suppliers based solely on a low price.

Table 9.1 shows an approach to rating a supplier for best value. Each category has a defined rating scale and a prescribed weight for importance. In Table 9.1, "weight value" is the importance of the element in the potential procurement. The values are ranked using a 1-3-5 scale. For the delivery, capacity, capability, and quality categories, the rankings are defined as:

1 = supplier's system does not meet the evaluation criteria

3 = supplier's system meets some of the evaluation criteria

5 = supplier's system meets most if not all of the evaluation criteria

In the case of price, the rankings are defined as:

1 = highest price

3 = medium price

5 = low prices

In this example, supplier C would provide the best value based on the scores for each of the categories.

Lead Time

Lead time is defined as the time between the initiation and completion of a process. Consistency in service delivery or manufacturing cycle/lead time, including preparation, queuing, run time, and shipping/service delivery, is important when evaluating existing suppliers. Increase in volume or output can impact current orders and deliveries or lead to costly expedited shipping in order to fulfill contract requirements.

Just-in-time (JIT) manufacturers need to factor lead time into the supplier selection decision to determine whether an existing supplier will continue to meet

Table 9.1 Best value ranking approach.

	Criteria										
	Price		Delivery		Capacity		Capability		Quality		Best value rating
Supplier	P	Weight value	D	Weight value	Ca	Weight value	Cb	Weight value	Q	Weight value	Total
A	1	1	1	0.75	3	0.5	1	1.25	3	1.5	9
B	1	1	5	0.75	3	0.5	1	1.25	5	1.5	15
C	3	1	3	0.75	5	0.5	5	1.25	5	1.5	21
D	5	1	3	0.75	3	0.5	3	1.25	3	1.5	17

the push triggers set up under existing contracts. In the service sector, lead time can indicate the amount of time a supplier commits to deploying or completing the service. Where lead time is a critical component in the selection, it can be included in the best value rating process (described in the previous section).

Responsiveness to Identified Requirements

Responsiveness to identified requirements includes the technical competency of a supplier to meet the customer's identified requirements. Customer organizations need to choose a method for getting feedback from the supplier on their understanding of technical requirements. Submission of manufacturing documentation such as batch records could be used to assess this. Verification and qualification of processes, products, and services are other ways of assessing responsiveness.

Assessing existing suppliers' ability to incorporate technical changes in a timely manner and to perform concurrent engineering or process qualification can also provide assurances that the existing supplier will be responsive to requirements. Another approach to responsiveness is a commitment to a time frame for responding to inquiries or incorporating requirement changes. This is usually used in customized designs or tailored service procurements/contracts.

POTENTIAL SUPPLIER EVALUATION

> Assess potential new suppliers against identified requirements using tools such as self-assessments, audits, financial analysis, and quality function deployment. Verify third-party certification status and regulatory compliance and analyze and report on results of assessments to support the supplier selection process. (Evaluate)
>
> Body of Knowledge III.B.2

The process for evaluating a potential supplier should be defined and should involve multiple functions or a cross-functional team. Those functions that are stakeholders in the supplier selection process or that will utilize the supplier's product or service should be involved in assembling a set of criteria for evaluation and activities for gathering relevant information to support the evaluation process. The evaluation criteria can be weighted based on the importance of each element of the part, process, or service and its impact on the end product or service. The following are considered typical activities in the evaluation of a potential supplier.

Self-Assessments

A self-assessment is typically a questionnaire sent to a potential supplier to gather information on capability, capacity, staffing, and resources available to support the procurement. The questionnaire can gather information on the following key parameters: organizational structure, infrastructure and facilities, major awarded contracts and active customers, key sub-tier suppliers, production processes, special processes, testing processes, etc. In addition, the self-assessment is used to obtain details on a supplier's QMS. Although a supplier self-assessment is part of a new supplier selection process, it is rarely used as the only element in the selection process. The self-assessment is a tool to determine next steps in the evaluation process or to identify activities or inquiries that will delve deeper into a supplier's infrastructure and ability to support the procurement.

Audits

An audit can be used to determine a potential supplier's preparedness for and potential compliance with the requirements of the procurement. There are several types of audits that can be utilized in the evaluation of a potential supplier. The following are examples of supplier audits:

- *Product/service.* Review of product or service requirements against the potential supplier's approach or documented procedures. Product/service audits can include assessing the impact of the procurement; for example:
  - —Catalogue/commodity parts/off the shelf versus build to print/customized designs and services
  - —Does the procurement include distributors or a direct relationship with the manufacturer?
- *Process.* Review of a particular process unique to the product or service to be provided by the supplier. Process audits provide an opportunity to assess a potential supplier's:
  - —Lead times for product/service delivery or release
  - —Ability and agreement to support end customer or agency audits/assessments
  - —Understanding of and responsiveness to flowed requirements
- *QMS.* Review of a supplier's QMS related to potential procurement or contract requirements. An example of the ranking of a potential supplier's QMS maturity is shown in Table 9.2.
- *Financial.* Review of a potential supplier's compliance with financial requirements (if included in the supplier evaluation criteria) and/or ability to remain financially viable during the procurement

The process of auditing a potential supplier may be the beginning of a customer–supplier relationship. Communicating openly and transparently creates an open environment for information exchange. When scheduling an audit of a potential supplier, the customer organization is highly encouraged to do the following:

- Ensure the supplier has been given the applicable technical and requirements documents for the potential procurement

Table 9.2 QMS maturity grid.

Detection	Criteria	1	2	3	Suggested method	Ranking
Almost impossible	QMS will not and/or cannot detect a potential cause/mechanism and subsequent failure mode.				None/no QS in place.	10
Low	QMS has a poor chance of detecting a potential cause/mechanism and subsequent failure mode.			X	Indirect/infrequent checks only.	8
Moderate	QMS may detect a potential cause/mechanism and subsequent failure mode.		X	X	Visual checks and measurements/gauging components. Periodic charting and tracking.	6
High	QMS has a good chance of detecting a potential cause/mechanism and subsequent failure mode.	X	X		Measurements/gauging or testing/sampling of components during production to prevent potential failures. Continuous charting and tracking.	4
Almost certain	QMS is almost certain to detect a potential cause/mechanism and subsequent failure mode.	X			Continuous monitoring predicts potential failures prior to occurrence.	2

Note: In this approach, a lower score indicates a more mature QMS.

- Send the supplier a copy of the checklist that will be used for the audit
- Communicate the expectations for documents and activities to be reviewed during the audit
- Collaborate on an agenda for the audit

It should be noted that results of an audit of a potential supplier would note the gaps in the supplier's process related to requirements in the new opportunity. These audits do not typically issue corrective action requests as there is formal purchase order or contract in place that flows the requirements to the supplier.

Financial Analysis

A financial analysis of a potential supplier can provide important information on the supplier's financial ability to meet the contractual obligations, such as:

- Financial stability for continued business through the contract phase
- Financial stability to support the manufacturing process and resources

- Elements that can be reviewed in a financial analysis of a potential supplier include:
  - —Supplier's credit rating, business financial profile ratings, business reviews including endorsements and complaints
- Supplier's primary customer and/or industry portfolio
  - —Percentage of business distributed across various customers
  - —Concentration of business in a particular industry or with a single customer
  - —An estimation of a percentage of business this new procurement or contract would represent for the supplier

Quality Function Deployment

Quality Function Deployment (QFD) is a tool that can identify key requirements for the customer organization and determine the degree of alignment with the supplier's product/process/service capabilities, technical capabilities, and quality management system capabilities, as shown in Figure 9.1. The QFD tool can compare the degree to which suppliers can comply with process/product/service requirements for a new or revised process/product or service. Inputs to the QFD can be obtained from the Self-Assessments and/or audits performed at the supplier. The results from the QFD would be analyzed by a cross-functional team to those suppliers that best meet the process/product/service requirements. These results can be technical inputs into a supplier comparison matrix.

Verification of Third-Party Certification Status and Regulatory Compliance

Third-party certifications such as ISO 9001, ISO 13485, ISO 17025, AS9100, IATF 16949) are used by many industries to confirm a QMS structure or compliance with industry regulatory requirements. Suppliers are often required to have certifications to bid on a procurement or contract, provided the certifications are relevant to the product or service to be provided. The status of a supplier's third-party certifications is pertinent. An active certification signifies a supplier's continued commitment to continuous improvement, the certification process, and industry requirements. A certification that has expired or lapsed indicates that the supplier may be exiting that particular industry segment and may not be maintaining the process rigor required by the certification. Third-party certifications can be used as the basis for approval as selection processes or procedures allow.

Analysis of Audit Results to Support the Supplier Selection Process

The results of audits and other supplier-related information should be assembled and compared with the supplier selection criteria to determine the degree to which a potential supplier can meet the requirements.

It is encouraged to review the results of supplier audits and supplier self-assessments with the supplier to ensure the information is accurate and contains sufficient detail with which to support the selection process. After the supplier information is assembled and reviewed for accuracy, the information should be

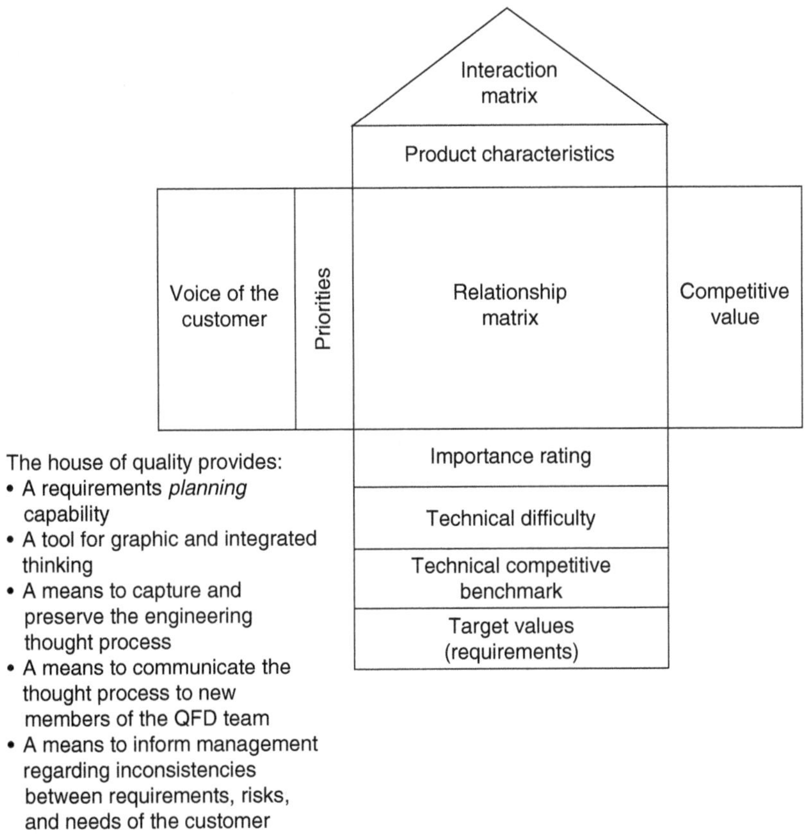

Figure 9.1 House of quality matrix.

Source: https://asq.org/quality-resources/qfd-quality-function-deployment.

reviewed by the functions supporting the supplier selection process. A multifunction review provides the opportunity for varying perspectives to be considered and confirms the alignment of supplier evaluation results to the selection criteria. The criteria for potential supplier evaluations can be used to create a best value supplier rating matrix with modifications to include the evaluation data discussed in this section.

SUPPLIER SELECTION

> Evaluate and select suppliers based on analysis of assessment reports and existing supplier evaluations using decision analysis tools such as weighted decision matrices and selection matrices. (Evaluate)
>
> **Body of Knowledge II.B.3**

The importance of the supplier selection process has been increasing across industries as supply chains have been widening and deepening. Supplier selection criteria must be developed and defined in the form of a procedure, process document, or a work instruction, to ensure consistency in the process over time. It is recommended to develop the selection criteria with a cross-functional team prior to identifying the suppliers to be evaluated. This reduces the likelihood of biases from a particular function or person during the selection process. A cross-functional team brings a variety of perspectives to identify and prioritize supplier related requirements. Items pertinent for supplier selection include, but are not limited to:

- Supplier survey/assessment form
- Consultant survey/assessment form
- Supplier/consultant approval form
- Audit report (on-site, desktop)
- Quality agreement
- Purchase order/quote documentation
- Product batch records
- Service satisfaction ratings
- Performance data/ratings
- Inspection results (service suppliers and manufacturers)
  —Receiving
  —Acceptance
  —Results of data analysis
  —Nonconforming material reports
  —Records of corrections

Table 9.3 displays an approach to calculating the risks of prospective suppliers. The list of potential suppliers should be narrowed down. The total risk factor could be included as an element in the best value rating to strengthen the evaluation criteria.

After completing the evaluation process activities, the customer organization should document the findings (i.e., whether the supplier can meet the requirements). A checklist can be used to ensure all the documentation required by standard operating procedure (SOP) for supplier selection—and, most importantly, the justification for approving a supplier—is complete and contains appropriate levels of detail. In the event a supplier is selected that does not meet the requirements, a justification for using the supplier should be documented. This documentation is a record of the decision and an acceptance of the associated risks.

Supplier selection is a process, not a single task. The current global supply chain environment is competitive, and suppliers should be selected based on

Table 9.3 Rating system example.

Supplier	Criteria															
	Severity (S)			Supplier QS maturity (D)			Financial stress factor (F)			Lead time (LT)			Order capacity (OC)			Total risk factor*
	S	Weight value	Risk	D	Weight value	Risk	F	Weight value	Risk	LT	Weight value	Risk	OC	Weight value	Risk	
A	8	1.5	12	6	0.75	4.5	2	1	2	1	0.5	0.5	3	0.25	0.75	19.75
B	8	1.5	12	4	0.75	3.0	5	1	5	3	0.5	1.5	1	0.25	0.25	21.75
C	8	1.5	12	10	0.75	7.5	5	1	5	5	0.5	2.5	5	0.25	1.25	28.25
D	8	1.5	12	2	0.75	1.5	1	1	1	1	0.5	0.5	1	0.25	0.25	15.25

*The higher the total risk factor, the higher the risk involved in utilizing the supplier.

objective evidence. Each supplier should be evaluated on its ability to meet specified requirements, including quality requirements. The scope and depth of the evaluation and the degree of control will vary based on the potential impact (i.e., risk) the outsourced product or service may have on the customer's own product or service.

Chapter 10
Part, Process, and Service Qualification

Part, process, and service qualification starts with a well-thought-out design based on defined user needs including regulatory requirements as applicable. The output of the design can take the form of engineering drawings, mathematical models, requirements, specifications, compounds, or formulas. Without these prerequisites, qualification activities will be ineffective and may require significant design modifications.

TECHNICAL REVIEW

> Interpret and evaluate technical specification requirements and characteristics such as views, title blocks, dimensioning and tolerancing, and apply GD&T symbols as they relate to the product and process. (Evaluate)
>
> Body of Knowledge III.C.1

Engineering drawings and specifications are outputs of the design and development process. They are necessary to communicate to the supplier what is to be made and, in some cases, how it is to be made.

Engineering drawings are sometimes called "the language of industry." Drawings (and computer files) are an important way of communicating customer requirements so the supplier can provide the products that meet the specifications and expectations. During design and development, the critical to quality characteristics (CTQs) are defined and documented. Engineering drawings are composed of various types of lines, dimensions, and notes that convey the design intent. Figure 10.1 depicts the types of lines typically seen on engineering drawings and their uses.

The lines are drawn in such a way as to convey the final shape of the part to be manufactured. Orthographic projection is a way of drawing a three-dimensional object from different directions, usually a front, side, and top view. There are

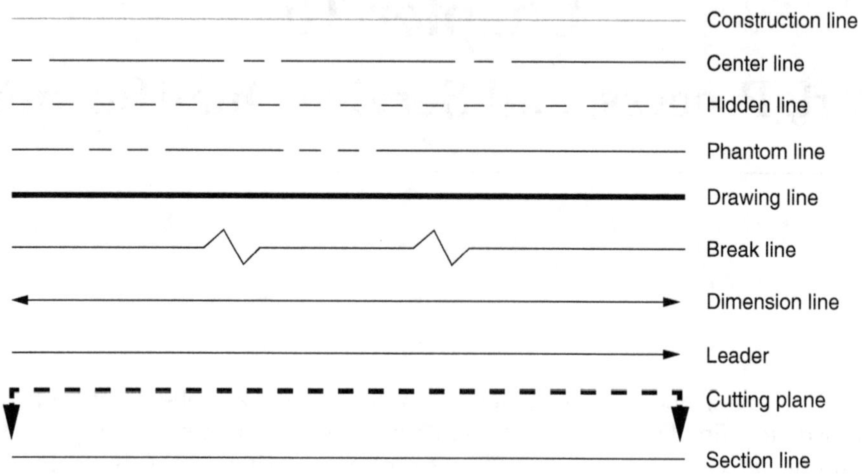

Figure 10.1 Drawing line types.

two primary ways to arrange the views: first angle projection and third angle projection (see Figure 10.2).

These methods differ only in the positions of the top, front, and side views, as shown in Figures 10.3 and 10.4.

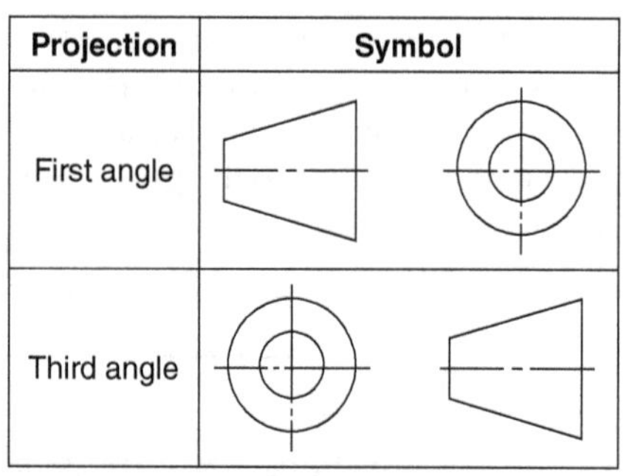

Figure 10.2 First and third angle projections.

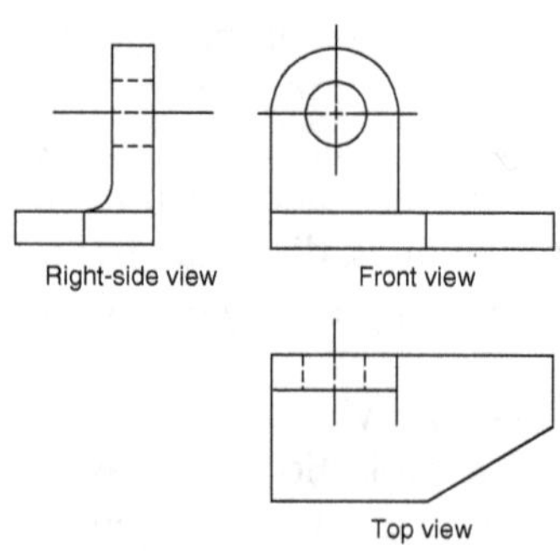

Figure 10.3 First angle projection.

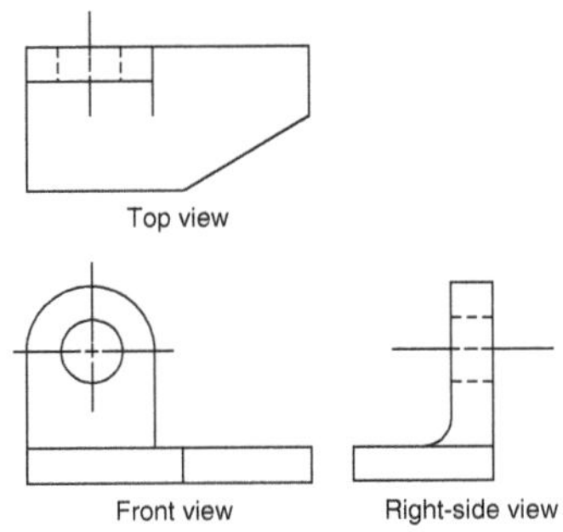

Figure 10.4 Third angle projection.

ISO and ANSI have defined standard drawing sheet sizes, as shown in Table 10.1.

Drawings usually include a title block that contains information necessary to manufacture the part, including drawing scale, material, part number, general tolerances, drawing number, and the initials of the individuals who created, checked, and approved the drawing (see Figure 10.5). Additionally, drawings usually contain a notes listing, which provides other types of information necessary to produce the part, and a change block, which documents the history of changes made to the part and/or drawing.

Table 10.1 ISO and ANSI standard drawing sizes.

ISO

Sheet designation	Sheet width (mm)	Sheet length (mm)
A0	841	1189
A1	594	841
A2	420	594
A3	297	420
A4	210	297

ANSI

Sheet designation	Sheet width (inches)	Sheet length (inches)
A	8.5	11
B	11	17
C	17	22
D	22	34
E	34	44

<table>
<tr><td>SCALE
1:1</td><td>Applicable specs.</td><td colspan="2">Next assembly</td><td colspan="2">Final application</td></tr>
<tr><td colspan="2">Material</td><td colspan="4">Drawing name</td></tr>
<tr><td colspan="2">Part number</td><td colspan="4" rowspan="2"></td></tr>
<tr><td colspan="2">Designer</td></tr>
<tr><td colspan="2" rowspan="4">Unless otherwise specified:
Dim. are in inches
Dim. in parentheses are (mm)
Dec: .XX ± .01 Frac: ±1/64
.XXX ± .005 Ang: ±1°
Remove all burrs and sharp corners</td><td>Date</td><td></td><td>Drawing no.</td><td>Alt.</td></tr>
<tr><td>Chk.</td><td></td><td rowspan="3">Drawing no.</td><td rowspan="3">A</td></tr>
<tr><td>Eng.</td><td></td></tr>
<tr><td>App.</td><td></td></tr>
</table>

Figure 10.5 Title block.

Dimensions are simply the size of an object. They can be linear, circular, or angular. Dimensions usually have tolerances associated with them—an acceptable deviation from the nominal dimension. There are two general types of tolerances on drawings: unilateral tolerance and bilateral tolerance. Unilateral tolerances vary above *or* below the nominal dimension of the feature. Bilateral tolerances vary above *and* below the nominal dimension of the feature. Note that the variance above or below the nominal dimension does not have to be the same. Figure 10.6 depicts examples of unilateral and bilateral tolerances.

Many different types of symbols are used in engineering drawings, including but not limited to welding, surface finish, electrical, architectural, and fluids/

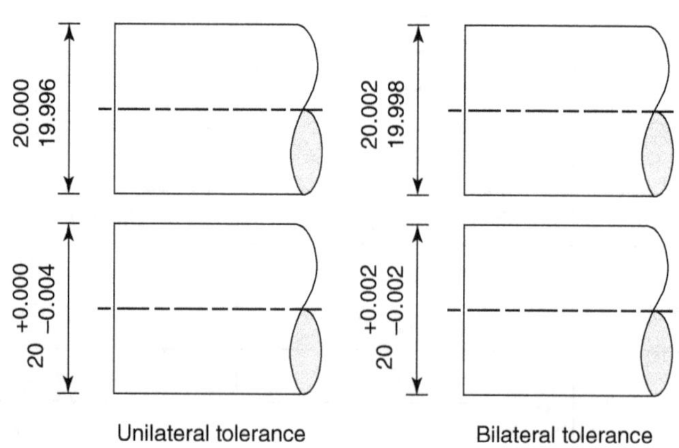

Figure 10.6 Unilateral and bilateral tolerances.

pneumatic/hydraulic symbols. The use of these symbols is necessary to communicate the complexities that can be present in engineering drawings.

Geometric dimensioning and tolerancing (GD&T) is another type of dimensioning. GD&T is a symbolic language used on engineering drawings to define the allowable deviation of feature geometry. GD&T consists of dimensions, tolerances, symbols, definitions, rules, and conventions that can be used to precisely communicate the functional requirements for the location, orientation, size, and form of each feature of the design model. GD&T symbols are shown in Figure 10.7. Geometric characteristics can have modifiers to the form, profile, orientation, location, and runout. Some of the more common modifiers are shown in Figure 10.8.

All of the symbols and modifiers are brought together in a feature control frame. The feature control frame is a rectangular box divided into compartments within which the geometric characteristic symbol, tolerance value, modifiers, and datum references are referenced. Figure 10.9 shows an example of a feature control frame.

In GD&T, Maximum material condition (MMC) is represented by Ⓜ and designates the condition of a feature where the maximum amount of material exists within its dimensional tolerance (see Figure 10.10). The MMC for a hole (internal

Symbol	Characteristic	Category
⏤	Straightness	Form
⏥	Flatness	
○	Circularity	
⌭	Cylindricity	
⌒	Profile of a line	Profile
⌓	Profile of a surface	
∠	Angularity	Orientation
⊥	Perpendicularity	
//	Parallelism	
⌖	Position	Location
◎	Concentricity	
⌯	Symmetry	
↗	Circular runout	Runout
⌰	Total runout	

Figure 10.7 GD&T symbols.

Geometric characteristic	Geometric characteristic symbol
Diameter or radius	Ø or R
Spherical diameter or radius	SØ or SR
Basic dimension	[.85]
Datum feature	[–A–]
Maximum material condition	Ⓜ
Least material condition	Ⓛ
Regardless of feature size	Ⓢ

Figure 10.8 Geometric characteristic modifiers.

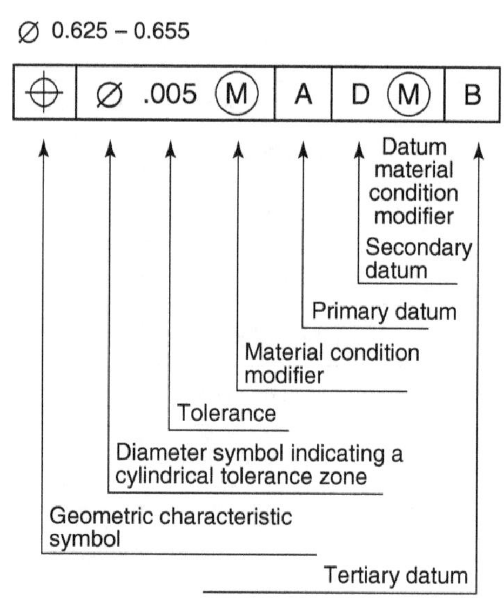

Figure 10.9 Feature control frame.

feature) is the smallest hole size. The MMC for a pin (external feature) is the largest pin size.

In GD&T, Least material condition (LMC) is represented by Ⓛ and designates the condition of a feature where the least amount of material exists within its dimensional tolerance. The LMC for a hole (internal feature) is the largest hole size. The LMC for a pin (external feature) is the smallest pin size. An example of MMC is the worst-case of a part that will still guarantee the assembly between a pin and hole when a hole is at its smallest size and a pin is at its largest diameter.

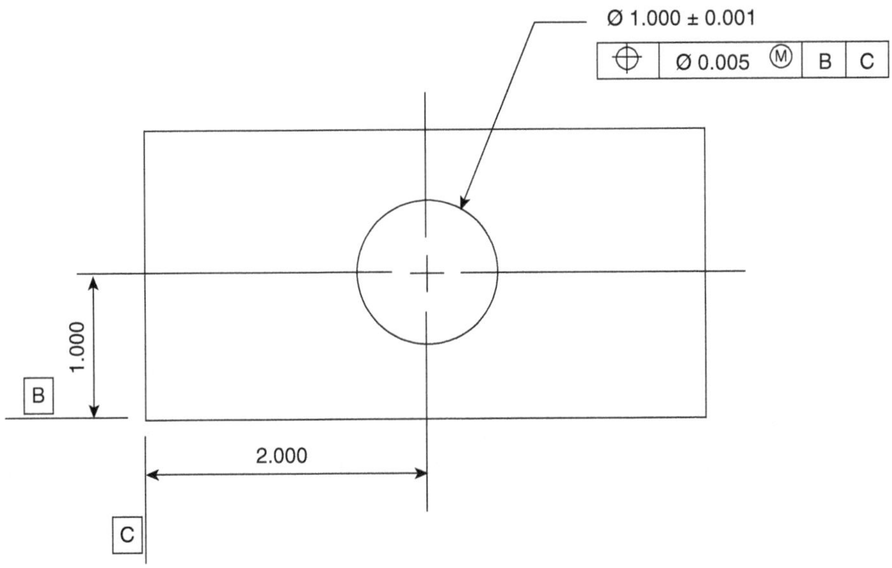

Figure 10.10 MMC location BT example.

Table 10.2 BT and location tolerance for Figure 10.10.

Condition	Ø	BT	Location Tolerance
MMC	0.999	0.000	0.005
Nominal	1.000	0.001	0.006
LMC	1.001	0.002	0.007

In GD&T, Bonus tolerance (BT) is the amount of extra tolerance that a feature is privileged to get when the feature of size departs from MMC to LMC (see Table 10.2).

BT is calculated using the following formula:

$$BT = \text{actual size} - MMC$$

For example, if the hole in Figure 10.10 is at MMC 0.999 there is no BT allowed.

SUPPLIER RELATIONS

Collaborate with suppliers to define, interpret, and classify quality characteristics for the part/process/service. (Evaluate)

Body of Knowledge III.C.2

AIAG has developed and successfully deployed what are known as the five core tools:

1. APQP
2. FMEA
3. SPC
4. Measurement system analysis (MSA)
5. Production part approval process (PPAP)

These tools are used to foster communication between the customer and the supplier. Although developed for the automotive industry, these tools can be readily adapted to virtually any industry, including the service industry. The purpose of the APQP process is to involve suppliers early in the design and development process and to produce a comprehensive quality plan that supports product and service development to meet customer requirements (see Figure 10.11).

The APQP process involves these major elements:

- *Understanding customer needs.* This is done by using voice of the customer techniques to determine customer needs and using QFD to

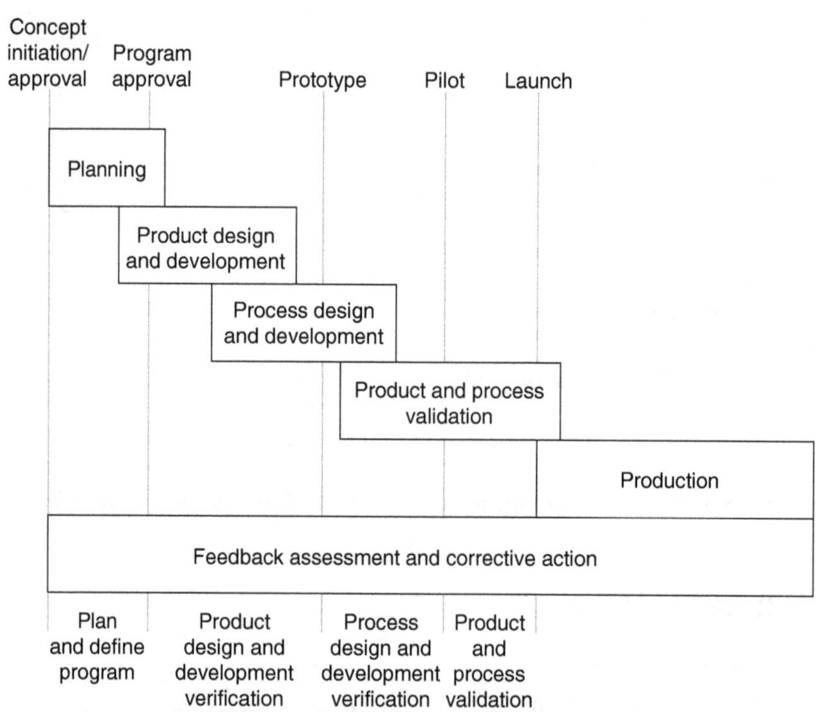

Figure 10.11 APQP process.

Source: Ken Crow, "Advanced Product Quality Planning," NPD Solutions, http://www.npd-solutions.com/apqp.html. Used with permission.

organize those needs and translate them into product characteristics/ requirements.

- *Providing proactive feedback and corrective action.* The advance quality planning process provides feedback from similar projects with the objective of developing countermeasures on the current project. Other mechanisms used with verification and validation, design reviews, analysis of customer feedback, and warranty data also satisfy this objective.
- *Designing within process capabilities.* This objective assumes that the company has brought processes under statistical control, has determined its process capability and has communicated its process capability to its development personnel. Once this is done, development personnel need to formally determine that critical or special characteristics are within the enterprise's process capability or initiate action to improve the process or acquire more capable equipment.
- *Analyzing and mitigating failure modes.* This is done using techniques such as FMEA and anticipatory failure determination.
- *Verification and validation.* Design verification is testing to ensure that the design outputs meet design input requirements. Design verification may include activities such as design reviews, performing alternate calculations, understanding tests and demonstrations, and review of design documents before release. Validation is the process of ensuring that the product conforms to defined user needs, requirements, and/ or specifications under defined operating conditions. Design validation is performed on the final product design with parts that meet design intent. Production validation is performed on the final product design with parts that meet design intent produced with production processes intended for normal production.
- *Design reviews.* Design reviews are formal reviews conducted during the development of a product to ensure that the requirements, concept, product, or process satisfies the requirements of that stage of development and that the issues are understood, the risks are being managed, and there is a good business case for development. Typical design reviews include requirements review, concept/preliminary design review, final design review, and a production readiness/launch review.
- *Control of special/critical characteristics.* Special/critical characteristics are identified through QFD or a similar structured method. Once these characteristics are understood and there is an assessment that the process is capable of meeting these characteristics (and their tolerances), the process must be controlled. A control plan is prepared to indicate how this will be achieved. Control plans provide a written description of systems used in minimizing product and process variation, including equipment, equipment setup, processing, tooling, fixtures, material, preventive maintenance, and methods.

PROCESS AND SERVICE QUALIFICATION PLANNING

> Develop a part/process/service (e.g., calibration, laboratory, software, and design) qualification plan with supplier and internal team that includes service provider audit, calibration requirements, sample size, first article inspection, measurement system analysis (MSA), process flow diagram (PFD), failure mode and effects analysis (FMEA), control plans, critical to quality (CTQ), inspection planning, capability studies, material and performance testing, appearance approval, and internal process validation. (Analyze)
>
> **Body of Knowledge III.C.3**

When the company decides to purchase materials, parts, services, and outsourced processes from a supplier, there are several commonly used tools that can be used to develop the requirements that a supplier must be able to comply. The tools used will vary on the types of materials, parts, services, and outsourced processes provided as well as the risks involved to the customer or end user. For example, a supplier providing machined parts may be required to provide a First Article Inspection (FAI) report whereas a supplier providing customized software bespoke may provide a verification and validation report.

First Article Inspection (FAI)

FAI, also known as first article layout (FAL), is a formal method of providing a measurement report for the part being manufactured. This technique requires measuring the properties and geometry of an initial sample (typically three pieces) against the dimensions and specifications shown on the engineering drawing. Dimensions and specifications to be checked are many and varied and may include distances between edges, positions of holes, diameter and shape of holes, weight, density, stiffness, color, reflectance, and surface finish. Despite the name, the inspected article is not necessarily the first produced. FAI is typically called for in a contract between the supplier and the manufacturer to ensure that the production process reliably produces what is intended. A FAI should be performed for each cavity in the case of a mold or each part from a stamping die in the case where multiple parts are made at the same time to ensure the tooling is correct. Additionally, if tooling or the process is changed a full or partial FAI may be required. Figure 10.12 provides an example of a ballooned drawing.

To properly document a FAI, a drawing is "ballooned," meaning that each dimension, feature, or specification is assigned a number which is enclosed in a circle. Each ballooned number is then placed on an inspection form where the results are recorded. Table 10.3 provides an example FAI report.

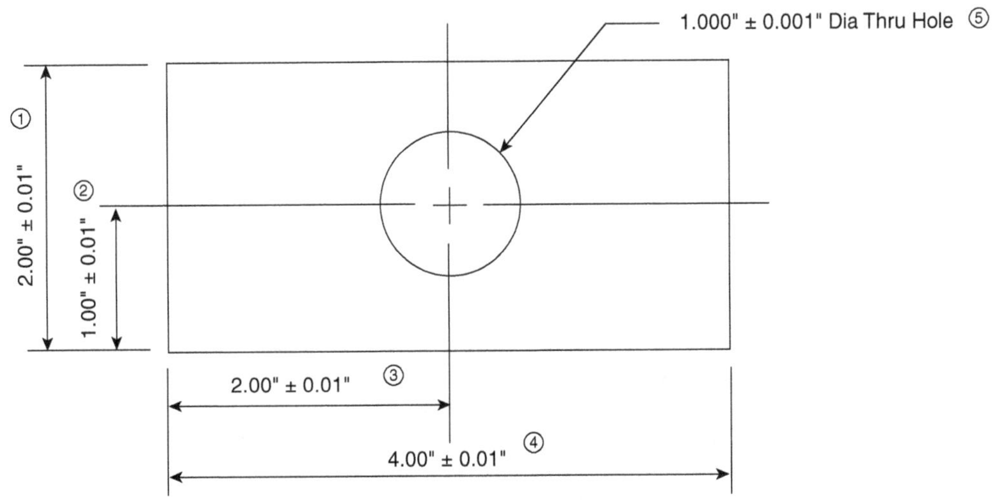

Figure 10.12 Ballooned drawing.

Table 10.3 FAI inspection report.

Feature	Dim	Tol	Actual	Result	Notes
1	2.00	± 0.01	2.005	Pass	
2	1.00	± 0.01	1.001	Pass	
3	2.00	± 0.01	2.015	Fail	Readjust set up
4	4.00	± 0.01	4.009	Pass	
5	1.000	± 0.001	1.0005	Pass	

If fail conditions are encountered during a FAI, the cause of the issue should be determined and a new FAI completed after addressing the issue.

It should be noted that a FAI is not the same as a first piece inspection. A first piece inspection is used to verify a set up for existing projects, where as a FAI is used for new projects or when changes are made to a process or supporting tooling.

Measurement System Analysis

MSA or gage repeatability and reproducibility (gage R&R) studies analyze the variation in measurements of a gage (repeatability) and the variation in measurements by operators (reproducibility). Repeatability is the variation observed when an operator measures the same sample using the same gage several times. Reproducibility is the variation observed when different operators measure the same sample. To understand why MSA is so important, the goal of process control is to reduce variation in the process and ultimately the product. To address actual process variability, the variation due to the measurement system must be identified and separated from that of the process. A prerequisite to this process is to be sure the gage is calibrated prior to beginning the study to ensure the reliability of the measurements.

Critical to Quality

CTQs are usually defined during the design and development phase, usually when design FMEA or process FMEA is being performed. A CTQ is a characteristic that has a direct effect on whether the overall process or product is perceived by the customer to be of acceptable quality or the process or product's fit, form, function, or safety. The identification of specific, measurable CTQs is critical in the planning and commercialization of goods and services.

Determining QTQ attributes will help develop inspection plans, process capability studies, material and performance testing requirements, and appearance approval standards for surface finish and color. These items are required as part of the part approval process. Inspection plans (in-process and final) may be linked to process capability. For instance, a process that is highly capable may require fewer in-process and final inspections than a process with a lower capability.

Sample Size Determination

Determination of sample sizes for incoming, in-process, and final inspection as well as for process validation can be accomplished in many ways and should be determined using a risk-based approach. One of the most common sample size determination methods is the use of an Acceptance Quality Limit (AQL). The two most common AQL sampling standards are summarized below:

ANSI/ASQ Z1.4-2003 (R2018): Sampling Procedures and Tables for Inspection by Attributes is an acceptance sampling system to be used with switching rules on a continuing stream of lots for Acceptance Quality Limit (AQL) specified. It provides tightened, normal, and reduced plans to be applied for attributes inspection for percent nonconforming or nonconformities per 100 units.

ANSI/ASQ Z1.9-2003 (R2018): Sampling Procedures and Tables for Inspection by Variables for Percent Nonconforming is an acceptance sampling system to be used on a continuing stream of lots for Acceptance Quality Limit (AQL) specified. It provides tightened, normal, and reduced plans to be used on measurements which are normally distributed. Variation may be measured by sample standard deviation, sample range, or known standard deviation. It is applicable only when the normality of the measurements is assured.

A third AQL-based sampling plan developed by Nicholas L. Squeglia is the Zero Acceptance Number Sampling Plan commonly known as the c=0 sampling plan, which is derived from the ANSI/ASQ Z1.4 sampling plan. The Zero Acceptance Number Sampling Plans are widely used in regulated industries for final acceptance activities as they provide greater protection for the consumer than the producer for a given AQL.

Table 10.4 is a partial c=0 sampling plan with three AQL indices provided, 0.010, 0.65, and 1.5. The smaller the AQL value, the larger the required sample. For

Table 10.4 c=0 Sampling Plan (partial).

	AQL		
Lot size	**0.10**	**0.65**	**1.5**
9–15	A	A	8
16–25	A	20	8
26–50	A	20	8
51–90	A	20	13
91–150	125	20	19
151–280	125	29	19
281–500	125	47	21

Note: A mean all parts in the lot must be inspected.

example, if a lot contains 200 pieces and we are using an AQL of 0.65, 29 samples are required to be inspected. For a c=0 sampling plan, we would accept the lot with zero failures and reject the lot if we detected one or more failures in the 29 samples.

The *ANSI/ASQ Z1.4* and *ANSI/ASQ Z1.9* sampling plans are somewhat more complicated. It is suggested to reference these standards for further guidance on their use and application.

Process Validation

Verification is confirmation by examination and provision of objective evidence that a product or service meets specified requirements. Process validation is establishing through objective evidence that a process consistently produces a result or product meeting its predetermined requirements. The requirements and methods for process validation can be driven by internal requirements or externally by customer or regulatory requirements. The flowchart in Figure 10.13 can be used to determine whether process validation is necessary.

Quality and manufacturing engineers should understand the value of process validation and the need to ensure that product features cannot be fully verified through inspection or testing are validated. There are multiple process validation prerequisites that need to be considered before an engineer starts the validation protocol, including:

- Using design of experiments (DOE) to characterize the process and determine the critical process parameters
- Understanding the importance of calibration and metrology
- Ensuring MSA activities have been completed
- Ensuring an adequate level of training has been pursued for process validation engineers, reviewers, approvers, technicians, operators, etc.

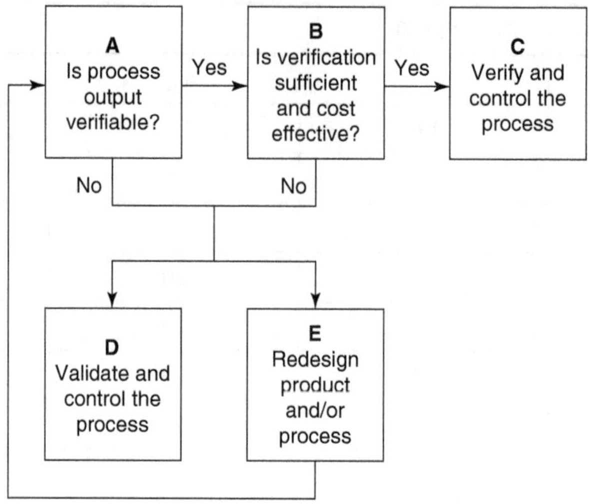

Figure 10.13 Process validation decision tree.

Source: Adapted from Global Harmonization Task Force (GHTF) Study Group 3, *Quality Management Systems—Process Validation Guidance,* January 2004.

The validation master plan (VMP), or master validation plan (MVP), serves as the validation road map. It is important to clearly define the intended purpose of the validation to determine whether revalidation and retrospective validation activities are necessary. The main output of a VMP is a compilation of the process to be validated and associated documents with a schedule for revalidation and retrospective validation activities. The VMP should be developed and approved prior to the start of validation activities. A VMP contains the following elements:

- Installation qualification (IQ) establishes, through objective evidence, that all key aspects of the process equipment and ancillary system installations adhere to the manufacturer's approved specification and that the recommendations of the supplier of the equipment are suitably considered. Additionally, IQ determines, through documented evidence, that all systems and equipment are installed correctly.
- Operational qualification (OQ) establishes, through objective evidence, that the equipment process control limits meet all predetermined requirements. For a process OQ, the process control limits are challenged to provide evidence that the predetermined process output requirements on the product meet the predetermined requirements. Additionally, OQ determines, through documented evidence, that the equipment process control limits meet all predetermined requirements. The OQ also challenges the process parameters (often referred to as "limits" or "worst-case" testing) to make sure they result in a product that meets all defined requirements under all anticipated conditions of manufacturing. For OQ, sample size selection is premised on the initial process risk index usually from a FMEA and data type (variable/attribute).

- Performance qualification (PQ) establishes, through objective evidence, that a process consistently produces a result and/or product that meets the predetermined requirements. Another objective of PQ is to demonstrate that the process will consistently produce acceptable product under normal operating conditions. PQ testing should always take place in nominal process conditions. Moreover, PQ samples should always be taken from product lots representative of production so that data generated from these samples will demonstrate reproducibility and provide an accurate measure of variability for the processes being qualified. Sample size selection is premised on the initial process risk index usually from a FMEA and data type (variable/attribute).
- Process performance qualification (PPQ) is the collection and evaluation of data, from the process design stage through commercialization, to prove that a process is capable of consistently delivering quality products. The object of PPQ is to demonstrate that all validated manufacturing processes produce finished product that meets the product specifications. PPQ looks at an entire manufacturing process versus a single process. Complete product configurations are built in accordance with nominal process conditions. In support of PPQ, samples should be taken from lots representative of production so that data generated in accordance with the PPQ protocol affirm that the entire manufacturing process being challenged reflects that all significant variables are in a state of control. Sample size selection is premised on the initial process risk index usually from a FMEA and data type (variable/attribute).

Once a process has been validated, it is important to ensure that the process does not drift or otherwise change. Control charts are an essential tool for satisfying continuous process monitoring requirements for validated processes. Control charts are decision-making tools that provide information for timely decisions concerning recently produced products.

Control Plans

A control plan is used to document the process characteristics and variables that must be controlled to ensure that the process remains stable ("in control"). Control plans can be used for training as well as for reference during process troubleshooting, as they list the existing control points in a process and can indicate which control point failed to prevent or detect an issue or where a control point is needed but does not exist. Figure 10.14 provides a partial example of a control plan.

Control plans are living documents that should be reviewed periodically to ensure that they remain current. Control plans should also be reviewed and updated whenever changes are made to a process, including changes made as a result of root cause analysis and corrective action.

If a control plan is not already available for the process that has been affected by corrective action activity, the team should create one. The first step is to define the scope of the process to be covered by the control plan, which is usually done by creating a high-level process flow diagram (if one does not already exist). Next,

Control Plan						
Fabricate cover						
	Process/product parameters	**Requirement(s)**	**Evaluation/ measurement technique**	**Sample size/ frequency**	**Responsible (who measures?)**	**Reaction plan (if abnormal or nonconforming conditions are found)**
Cut material	Piece dimensions	As specified in part drawing	Measure with calipers	Each piece	Cutting operator	Stop equipment; invoke nonconforming material process (work instruction WI-01-0004)
Bend material (press brake)	Piece dimensions after bend	As specified in part drawing	Measure with calipers	First piece, then every 5th piece	Brake operator	Adjust equipment as necessary; invoke nonconforming material process
	Bend angles	As specified in part drawing	Measure with digital protractor	First piece, then every 5th piece	Brake operator	Adjust equipment as necessary; invoke nonconforming material process

Figure 10.14 Partial control plan.

identify the process and/or product characteristics that are to be evaluated, as well as the requirements or specifications for those characteristics. More than one characteristic may need to be evaluated at each step of the process.

List the technique to be used to measure or evaluate the process output to ensure that the requirements are met, as well as the sample size and sampling frequency and the person responsible for taking the measurement or performing the evaluation.

Finally, indicate what to do if an abnormal or nonconforming result is detected. While most companies have a defined process for segregating and dispositioning nonconforming material, the person performing the process may also be able to make some process adjustments, which should also be listed.

Hazard Analysis and Critical Control Points

Hazard analysis and critical control points (HACCP) is used to manage risks with physical, chemical, biological, allergenic, and radiologic hazards. HACCP can and is used in many industries including food, drugs, cosmetics, medical devices, and biologics. HACCP has seven steps:

1. Conduct a hazard analysis and identify preventive measures for each step of the process
2. Determine the critical control points
3. Establish critical limits
4. Establish a system to monitor the critical control points
5. Establish the corrective actions to be taken when monitoring indicates that the critical control points are not in a state of control
6. Establish a system to verify that the HACCP system is working effectively
7. Establish a record-keeping system

Potential areas of use for HACCP might be to identify and manage risks associated with physical, chemical, biological (including microbiological contamination), allergenic, and radiological hazards. HACCP is useful when product and process understanding is sufficiently comprehensive to support identification of critical control points. The output of an HACCP analysis is risk management information that facilitates monitoring of critical points not only in the manufacturing process but also in other lifecycle phases.

A good HACCP plan much like a FMEA begins with a process flow chart and looks to identify the potential sources of contamination at each step of the process.

The foundation of an effective HACCP program is a rigorous prerequisite program. Prerequisite programs generally consist of the following elements:

- Facilities: The establishment should be located, constructed, and maintained according to sanitary design principles. There should be linear product flow and traffic control to minimize cross-contamination from raw to cooked materials.

- Supplier Control: Each facility should assure that its suppliers have in place effective GMP and food safety programs. These may be the subject of continuing supplier guarantee and supplier HACCP system verification.
- Specifications: There should be written specifications for all raw matertials, products and packaging materials.
- Production Equipment: All equipment should be constructed and installed according to sanitary design principles. Preventive maintenance and calibration schedules should be established and documented.
- Cleaning and Sanitation: All procedures for cleaning and sanitation of the equipment and the facility should be written and followed. A master sanitation schedule should be in place.
- Personal Hygiene: All employees and other persons who enter the manufacturing plant should follow the requirements for personal hygiene.
- Training: All employees should receive documented training in personal hygiene, GMP, cleaning and sanitation procedures, personal safety, and their role in the HACCP program.
- Chemical Control: Documented procedures must be in place to assure the segregation and proper use chemicals in the plant. These include cleaning chemicals, fumigants, and pesticides or baits used in or around the plant.
- Receiving, Storage and Shipping: All raw materials and products should be stored under sanitary conditions and the proper environmental conditions such as temperature and humidity
- Traceability and Recall: All raw materials and products should be lot-coded and a recall system in place so that rapid and complete traces and recalls can be done when a product retrieval is necessary.
- Pest Control: Effective pest control programs should be in place.

For more information about HACCP, refer to 21 CFR Part 120 Hazard Analysis and Critical Control Point (HACCP) Systems or ISO 22000 Food Safety Management Systems - Requirements for Any Organization in the Food Chain.

Process Flow Diagrams

Process Flow Diagrams (PFDs) otherwise known as flowcharts are used to graphically depict the structure of a process. PFDs help in understanding how a process works by showing the individual tasks, their interactions, decision points, and when inspections are required in a logical flow. PFDs can help with problem solving (8-D and CAPA) and provide the foundation for a pFMEA, HACCP, and control plan. Figure 10.15 is an example of a simple flowchart with one decision point followed by one process step.

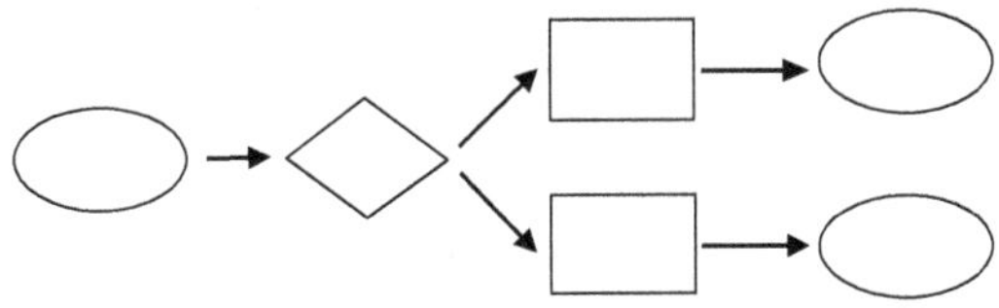

Figure 10.15 Example flowchart.

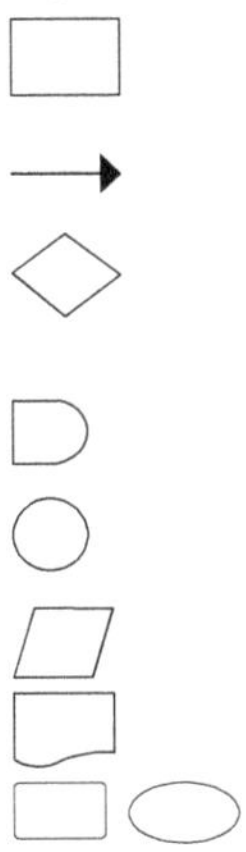

One step in the process. The step is written inside the box. Usually, only one arrow goes out of the box.

Direction of flow from one step or decision to another.

Decision based on a question. The question is written in the diamond. More than one arrow goes out of the diamond, each one showing the direction the process takes for a given answer to the question. (Often the answers are "yes" and "no.")

Delay or wait

Link to another page or another flowchart. The same symbol on the other page indicates that the flow continues there.

Input or output

Document

Alternate symbols for start and end points

Source: https://asq.org/quality-resources/flowchart.

Failure Modes and Effects Analysis

FMEA involves the evaluation of potential failure modes and their effect on outcomes. It methodically breaks down the analysis of complex processes into manageable steps; risk reduction can then be used to eliminate, contain, reduce, or control failures.

Potential areas of use for FMEA include prioritizing risks and monitoring the effectiveness of risk control activities. It can be applied to equipment or facilities and could be used to analyze a manufacturing operation and its effect on product or process. It identifies elements or operations within the system that render it vulnerable. The output or results of FMEA can be used as a basis for design or further analysis or to guide resource deployment.

Table 10.5 is an example of an FMEA rating scheme. Severity is a measure of the seriousness of the possible consequences of a hazard. Occurrence is the likelihood that a cause will occur and that it will induce the detrimental consequences of the associated hazard. Detection is the probability that control will catch the defect. Multiplying the ratings for severity, occurrence, and detection yields the RPN. The RPN is a numerical assessment of the risk for each failure mode and is used to relatively rank the modes.

The RPN is calculated by multiplying the three ratings:

$$\text{RPN} = \text{severity} \times \text{occurrence} \times \text{detection}$$

For example, say a process FMEA has been developed for a new drug product. A certain characteristic has a severity rating of 3, a probability of occurrence rating of 3, and a probability of detection rating of 3. The RPN calculation is:

$$\text{RPN} = 3 \times 3 \times 3 = 27$$

Table 10.5 Five-point FMEA rating scheme.

	Severity
5	Catastrophic. Failure affects safety and/or involves noncompliance with government regulations.
4	Critical. Customer dissatisfaction due to the nature of the failure, such as unusable product. May cause disruption to future processes. Process changes to be examined.
3	Serious. Customer is made uncomfortable or is bothered by failure. May require change in controls.
2	Minor. Customer may notice a slight deterioration of the product.
1	Negligible. Customer will probably not notice the failure.
	Occurrence
5	Certain to fail (1 in 10 or 0.1)
4	High number of failures likely (1 in 100 or 0.01)
3	Occasional failure likely (1 in 1000 or 0.001)
2	Few failures likely (1 in 10,000 or 0.0001)
1	Failure unlikely (1 in 100,000 or 0.00001)
	Detection
5	Undetectable. Controls very likely will *not* detect the existence of a failure. Very high risk that the product will be shipped with this defect.
4	Poor. Controls likely will *not* detect the existence of a failure. Likely that the product will be shipped with the defect.
3	Moderate. Controls may detect the existence of a failure. Moderate risk that the product will be shipped with this defect.
2	Good. Controls have a good chance of detecting the existence of a failure. Low risk that the product will be shipped with this defect.
1	Excellent. Controls almost certainly will detect the existence of a failure. Remote risk that the product will be shipped with this defect.

According to Table 10.6, an RPN of 27 is considered tolerable. However, the criticality and probability of occurrence for this characteristic should also be evaluated.

A drawback of RPN scores is that there are multiple ways the same RPN can be achieved. For example, scenario 1, an RPN of 100 could have a Severity (S) rating of 10, Occurrence (O) rating of 10, and Detection (D) rating of 1

or

scenario 2, an RPN of 100 could have a Severity (S) rating of 1, Occurrence (O) rating of 10, and Detection (D) rating of 10

or

scenario 3, an RPN of 100 could have a Severity (S) rating of 10, Occurrence (O) rating of 1, and Detection (D) rating of 10.

Table 10.6 RPN action requirements for a five-point scale.

RPN	Risk acceptability	Action
1 to 14	Low	Although risk is low, continue mitigation process as far as possible.
15 to 29	Tolerable	This should only be revisited in future design review if corrective action enhances the reliability or product appeal.
30 to 49	Undesirable	Risk is acceptable only if it cannot be further mitigated by organizational or technological solutions that do not reduce the clinical/functional utility of the product.
Above 49	Intolerable	Risk should be eliminated or reduced by protective measures. Justification required for risk that is accepted.

Table 10.7 Criticality and occurrence action requirements for a five-point scale.

		Severity				
		Catastrophic	Critical	Serious	Minor	Negligible
Occurrence		5	4	3	2	1
Certain	5	Unacceptable	Unacceptable	Unacceptable	Unacceptable	ALARP
High	4	Unacceptable	Unacceptable	Unacceptable	ALARP	ALARP
Occasional	3	Unacceptable	Unacceptable	ALARP	ALARP	ALARP
Few	2	Unacceptable	ALARP	ALARP	ALARP	Acceptable
Unlikely	1	ALARP	ALARP	ALARP	Acceptable	Acceptable

ALARP = as low as reasonably possible

Although the Risk Priority Number (RPN) is 100 for each of these scenarios, which of these scenarios pose the greatest risk? To answer this question, we must first look at severity. Scenarios 1 and 3 both have a Severity (S) rating of 10, however scenario 1 has an Occurrence (O) rating of 10 making it higher risk. Scenario 2 has a Severity (S) rating of 1 making lower risk than scenarios 1 and 3.

Ranking the order RPN scenario risks in this example, the order 1, 3, 2 is the obvious answer.

Figure 10.16, Partial FMEA, provides an example of where piece dimensions after bend required further mitigations to reduce the risk as the initial RPN of 120 was deemed to undesirable according to Table 10.6, RPN action requirements for a five-point scale. With the improved gaging and vision system the RPN was reduced to 12 which is low.

Table 10.7 indicates that for a characteristic with a severity of 3 and a probability of occurrence of 3, the risk is considered to be as low as reasonably possible.

The examples shown in Tables 10.8, 10.9, and 10.10 are based on a 10-point scale. The 5-point and 10-point rating scales are provided as examples only. Each

Process FMEA													
Fabricate cover													
Process step	**Process/product parameters**	**SEV**	**OCC**	**DET**	**RPN**	**Evaluation**	**Actions Taken**	**SEV**	**OCC**	**DET**	**OCC**	**Evaluation**
Cut material	Piece dimensions	3	3	3	9	risk acceptability low						
Bend material (press brake)	Piece dimensions after bend	4	5	3	120	risk acceptability is undesirable	Improve gaging and add vision system	4	3	1	12	risk acceptability low
	Bend angles	1	3	3	6	risk acceptability low						

Figure 10.16 Partial FMEA.

Table 10.8 Ten-point FMEA rating scheme.

	Severity
10	Failure could injure the customer or an employee
9	Failure creates noncompliance with federal regulations
8	Failure renders the unit inoperable or unfit for use
7	Failure causes a high degree of customer dissatisfaction
6	Failure results in a subsystem or partial malfunction of the product
5	Failure creates enough of a performance loss to cause the customer to complain
4	Failure can be overcome with modifications to the customer's process or product, but there is minor performance loss
3	Failure creates a minor nuisance for the customer, but the customer can overcome it without performance loss
2	Failure may not be readily apparent to the customer but has minor effects on the customer's process or product
1	Failure is not noticeable to the customer and does not affect the customer's process or product
	Occurrence
10	More than one occurrence per day or a probability of more than three occurrences in 10 events ($C_{pk} < 0.33$)
9	One occurrence every three to four days or a probability of three occurrences in 10 events ($C_{pk} \approx 0.33$)
8	One occurrence per week or a probability of five occurrences in 100 events ($C_{pk} \approx 0.67$)
7	One occurrence every month or one occurrence in 100 events ($C_{pk} \approx 0.83$)
6	One occurrence every three months or three occurrences in 1000 events ($C_{pk} \approx 1.00$)
5	One occurrence every six months to one year or five occurrences in 10,000 events ($C_{pk} \approx 1.17$)
4	One occurrence per year or six occurrences in 100,000 events ($C_{pk} \approx 1.33$)
3	One occurrence every one to three years or six occurrences in 10 million events ($C_{pk} \approx 1.67$)
2	One occurrence every three to five years or two occurrences in 1 billion events ($C_{pk} \approx 2.00$)
1	One occurrence in greater than five years or less than two occurrences in 1 billion events ($C_{pk} > 2.00$)

(continued)

Table 10.8 Ten-point FMEA rating scheme. (continued)

Detection	
10	The product is not inspected or the defect caused by failure is not detectable
9	Product is sampled, inspected, and released based on acceptable quality level (AQL) sampling plans
8	Product is accepted based on no defectives in a sample
7	Product is 100% manually inspected in the process
6	Product is 100% manually inspected using go/no-go or other mistake-proofing gages
5	Some SPC is used in process and product is final inspected off-line
4	SPC is used and there is immediate reaction to out-of-control conditions
3	An effective SPC program is in place with process capabilities (C_{pk}) greater than 1.33
2	All product is 100% automatically inspected
1	The defect is obvious or there is 100% automatic inspection with regular calibration and preventive maintenance of the inspection equipment

Table 10.9 RPN action requirements for a 10-point scale.

RPN	Risk acceptability	Action
1 to 65	Low	Although risk is low, continue mitigation process as far as possible.
66 to 125	Tolerable	This should only be revisited in future design review if corrective action enhances the reliability or product appeal.
126 to 300	Undesirable	Risk is acceptable only if it cannot be further mitigated by organizational or technological solutions that do not reduce the clinical/functional utility of the product.
Above 300	Intolerable	Risk should be eliminated or reduced by protective measures. Justification required for risk that is accepted.

company should develop its own rating scales based on its risk acceptance threshold, industry practice, guidance documents, and regulatory requirements.

As an alternative to the traditional FMEA method of using an RPN, the Automotive Industry Action Group (AIAG) and the Verband der Automobilindustrie (VDA) jointly released an updated FMEA handbook. One of the major changes with the new AIAG-VDA FMEA process is the use of the RPN has been eliminated. The RPN has been replaced by an action priority (AP) table. APs place the emphasis on severity, followed by occurrence, and finally detection. The AP tables assign

Table 10.10 Criticality and occurrence action requirements for a 10-point scale.

		Severity				
		Catastrophic	**Critical**	**Serious**	**Minor**	**Negligible**
Occurrence		**9–10**	**7–8**	**5–6**	**3–4**	**1–2**
Certain	10	Unacceptable	Unacceptable	Unacceptable	Unacceptable	ALARP
	9	Unacceptable	Unacceptable	Unacceptable	Unacceptable	ALARP
High	8	Unacceptable	Unacceptable	Unacceptable	ALARP	ALARP
	7	Unacceptable	Unacceptable	Unacceptable	ALARP	ALARP
Occasional	6	Unacceptable	Unacceptable	ALARP	ALARP	ALARP
	5	Unacceptable	Unacceptable	ALARP	ALARP	ALARP
Few	4	Unacceptable	ALARP	ALARP	ALARP	Acceptable
	3	Unacceptable	ALARP	ALARP	ALARP	Acceptable
Unlikely	2	ALARP	ALARP	ALARP	Acceptable	Acceptable
	1	ALARP	ALARP	ALARP	Acceptable	Acceptable

ALARP as low as reasonably possible/practical or AFAP as far as possible/practical. These terms both suggest that mitigations should be investigated and implemented to prevent or minimize risks to ensure safety and performance.

one of three suggested rankings for each action based upon the effect (severity), prediction of failure cause occurring (occurrence), and ability to detect (detection) values. The resulting action priorities are defined as Priority High (H), Priority Medium (M), and Priority Low (L). Table 10.11 depicts an example Action Priority Table and required mitigations.

As an example, an effect on the product may be very high, a prediction of failure cause occurring of low, and with a moderate ability to detect would yield an AP of Priority Medium (M). Thus, the FMEA team would collectively decide if the additional mitigations are required or justify the adequacy of current controls. See FMEA Handbook 1st Ed., 2019, Automotive Industry Group (AIAG) and the Verband der Automobilindustrie (VDA), Southfield, MI page 116.

Table 10.11 Example action priority table and required mitigations.

Priority High (H)	Priority Medium (M)	Priority Low (L)
Highest priority for evaluation. Requires mitigation.	Medium priority for evaluation. May require mitigation, justify adequacy of current controls.	Low priority for evaluation. Further mitigations could be implemented but may not be necessary.

PART APPROVAL

> Understand production part approval process (PPAP) requirements and ensure suppliers understand the processes required to produce parts with consistent quality during an actual production run at production rates. (Understand)
>
> **Body of Knowledge III.C.4**

PPAP, developed by AIAG, is a process that helps manufacturers and suppliers communicate and approve production designs and processes before, during, and after manufacture. PPAP helps ensure that the processes used to manufacture parts can consistently reproduce the parts at stated production rates during routine production runs. The PPAP process is used to document:

- New parts
- Engineering changes
- Durable tooling: transfer, replacement, refurbishment, or additional
- Tooling inactive for more than one year
- Correction of discrepancies
- Changes to optional construction or material
- Sub-supplier or material source change
- Changes in part processing
- Parts produced at a new or additional location

The PPAP process has 18 requirements, shown in Table 10.12.

The part submission warrant (PSW) is a summary of the PPAP process. The PSW indicates the reason for submission (changes in the design, tooling, materials, etc.) and the level of documents submitted to the customer (see Table 10.13). The PSW includes a simple yes/no question asking whether the results meet all drawing and specification requirements. If there are any deviations, the supplier should note this on the PSW or inform the customer according to prearranged requirements that PPAP cannot be submitted until the issues are adequately resolved. PPAP Levels for the PSW are generally driven by the customer. A Level 3 PPAP is usually accepted as the default for most submissions.

Safe Launch is a process of elevated inspection and monitoring activity (more frequent and/or larger sample sizes) until Safe Launch exit. Safe Launch helps ensure defect free supplies, detect and address failure modes, identify potential issues in new or modified production processes for a period of time after PPAP submission and approval.

Table 10.12 PPAP requirements based on levels.

	Requirement	Level 1	Level 2	Level 3	Level 4	Level 5
1.	Design record	S	X	X	R	X
2.	Engineering change documents	S	S	X	R	X
3.	Customer engineering approval	S	S	X	R	X
4.	Design FMEA	S	S	X	R	X
5.	Process flow diagrams	S	S	X	R	X
6.	Process FMEA	S	S	X	R	X
7.	Control plan	S	S	X	R	X
8.	MSA studies	S	S	X	R	X
9.	Dimensional results	S	X	X	R	X
10.	Material, performance test results	S	X	X	R	X
11.	Initial process studies	S	S	X	R	X
12.	Qualified laboratory documentation	S	X	X	R	X
13.	Appearance approval report (AAR)	S	X	X	R	X
14.	Sample products	S	X	X	R	X
15.	Master samples	S	S	S	R	S
16.	Checking aids	S	S	X	R	X
17.	Records of compliance	S	S	X	R	X
18.	Part submission warrant (PSW)	X	X	X	X	X

X = submit and retain a copy
R = retain and make available upon request
S = recommended to retain and make available

Table 10.13 PPAP levels.

Level 1	PSW only (and for designated appearance items, an AAR) submitted to the customer
Level 2	PSW with product samples and limited supporting data submitted to the customer
Level 3	PSW with product samples and complete supporting data submitted to the customer
Level 4	PSW and other requirements as defined by the customer
Level 5	PSW with product samples and complete supporting data reviewed at the organization's manufacturing location

VALIDATE REQUIREMENTS

> Collaborate with internal team to interpret the results of the executed qualification plan for the part/process/service, including reviewing Certificate of Compliance (CoC), Certificate of Analysis (CoA), and production readiness reviews (PRR). (Evaluate)
>
> **Body of Knowledge III.C.4**

Certificates of Compliance and Conformance

A Certificate of Conformance (COC) or Certificate of Compliance is a document provided by a supplier verifying the products or services were performed meet required specifications. A COC usually does not include inspection, measuring, and test results. The organization's Quality Assurance function is generally responsible for generating a COC after a thorough review of production and inspection records including any deviations or nonconformances that may have been encountered. COCs are typically seen for parts that are machined and for services such as coating and heat treating.

A Certificate of Analysis (COA) is a document normally issued by the organization's Quality Assurance function that confirms a product meets predetermined specifications. A COA includes actual inspection, measuring, and test results. A COA is generated after a thorough review of production and inspection records including any deviations or nonconformances that may have been encountered. Common materials for which a COA may be used include chemicals, resins, and metals. Often suppliers will charge an extra fee for providing a COA due to the extra work involved in providing the actual inspection, measuring, and test results.

When using a COC or COA as the sole basis for incoming acceptance activities, a representative sample of receipts should be independently verified to ensure the accuracy of the information provided to help qualify the supplier. Additionally, once the supplier is qualified, a COC or COA should be independently verified at least annually as part of the supplier monitoring and requalification program.

While reviewing COAs, pay special attention to the reported inspection, measuring, and test results. If there is no variability in the reported results, it may indicate the actual inspection, measuring, and test results are not being reported.

Production Readiness Review

IEEE 15288.2 Standard for Technical Reviews and Audits on Defense Programs Department of Defense (DoD) and other defense agencies guides the DoD and contractor on what is required during a Production Readiness Review (PRR) and establishes the requirements for technical reviews and audits to be performed throughout the acquisition life cycle for the US Department of Defense (DoD) and other defense agencies. This standard provides the definition, description, and intent, as well as the entry, exit and success criteria, for each technical review and audit. This standard is to be used to establish agreement between acquirers and suppliers on the technical reviews and audits that are needed for the project, as well as the focus and expectations of each technical review and audit.

The Production Readiness Review (PRR) process is used to determine if a systems design is production ready and if the developer of the system has adequate production planning to enter Low-Rate Initial Production (LRIP) and Full-Rate Production (FRP). The PRR assesses the program to determine if the design is ready for production and focuses on:

- manufacturing readiness
- quality management system
- production planning
- compliance of system requirements
- inventory management
- supplier management

Part IV

Supplier Performance Monitoring and Improvement

Chapter 11
Supplier Performance Monitoring

SUPPLIER METRICS

> Define, implement, and monitor supplier performance metrics such as quality, delivery (e.g., on-time delivery [OTD] and on-time in full delivery [OTIF]), cost, and responsiveness. (Evaluate)
>
> **Body of Knowledge IV.A.1**

Without data you're just another person with an opinion.

—W. Edwards Deming

What's measured improves.

—Peter F. Drucker

Why Metrics?

The reasons for using metrics are many and varied. ISO standards such as ISO 9001 require measurements that assess an organization's performance on key business and process metrics. Processes cannot be improved if one cannot measure them first. Processes that are monitored get attention, and the majority of those processes are improved.

Metrics are also used for financial reasons. It is important to understand the value of the products and services we receive. In addition to the direct cost of buying products or services, anytime a product is recalled or removed from the market, there is an associated cost. It is very difficult for a company to accept the expenses resulting from a supplier-related failure.

Currently, we place a strong emphasis on the part cost but rarely look at the TCO (again, total cost of ownership). The TCO model better defines risk and creates mitigation strategies.

Data are numbers or letters that are the output of a process or query. They do not provide very much in the way of actionable activities on their own. Data should be collected and summarized and then used with metrics to produce information that becomes actionable. Those metrics should be used for making decisions for the business.

What Metrics?

If you can't describe what you are doing as a process, you don't know what you are doing.

—W. Edwards Deming

There are many metrics, and it's critical to know why you are using the ones you are using. Metrics should drive you to focus on the critical few issues and drive activities that make improvements. The supplier-related metrics that are important and easy to create are:

- Incoming acceptance performance
- In-process acceptance
- Field returns that result from supplier-related failures
- Cost impact of failures or rejections
- On-time delivery
- SCAR response time
- Weighted calculations of metrics
- TCO

Metrics should be reliable and readily obtainable, meaning the data can be extracted from a single source, such as a database that captures the information.

If the system and methodology for flagging the rejections is not well defined, the report integrity will be compromised.

For example, here are some scenarios where the disposition may be clearer:

Supplier responsibility	NOT supplier responsibility
Shipped parts that don't meet the dimensional requirements of the drawing	Supplier made the parts per the drawing, but the critical dimension is incorrect so the parts do not fit
Parts were not wrapped to protect them during shipment	Shipper accidentally ran the forks of the forklift into the box

Each rejection report or tracking mechanism assigns the responsibility to the supplier or not to the supplier. A simple way to differentiate this disposition is a block on the rejection form, as in the following example:

Disposition responsibility (select one):
☐ Supplier ☐ Company

Incoming Acceptance Performance

Incoming acceptance can be done through either lot or part acceptance. Lot acceptance measurement is the most common method since the calculations are easy—the lot is either all good or all rejected. Part acceptance can be a challenge, since the inspection is usually done on a sample.

> Incoming inspection = lots accepted / lots received
>
> Or = parts accepted / total parts received
>
> *(This can be done for any period of time, e.g., monthly, quarterly)*

Look at the monthly as well as rolling three-month averages. This can indicate if there was a recent spike or a trend. Also, it's important to provide the source data under each chart. For suppliers that ship one lot per month or one or two lots per year, the data can be skewed. For example, in the following table, the supplier that provides multiple lots has different performance data than a supplier that ships one or two lots per year. Also, when questions are asked regarding performance, the data can be reviewed to answer those questions.

	Month: March			1st quarter (Jan–Mar)		
Supplier	**Lots accepted**	**Lots received**	**%**	**Lots accepted**	**Lots received**	**%**
A	1	1	100%	3	5	60%
B	19	20	95%	40	45	89%

In-Process Acceptance Performance

In-process acceptance can be done by identifying the count of rejected parts found after the incoming inspection. Unless there is lot traceability, it may be difficult to identify the original lot quantity. Most companies have some traceability back to the receiving lot information, but gathering the data for reporting metrics might be a challenge.

> In-process acceptance = (total parts – rejected parts) / total parts from the initial lot

Using the total parts received for the month along with three-month historical data is the best method. This method is based on the lag that usually occurs between the time the parts are received and the time they are put into production and actually used. It's not uncommon to not receive parts every month, so the total number of parts could be zero. Most companies do not utilize lean inventory practices such as kanban. The three-month historical data is used because it's more likely that the parts were received during that longer period. Also, it's a way to normalize the data to avoid spikes in inventory levels.

Field Returns That Result from Supplier-Related Failures

Field returns that result from supplier-related failures refers to the quantity of parts that are components of products returned by customers. This metric is very important to track as there are additional costs associated with this type of failure, especially since the warranty/repair costs of the finished product are the result of one supplier component.

Having the "disposition responsibility" selection is critical when using this metric, since not all field failures are directly related to a supplier issue.

Since there could be a lag between receiving the parts from the supplier and the actual failure occurring, we recommend using the three-month rolling window of parts received. This should allow for the data to be normalized as well as show any trends.

This is a difficult metric to track, since not all field failures can be traced back to the supplier. However, there are cases where the supplier provides a distinct finished device or accessory and the failure is specific to that item. Either way, it's critical that the root cause be clearly identified and understood before the disposition is assigned.

Repeat Failures by Failure Mode or Specific Item

As part of the rejection analysis, the repetition of a failure mode or of a specific item should be noted. The trending of data should help you see whether a specific failure mode is recurring with the same supplier. For example, a trend showing that parts are not being marked properly can tell you something is systemically incorrect at the supplier and requires some form of escalation.

On-Time Delivery (OTD)/On-Time In Full Delivery (OTIF)

On-time delivery is a straightforward metric in the world of expedited delivery platforms such as Amazon Prime. We expect products to be received on time without excuses. If parts are critical to a production line, then any late parts could impact the entire factory.

On-time delivery means the right product is received in the right quantity. If multiple delivery dates and quantities are required, the purchase order should include specific line items to track this information. Otherwise, the system may not be able to distinguish whether the order is on time. Some MRP systems may not be able to accommodate this type of situation.

Some factors, such as weather, cannot be controlled, and shipments can be late. If the supplier contacts you ahead of time in this type of situation, you should allow for the delivery dates to be changed on the purchase order. There should be some way to document these exceptions. These situations should be recorded as they occur or a different set of expectations will be created and the integrity of the data will be questioned.

On-Time in Full (OTIF) is a critical metric because not having the full amount may cause further delays in manufacturing. With more companies having tighter controls on inventory, having too much is considered a carrying cost that affects the profitability of a company. If you can't turn inventory into profit, then you usually end up with an obsolete product that results in scrap costs. However, if you don't have the right quantity when you need it the net effect can be a delivering product to customers. Refer to "Cost Associated with Inventory" later in this chapter for more details.

Responsive Metrics

Audit Results and Audit CAPA Response Time

Let's start with the easiest metric to monitor because these audit findings can be identified while at the supplier. The specific resolution may vary based on the complexity of the issue and the time required to implement the solution and to test its effectiveness.

Supplier Corrective Action Requests

A SCAR can be issued for audit findings and other supplier performance issues, but it should not come as a surprise to the supplier. Communicating the reasons and expectations will go a long way toward building a relationship with your supplier. The name itself—SCAR—has negative connotations. However, the industry standard is to use this acronym, and since it's something that everyone understands, we'll keep it.

When should a SCAR be issued? When the item is rejected, either at incoming inspection, in production, or because of a customer complaint? The answer is, it depends.

Many questions must be considered before issuing a SCAR:

- How critical is the item?
- What type of supplier is involved?
- Is this the result of a recall or product correction?
- What is the impact to the end customer?
- Is this a repeat occurrence? When was the last time this happened?
- What is the total cost of this rejection?
- Is this a long lead item?

These questions should help address 90% of the scenarios encountered, but there may be other company-specific circumstances that need to be addressed. The answers to the above questions may not be useful in deciding whether to issue a SCAR but are important to know.

Risk-Based SCAR Decisions

There are several ways to decide whether to issue a SCAR depending on the situation and risk levels. Table 11.1 shows a risk-based SCAR decision matrix, which can be used as the foundation for making the decision while exceptions are documented.

Table 11.1 Risk-based SCAR decision matrix.

Risk matrix table

	Severity			
Frequency	**Negligible**	**Low**	**Moderate**	**Critical**
Continually	SCAR	SCAR	SCAR and escalation	SCAR and escalation
Frequently	SCAR	SCAR	SCAR	SCAR and escalation
Occasional	Notification	Notification	SCAR	SCAR
Rarely	Notification	Notification	SCAR	SCAR

Risk severity table

Description	**Definition**
Critical	Performance impact, up to and including product failure with the potential for serious injury to the operator or patient* due to failure of product to meet performance specifications Potential for death of the operator and/or patient* as a result of product failure Impact to customer's ability to fulfill production requirements Critical risk to compliance with regulatory requirements including but not limited to product recall Product returned by the customer due to a rejection at its location Product from a critical or category 1 supplier Recurrence after issue was identified previously and corrective action was not effective
Moderate	Performance impact, up to and including product failure with the potential for injury to the operator and/or patient* that would not require medical intervention Moderate risk to product or process and/or compromised product Moderate risk to compliance with regulatory requirements including but not limited to internal audit nonconformance and supplier quality audit nonconformance Significant scrap determined by supply chain Significant process downtime, long lead time items Product from a category 2 supplier
Low	Little to no impact on production Low scrap cost determined by supply chain Product from a category 3 supplier

(continued)

Table 11.1 Risk-based SCAR decision matrix. (continued)

Negligible	Minimal risk of injury to the operator and/or patient Minimal risk to the product or process No impact on product or process performance or safety No impact on regulatory compliance

Frequency table	
Description	**Quantitative likelihood**
Continually	Nonconformance with other products being provided by the same supplier
Frequently	Has occurred several times in the past 12 months
Occasionally	Has not occurred in the past 12 months
Rarely	First-time occurrence

*Patient impact is applicable only where the information is provided by the customer or where sites that manufacture finished products have the regulatory and clinical expertise.

How does this work?

1. First identify the risk severity from the choices listed at the top of the matrix, which are based on the criteria outlined in the risk severity table. This information may be modified to best fit your company.
2. Then, select the frequency of this issue with this supplier using the definitions in the frequency table.
3. Find the intersection of the severity and frequency to determine the action.

For example, you may have a rejection that has a low risk severity and a one-time frequency. In this case the supplier should be notified so they are aware of the issue. A severity of critical and a one-time frequency, however, would warrant issuing a SCAR. A SCAR is issued in this case to make sure the supplier is taking corrective action because the severity of this failure could result in serious injury or death to a customer.

Actions and Escalations

The risk matrix outlines the recommended actions that should be taken. Communication can be done via e-mail, phone calls, Skype, text, and so forth.

Notification

- The supplier is notified about the nonconforming material by e-mail and a call is made to discuss the supplier's containment plan.
- The supplier should confirm the issue within 48 hours of it being dispositioned as its responsibility.

SCAR

- The supplier is notified about the nonconforming material by e-mail and a call is made to discuss the supplier's containment plan.
- The supplier should confirm the issue within 48 hours of it being dispositioned as its responsibility.
- The SCAR is e-mailed to the supplier to document the investigation and action plan; the supplier should send a formal response within 10 business days.

SCAR and Escalation

- The same steps listed in the SCAR action section should be followed.
- The quality and purchasing team should meet to determine whether any of the following actions need to be taken:
  - —Require source inspection at supplier prior to shipment
  - —Have supplier sort material on-site
  - —Increase sampling plan levels at incoming inspection
  - —Review quality performance of all other items from this supplier
  - —Review quality performance of this supplier at other sites
  - —Change supplier status on approved supplier list (e.g., conditional) and reconsider supplier for future business
  - —Perform for-cause audit at supplier
  - —Include other team members (e.g., production management, finance, regulatory)

The results of these investigations should be reviewed, verified, and shared within the organization. This information is a feeder to the audit process, inspection process, and scorecard/performance reporting.

SCAR Response Time

The SCAR process should specify what information the supplier should provide when a SCAR is issued and when the requested information should be sent. For example, "The supplier shall report on the root cause and containment plan for any SCAR within five days of being received." The number of days depends on the industry and the risk involved with failures.

Some quality professionals believe that the root cause and implementation of the solution should be done in five days. The reality is that some problems are difficult to reproduce and it may take longer to get to the root cause of such failures. Therefore, the number of days should be discussed within the company before a final decision is made.

Here are some examples of typical expectations segmented into milestones:

- Notification—confirmation within 48 hours of notification
- SCAR—confirmation within 48 hours of notification *and* a formal response within 10 business days. The formal response should include:
  - —The containment plan
  - —The results of the initial investigation into the root cause
  - —An action plan with dates for the complete investigation and implementation. The customer organization should follow up to confirm the supplier meets these deadlines.

Since this may take some time, track the confirmation of the issue, the containment plan, and the implementation plan date. The justification behind picking these three dates is based on acknowledging the issue.

SUPPLIER PERFORMANCE

> Analyze supplier performance data (e.g., warranty analysis/field returns and defect rates) and develop periodic reports (e.g., scorecard and dashboards). (Analyze)
>
> **Body of Knowledge IV.A.2**

Weighted Calculations of Metrics

If everything is a priority, then nothing is a priority.

—Stephen Perez

Prioritization is something we need to do every day. The need to have a weighted calculation with supplier metrics will be explained in this section.

Supplier performance can be assessed in many different ways, but the following are the two most common methods:

1. Each metric is reported separately.
2. Each metric represents a percentage of the total so that the final number is 100%.

Supplier Metrics

Figure 11.1 outlines the supplier lifecycle, from the identification of the need all the way through to the end of the relationship. Supplier metrics should only be collected and used for making good business decisions and improvements.

We recommend that you start collecting data for the metrics once the production of material has begun. In Figure 11.1, this is the sustaining, monitoring, and feedback step. You could start collecting data earlier, but that would require additional resources and time and might not add value or drive improvement.

Determining Which Suppliers Should Be Part of the Review

Current ISO and industry expectations assume that companies make decisions based on risk. Therefore, the best way to determine which requirements to apply to which suppliers is to create a risk classification. Based on the risk assessment, a supplier is assigned a risk classification, which dictates what actions are required (see Table 11.2).

The supplier may later need to be classified at a higher or lower level, but this is a good way to start, and justification of any changes in risk level can be documented. If your focus is priority or high-risk suppliers, this is a good tool to help justify your actions. Any auditor will ask why you are only focusing on certain suppliers, and with this classification and matrix, you can show them the documented method for addressing risk levels.

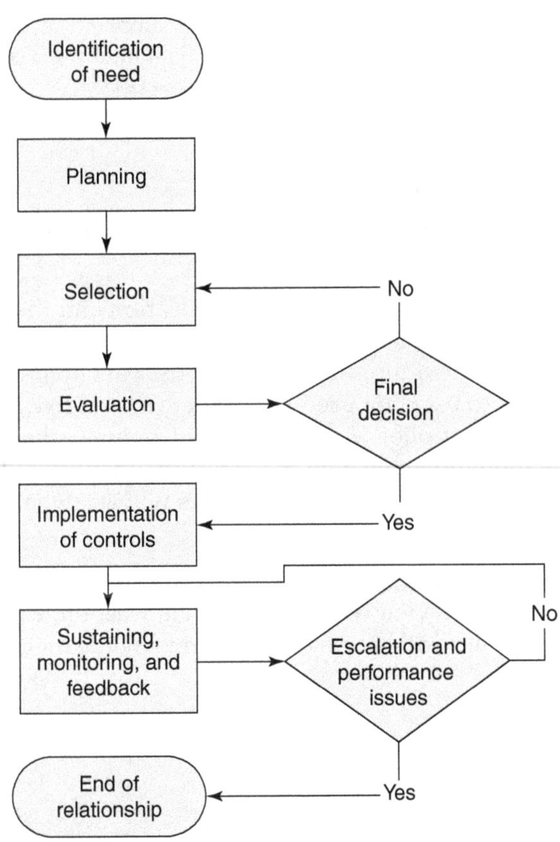

Figure 11.1 Supplier lifecycle.

Table 11.2 Supplier risk classification.

Category	Risk level	Commodities	Controls
1	Critical or high	Finished medical devices Contract manufacturers Outsourcing of software used in manufacturing	Quarterly/annual assessments Dun & Bradstreet reporting Process control plans Incoming inspection plans Monthly reporting
2	Medium	Precision components Printed circuit board assemblies	Annual assessments Dun & Bradstreet reporting Process control plans Incoming inspection plans Quarterly reporting
3	Low	Noncritical components Off-the-shelf items, noncritical	On-site assessments every two to three years (if necessary) Incoming inspection plans Supplier survey provided by the supplier Annual reporting
4	Negligible	Janitorial services for nonmanufacturing areas Office supplies	Paper survey provided by the supplier Monitoring based on areas that are serviced No reporting required

Before starting the supplier performance review process, take the following steps:

- Decide on the format of the communication and the information that will be sent to the supplier. We recommend sending the actual monthly metrics, the three-month trends, and the backup data. There's no reason not to use data used to populate a total risk factor (TRF) matrix to show the supplier where they stand against other suppliers, as long as you anonymize the data to protect the identity of the other suppliers (Shore & Freije, 2016). Good suppliers will ask what it would take to improve.
- Determine the time interval between sending the report to the supplier and the actual review. We recommend at least three to five business days so the supplier will have a chance to review the revised report prior to the meeting.

Table 11.3 OIL.

Date opened	Action item	Assigned to	Current status	Status*

*Green = on schedule, Yellow = may be late, Red = late

- Decide who will send out the reports and be the point of contact for setting up meetings and receiving any feedback. This will prevent any confusion and duplication of effort on the part of both companies.
- Decide on the agenda for these meetings. We recommend using the following as a standard agenda and modifying it to suit your needs:
  - —Confirm that information was received and address any discrepancies
  - —Discuss any open concerns that have arisen since the last review
  - —Discuss any open actions, e.g., SCAR responses

Once the suppliers are identified and the frequency of the information transfer has been determined, you need to choose the method of review by doing the following:

- Determine the day of the month that reports will be created and reviewed by key people, e.g., supplier quality, supply chain
- Determine the day of the month that the reports will be sent to the supplier along with the supporting data
- Schedule reviews with the critical or high-risk suppliers first, then the next level of suppliers (medium risk) based on the required intervals
- We recommend using videoconferencing to conduct meetings. Most laptops and monitors have video cameras installed, and videoconferencing software is relatively inexpensive compared with the cost of face-to-face meetings.

For the majority of these meetings, you won't need more than 15 minutes to check in with your supplier and verify the accuracy of the information. Create an open issues list (OIL) to track any follow-up items (see Table 11.3). After the meeting, be sure to e-mail everyone the notes and OIL. Also, document the meeting in your QMS records. Remember, if the meeting isn't documented, it didn't happen.

SUPPLIER PROCESS PERFORMANCE

> Define and implement lean principles and applications such as 5S, kaizen, value stream mapping, supplier process capabilities and controls, 8 wastes, single minute exchange of dies (SMED), kanban, muda, standardized work, takt time, and error-proofing to reduce waste and increase performance. (Evaluate)
>
> **Body of Knowledge IV.A.3**

Improvement usually means doing something that we have never done before.

—Shigeo Shingo

There are many lean principles that can be applied when evaluating supplier process performance. Of course, process control plans, which were discussed in Chapter 10, are critical in demonstrating the proper controls. Process control plans identify the critical inspection points, which can generate information that can be converted to usable metrics.

For example, say a company uses a pull test to determine whether a certain welded joint meets the minimum strength value. As this is a minimum value, the result is only a pass or fail. However, the test provides an opportunity to track the actual variable data, which shows how inconsistent the actual values are from lot to lot. When the company reviews the data from the past six weeks, a significant drop in the pull strength is discovered. Research reveals that a materials supplier made a material change in the past month. The current material is immediately contained until a determination can be made that the material can continue to meet the specification.

Implementation of Lean Principles

Lean manufacturing principles are the foundation of any successful manufacturing company. The implementation of these principles can be done in any order, as long as the implementation is effective and sustainable. Figure 11.2 outlines the lean manufacturing principles.

The following quality tools are commonly used by suppliers:

- 5S
- Kanban
- Standard work
- Single-minute exchange of dies (SMED)
- 8 Wastes

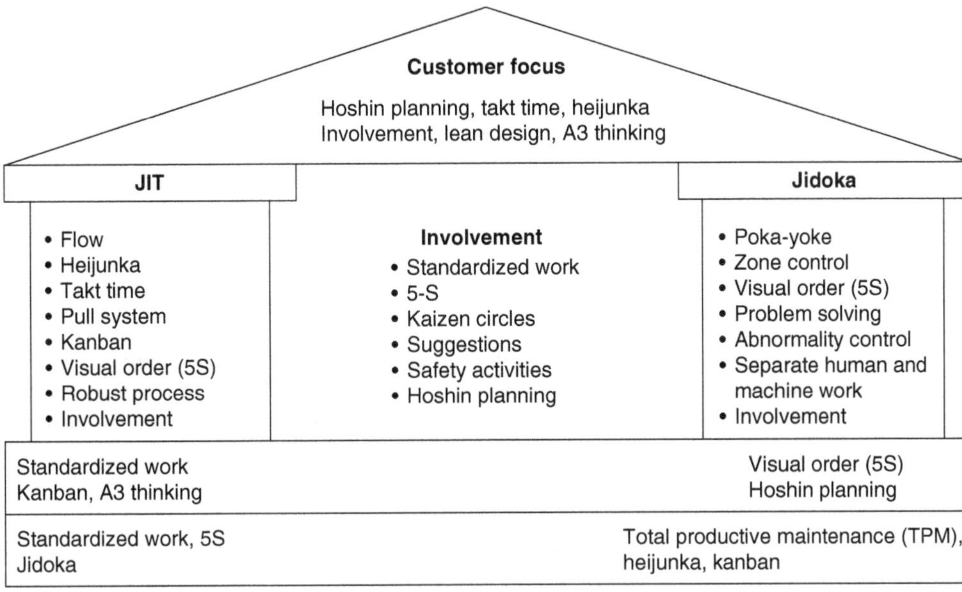

Figure 11.2 Lean manufacturing principles.

- Value stream mapping
- Takt time
- Mistake proofing

5S

A simple way to explain the 5S methodology is "a place for everything and everything in its place." The value of implementing a cultural tool such as 5S is the creation of a quality environment that promotes safety and efficiency.

Some of the benefits that can be realized are:

- Improved safety—e.g., trip hazards are removed and guards are implemented
- Lower defect rates—e.g., the right tools are available
- Better asset utilization—e.g., people are not looking for tools, which can waste time

The 5S methodology is broken down into five steps:

1. Sort—Eliminate whatever is not needed
2. Straighten—Organize whatever remains
3. Shine—Clean the work area
4. Standardize—Schedule regular cleaning and maintenance
5. Sustain—Make 5S a way of life

Safety should be regarded as the sixth "S." It should be considered throughout the 5S program.

Kanban

Let the flow manage the process, and let not the management manage the flow.

Taiichi Ohno, Toyota Production System:
Beyond Large-Scale Production

Kanban is a Japanese word that means "visual signal" or "card." The signal is to perform a specific task or activity in a certain quantity. This technique is used in production and business processes that rely on a pull or signal to do the work. In normal production settings, the historical approach has been batch building or lot production. Even in business processes, paperwork does not go to the next person until the first person is done with it. This creates bottlenecks in the process and causes delays. The end result can be other process steps waiting for work to be sent to them (loss of utilization), overtime to get orders done on time (overutilization and loss of revenue), or late orders (unhappy customers and potential loss of customers).

Kanban provides an even tempo or level load across all of the steps in a process. No one step in the process creates more than the next step can handle, and everyone is utilized. The other important aspect of kanban is the control of inventory from your suppliers as well as your own inventory.

The major benefits of kanban are:

- Less expediting with your suppliers
- Right parts and quantity when you need them
- Less inventory to count
- Reduction in overall costs
- Increase in inventory turns
- Better use of resources

Cost Associated with Inventory

The discussion of TCO earlier in the chapter explained the cost associated with holding inventory that is not turned into revenue. The true cost of inventory can be between 10% and 20% of the actual total dollar value at the component or piece part level and up to 30% at the subassembly level.

There are many metrics that can be used to analyze inventory levels, but the most common is called inventory turnover, or inventory turns. This measure looks at how much inventory is on hand versus how much is being used to create sales. The lower the number, the less inventory is being used to create sales, meaning inventory is higher than necessary. For general manufacturing, average inventory turns range from five to seven. Inventory on hand that is not being converted to sales is tying up cash that could be used for capital improvements or quality initiatives.

Kanban uses the production consumption rate to determine the right amount of inventory. Once that quantity is consumed, the signal is sent to the supplier to replenish the quantity required.

Risk Associated with Inventory

If a quality issue occurs, there may be problems with screening and sorting large amounts of inventory, which could impact the current scheduled deliveries. If this is the case, the entire lot must be quarantined; if the issue affects several production lots, this could stop all production.

Other risks associated with inventory include the following:

- Obsolescence of inventory due to technology or engineering design changes
- Damage caused by handling during general stockroom activities
- Damage caused by handling during inventory counts

Using the kanban method facilitates the traceability of components through a kanban replenishment system. Each kanban shipment has a unique number, so if a quality issue occurs, tracking back to the shipment and manufacturing lot by the supplier is easy (see Figure 11.3). If segregation of discrepant material is needed, the individual kanbans can be identified quickly. The quantity of each kanban shipment is much smaller than the total production order, so there is less inventory to identify and quarantine. Smaller quantities usually have less damage due to handling and can be counted very easily without opening the package, since the label lists the quantity.

Following are a few things to note about Figure 11.3:

- The purchase order number, part number, revision, and quantity are listed
- The kanban card number is a unique number

Figure 11.3 Kanban card.

- The bin location is the point of consumption, so when the parts are received the items can go directly into the work area

The basic process of kanban can be described with the following process flow:

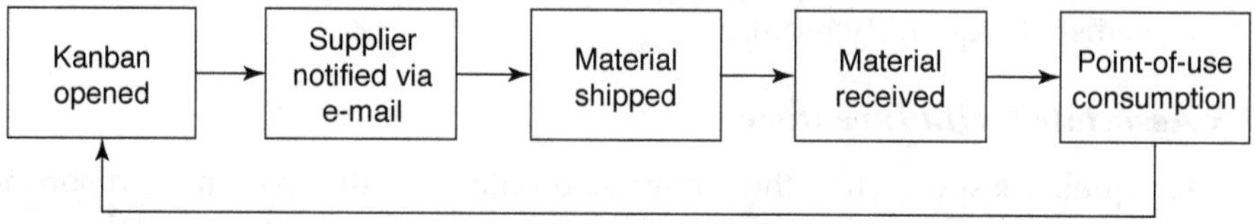

The specific steps in the process are easy to understand. Once the kanban is opened, the signal is sent to the supplier, e.g., via e-mail. The supplier sends the product and the material is received by the customer. The product is brought to the work cell or the point of consumption. The process starts by determining the number of kanbans required in this loop.

Standard Work

The most dangerous kind of waste is the waste we do not recognize.

—Shigeo Shingo

One of the most powerful tools in the lean toolbox, but one that is not always used to its full potential, is standard work. This tool allows best practices to be documented so that specific work can be performed by all workers within a certain time frame. Standard work is similar to kanban in that it is based on a signal from the process to perform a specific task.

Standard work is common in heavily regulated or customer-critical industries such as aerospace, medical devices, and automotive. In these industries, not following standard work can result in the catastrophic failure of equipment and potential harm to a customer, patient, or care provider.

Procedures and work instructions require proper revision control as well as change management. No one should make any changes without the team agreeing to the change, and the change should be properly tested to make sure it doesn't cause further issues.

The process to create procedures and work instructions can be started early in product development, but it can also be used to help standardize current production methods. The basic steps to create standard work are:

1. Identify the scope of work
2. Determine the process and draft the initial document
3. Determine what tools are required
4. Determine the amount of time required to make the part or perform the task
5. Test assumptions and document the work

There are many templates and documents that can be used. Figure 11.4 shows a recommended template that can be found on the Lean Enterprise Institute website.

Process Study Sheet														
Process study	Process:	Product:					Observer:					Date/Time:		Page /
	OPERATOR											MACHINE		
		Observed times												
Process steps	Work element	1	2	3	4	5	6	7	8	9	10	Repeatable	Cycle time	Notes

Figure 11.4 Standard work template.

Source: Lean Enterprise Institute, "Standard Work Process Study Sheet," December 7, 2012, http://www.lean.org/common/display/?o=2192.

The scope of work should be outlined or documented on a flowchart. The work content should be balanced and defined in a specific manner that flows well within the work cell and reaches a specific test or acceptance point. We recommend that there be some form of test or acceptance at the end of the process. That way, you can be assured that the product is correct before advancing it to the next step in the process.

The workers in the work cell should follow the same steps. If a worker has an idea for an improvement, that idea should be vetted by the team and the document should be updated. It should not be a surprise to hear that the product has been built in more than one way up to that point, which is the reason for this exercise. At this point in the exercise, the best method is to have the steps written out with critical points to follow. A worker should be selected to build the product following the process outlined in the document while someone takes pictures and everyone else observes (their job is to document anything they see during the build process). After completion, the team should debrief and get feedback from everyone. The team should vet the changes and add any critical pictures to help explain the instructions.

Once the process scope and steps are defined, the amount of time needed to produce the part or complete the task should be estimated. Again, a worker should be selected to determine the initial baseline against the standard. If possible, have at least two or three workers perform the same work and record their times. This should help identify any steps that take longer and that might be improved with additional fixtures or training.

The best standard work has pictures to accompany the written instructions. Visual instructions are often easier to understand and can help with training new people.

Single-Minute Exchange of Dies

When lean principles are implemented, production lot sizes will become smaller, which may necessitate more setup changes. However, with proper planning these changes can be minimized and done quickly. The goal of SMED is to complete these changeovers in less than 10 minutes. Before using this tool, it is very important to understand the process and tooling. Setup time can be greatly reduced by using standard tooling or fixturing.

The benefits of using SMED methods include:

- Overall lower costs, since faster changeovers mean higher equipment utilization
- Fewer overages, since the lot sizes are smaller
- Smaller lot sizes, which allows for flexibility in meeting various customer requirements
- Lower inventory, since the quantity made is what will be shipped as needed
- Less inventory tied up in production, which increases cash flow that can be used for business improvements

Document the Standard Work

When the initial work has been done, the standard work must be documented. As discussed earlier in the chapter, you need to write down each step, what tools are needed, and how long each step takes. Figure 11.5 shows a template that can be used. This is another instance where the power of observation is very important. The idea is to have people just observe and take notes. Another approach to documenting the work is to film someone doing the steps.

Identify which tasks can be done before and after the change (external activities). Go over the list of tasks and ask whether any need to be done while the machine is still running. For example, getting the tooling and needed tools staged next to the machine can be done off-line. These activities usually save 30% of the overall time of the SMED during initial improvements.

Examples of SMED include prewarming a die, having quality assurance (QA) perform the final inspection while the machine is still running, and using standard tooling points so tools can be switched over without any alignment time. These activities usually save 30%–50% of the SMED during initial improvements.

Once the improvements have been made, standardize the work and ensure everyone follows the new method. After the process has become the new standard and everyone is comfortable with it, conduct another observation session to see what other improvements can be made.

Kaizen

One of the most powerful improvement tools with the least amount of cost is the use of Kaizen. The waste of resources that are usually overlooked by most people can be easily seen by those trained in this tool. The only things that you need are a Senior Sponsor, a small group of employees, and a goal: improvement. The Kaizen tool can be used not only in manufacturing but Quality and Business processes as well. These improvements are not enormous changes but incremental changes that have very powerful positive results. This comes from the fact that the employees are part of the solution.

The ideas for improvement can come from many different inputs, but the best ideas come from the actual employees. The Senior Sponsor is usually the final decision maker regarding which idea will be worked on first. The facilitator of this improvement will hold what is called a "Kaizen Event." First, the facilitator defines the rules and the expectations of the team members and provides some basic training on Kaizen. Then, the problem that is to be solved is identified. The only restrictions are time and budget; these are usually a couple of days and limited money.

Some of the elements that are part of a Kaizen event are mapping out the process, identification of the issues, and brainstorming improvement ideas. The changes are incremental, which allows them to be done and tested quickly. Once the team agrees on a change, they try it on a larger scale, and the team evaluates the change.

An example is the process of receiving and shipping products by the same dock entrance. Due to the size of the area, it became very unorganized and difficult to accomplish either task effectively, resulting in late shipments and delays in receiving inventory. The Senior Operations Leader agreed to hold this two-day

Line	
Machine	
Operators	

Required setup tools	

Standard setup time	

No.	**Task/operation**	**Actual time**		**Improvement**	**Target time**		**Necessary activities**
		Internal	**External**		**Internal**	**External**	
1							
2							
3							
4							
5							
6							
7							
8							
9							
10							
11							
12							
13							
14							
15							

Figure 11.5 SMED standard work template.

Kaizen event with the goal of reducing the time for receiving a package within two hours of receipt and within the same day for delivery once it was put in the shipping area.

The team included the shipper/receiver, a manufacturing technician, a quality engineer, and a maintenance worker. The shipper/receiver explained to the team the process for each activity, and the team created the process flowcharts. Then each team member was given three red stickers to place next to the process step they felt would have impact if changed. The team went over each idea, explained why it would be better, and determined the best suggestions. All the items were removed from the area, and the process for receiving was observed. Only the items that were needed were returned to the room. The same process was followed for shipping, and again only the items that were needed were brought back into the room. Using this process, they eliminated all the unnecessary items that were taking up space.

To help improve the organization of the room, the maintenance worker built shelves to hold the boxes, which created more floor space and better organization. The manufacturing technician created a process for identification of priority parts when they were received so the delay was shortened. The quality engineer found a software package that allowed the shipment information to be pulled from the computer system to make shipping labels upon demand. After two days, the team presented the solutions to the Senior Sponsor that reduced the time to receive or ship to within one hour of receipt in the area. The total cost of materials was offset by the items that were removed and sold to a local company, such as racks and an old forklift.

If you integrate the kaizen principles into your workplace, you will see noticeable improvements right away and great improvements in the long term. You will see improvements at the workstations, in the office, and in your employees. And, you will get these results without spending a ton of money on new equipment. This is because the results come not from fancy innovation, but from the creation of a workforce that is actively involved in making small improvements to the workflow. It doesn't get any better than that.

Value-Stream Mapping and Takt Time

Value Stream Mapping (VSM) is very similar to processing mapping but provides more valuable information that can be used in conjunction with the other process improvement tools. The VSM tool can identify waste, reduce cycle time at each step as well as the overall process, and identify process improvements (see Table 11.4). As with any quality tool, team effort is usually the best way to

Table 11.4 Value Stream Mapping, The Kaizen Kick-Off (adapted from https://asq.org/quality-resources/lean/value-stream-mapping).

Day 1	Day 2	Day 3
• Introduce VSM. • Determine process families. • Walk the flow and gather information.	• Draw the current state map. • Perform lean concepts training.	• Create the future state map. • Develop the draft plan.

create the most powerful improvements. The cross-functional team is formed that includes a representative for each functional role. Sometimes, the inclusion of experts and suppliers may be needed to help the team be effective.

Once the team is formed, it holds a kaizen event, usually for three days. Note: It's important to include an individual with experience creating and drawing VSMs to help facilitate the kaizen event.

There are four important steps that need to be accomplished:

1. Determine the process family or process
2. Draw the current state map
3. Determine and draw the future state map
4. Draft a plan to arrive at the future state

The process family or process is the gathering of the group of products or processes that go through the same or similar processes. That allows the identification of the process that affects the most items so that increasing its efficiency will result in a large impact. In addition, the smaller items can be identified to determine if they are still needed or can be done in some other manner, such as using a different production line or through an alternative process.

The next step is to create a process map that reflects the "as is" process by walking the actual process, observing the steps, and interviewing the people. This is usually done independently of each team member so that they don't bias each other regarding what they observe. In addition, this gets the employees of the process to share their ideas and gain involvement. Figure 11.6 displays common VSM icons, which should be used to describe what the team members saw in the same manner as completing a flow chart. However, VSM goes deeper into the details of how the process flow actually works and identifies where delays occur or where actual excessive inventory might show up in a certain work step.

As we see in Figure 11.7, the VSM example shows the details of the process in detail as well as the value-added and non-value added time. The "gutters" on the top and bottom show those different times, which help identify not only the areas for improvement but also the potential process cycle time that needs to be improved. Those improvements can be achieved by changing the process steps, breaking the work up into smaller or sequential steps, or adding more or more efficient equipment.

The next step is probably the most critical because we are creating the future state of the process. Knowing how it works now, we have the insight to identify the changes that address the inefficiency and non-value-added activities, known as the 8 Wastes (to be discussed in the next section).

In order to make improvements, we first need to know how fast we need to generate output from the process. In lean terminology, we call this "Takt Time." Takt is a German word for the baton that a conductor uses to control the speed or tempo of an orchestra. So, we use takt time to determine how fast we need to make product to meet customer demand. The formula is very simple:

$$\text{Takt} = \frac{\text{Time available}}{\text{Demand for the product}}$$

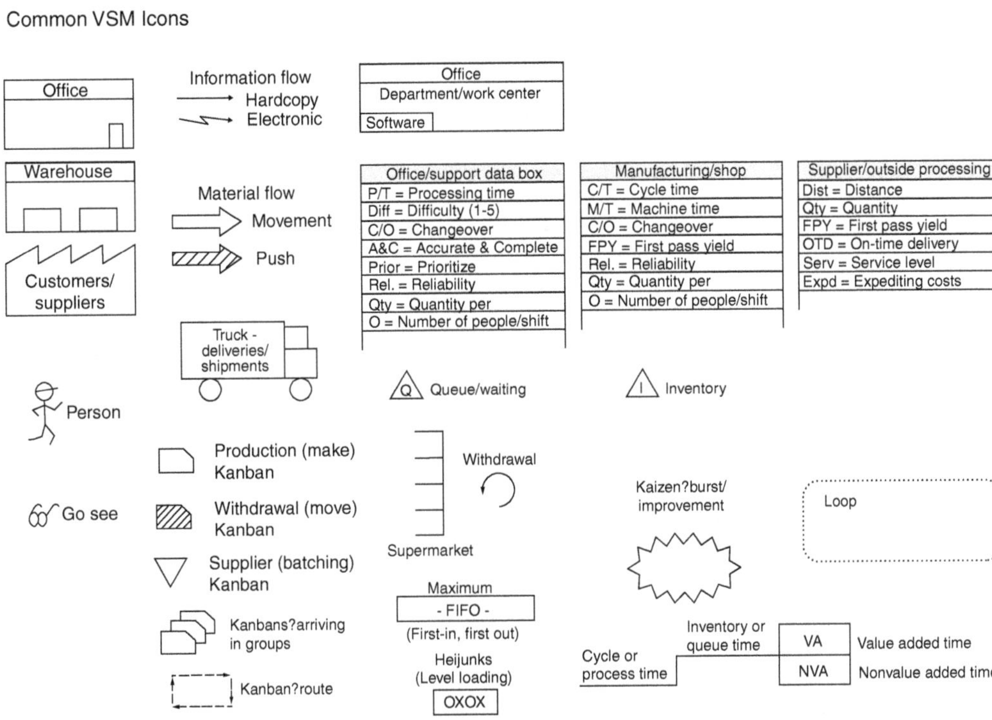

Figure 11.6 Common VSM symbols (adapted from https://asq.org/quality-resources/lean/value-stream-mapping).

So, if we have 40 hours in the week (2,400 minutes)/100 parts for each week = 24 minutes/piece. This means that a finished product has to be coming off the production line every 24 minutes.

Now that we know what the current process is to make the product and what we are trying to reach, we can use the VSM to identify the areas for improvement. Some of the steps that we can look at to make these improvements are:

- Bottlenecks or constraints – are there steps that take longer than others and material queues build up waiting to be processed? Are there places in the process where there are delays waiting for products to come to that station? The constant push/pull and tug of war only cause delays and overtime costs. The process should not be "choppy"; it should flow smoothly.
- Can we create buffer stock or safety stock inventory to help the flow of the process? These in-process queues may be needed where process steps cannot be done faster or the equipment is limited. For example, a heat treating step may take longer to complete, so in this case, we should consider looking at smaller furnaces that can process smaller batches or even create a continuous flow line so the next process step is not waiting.
- Other process improvements – are there steps that can be shortened or work cells that can be moved next to each other to eliminate any delay

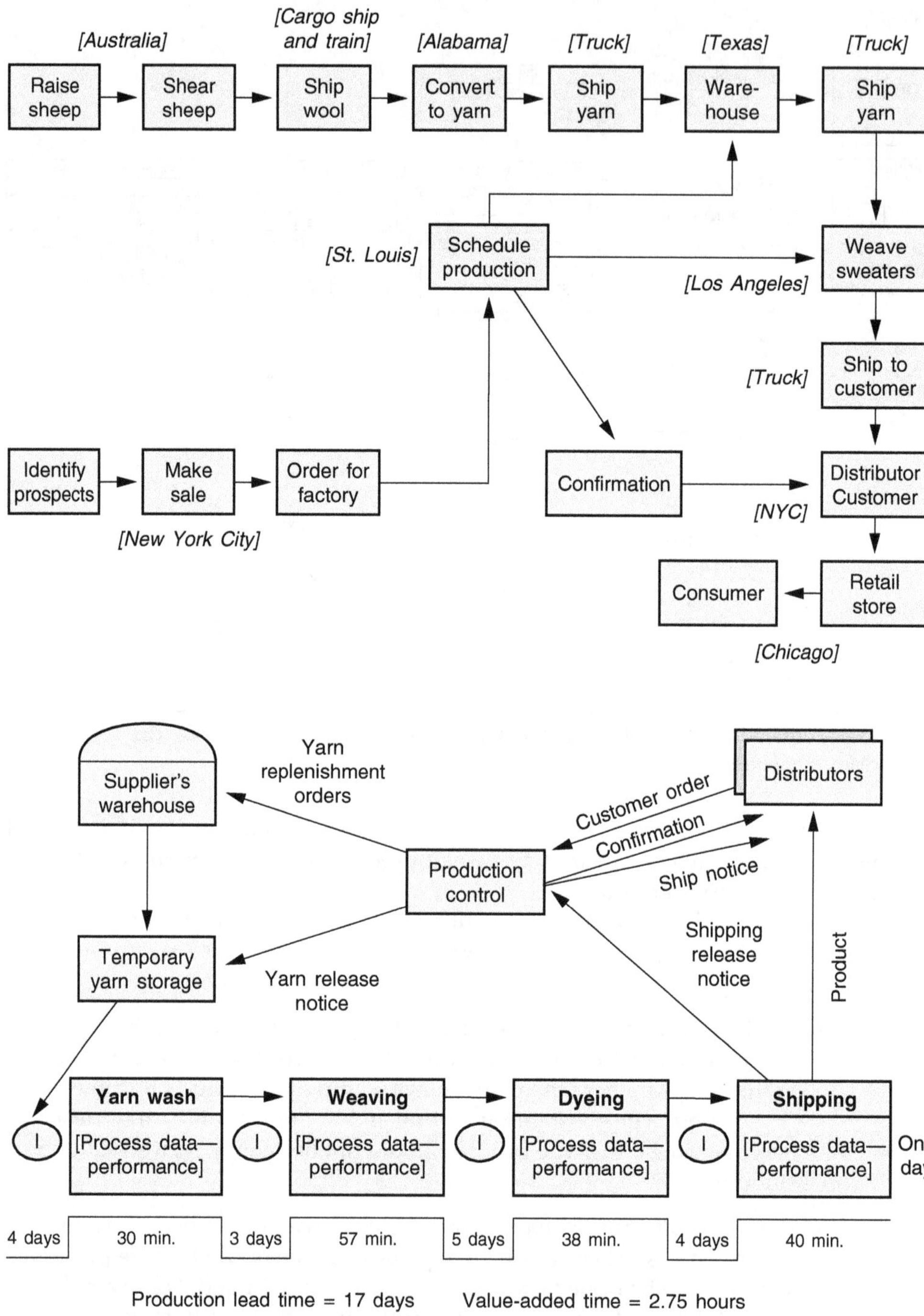

Figure 11.7 Example VSM, macro (top) and partial (bottom) (adapted from *The ASQ Certified Quality Manager of Quality/Organizational Excellence Handbook*, 5th Ed., ASQ Quality Press).

in moving the material? Another idea can be to look at what tools are needed in the work areas to see if the right tools are present at the right locations. The elimination of the need to search for tools or leave the work center to find tools can make improvements as well.

8 Wastes (Muda)

Waste is defined as any action or step in a process that does not add value to the customer or that the customer does not want to pay for. The original Seven Wastes were developed by Taiichi Ohno, the Chief Engineer at Toyota, as part of the Toyota Production System (TPS). The 8th waste of non-utilized talent or skills of workers was later introduced in the 1990s when the Toyota Production System was adopted in the Western world.

A good way to remember these wastes is through the acronym DOWNTIME:

- Defects - products or services that don't meet company standards. By using Quality tools, you can identify where and how many defects are made and any needed changes. If the process is not capable of making 100% good parts, identify methods that can quickly identify and remove defects.
- Overproduction - producing more of a product than customer demand requires. This can be identified via the Value Stream Mapping activity as well as by understanding customer demand. Having more inventory does not always add value.
- Waiting - the downtime between steps in a production process. These situations can be found using the VSM where one process step can't produce enough to keep up with the next step. We identified ways to address this in the previous section.
- Non-utilized talent - not fully using the time and skills of employees. By employing better manufacturing methods, you can better utilize the people in the production process. For example, providing better equipment training may reduce downtime of the equipment since the employees are trained to look for issues and fix them before they happen.
- Transportation - unnecessary movement of materials in production processes. These can quickly be eliminated by moving the work cells closer together so that the time between cells is minimized. There may be cases where you have to move the people closer to the equipment due to constraints and costs of moving the equipment.
- Inventory - making/storing more than required to meet customer demand. This is very similar to overproduction and can be addressed by using takt time or implementing a supermarket or single-piece flow process.
- Motion - people or equipment moving more than necessary. While transportation waste involves the movement of product, this involves people moving to different stations or areas in a work cell, leaving to

find tools, going to the stockroom to get more parts, and so on. Simple observations identified in the VSM should help identify and eliminate these issues.

- Excessive processing - steps that don't add value for customers but cost resources. For instance, consider a case where employees make the finish on a product better than the stated drawing requirements. They take pride in their work, but in some cases, the finish may not be even visible to the end customer, e.g., a bracket that holds up the exhaust in a car. The extra time does not add value, and the customer does not need it.

Error Proofing

Error Proofing or "Poka-yoke" in Japanese is the use of tools and methods to make it impossible for an error to occur or to make an error immediately obvious when it has occurred. The use of mistake proofing is very common in today's world, such as the USB connector that can only be connected one way or color coding of wires to aid in correct connection to prevent short-circuiting.

In manufacturing, parts and jigs can be designed so that parts can only go in one way. In some cases where metal parts are over-molded with plastic, the part must go into the drilling jig with the metal part first. The magnet on the other side of the jig allows the power connection so that the drill will work. If the part is placed backward with the plastic part next to the magnet, it will not complete the circuit and the drill will not work.

Chapter 12
Assess Nonconforming Product/Process/Service

> Segregate, control, and evaluate nonconforming materials to determine whether a material review board (MRB) requires disposition. Conduct risk assessments to prevent future discrepancies. (Evaluate)
>
> **Body of Knowledge IV.B**

There will be times when the processes at your facility create nonconforming products, and depending on your organization, those issues may fall into your area of responsibility. Some organizations have minimally staffed quality departments, so the quality professional may need to wear many hats. Here we are looking specifically at the role of the Supplier Quality Professional (SQP), in particular the SQP handling the of outsourced suppliers in your process. These suppliers may be an exceptionally large group of diversified suppliers depending on the processes performed at your facility. Manufacturing facilities have raw material suppliers providing everything from sheet aluminum to agricultural products, chemicals, and even water as an initial ingredient. Packaging facilities receive final customer packaging, cans, boxes, or bags already printed or decorated for final packaging. Some of your suppliers may provide services on your in-process products that are processes your facility does not specialize in. Heat treating of metals, specialty finishing such as electroplating, and even in-process material testing might be better suited to being performed by certified facilities. If your facility assembles complex products, your suppliers may provide intricate subassemblies or components. Keep in mind that, as the SQP, you hold your suppliers responsible for the products, processes, and services they provide your organization. It is their task to do the same with their suppliers. These products, processes, and services, as supplied to you, are the topics of concern in this chapter.

Even with the most stringent supplier qualification programs, nonconforming material will occur. What then? The first step is to have a formalized definition of nonconforming material. Nonconformance to what? To some type of inspection, verification, or test activity. If no inspection criteria, verification processes, or testing activities have been established, by definition, there cannot be nonconforming material. The material requirements are usually defined well before production starts. Blueprints, specifications, and the customer's needs (not necessarily their desires) dictate the parameters to be inspected, verified, and tested.

IDENTIFYING NONCONFORMING MATERIAL

Inspection protocols should be established for incoming materials. Generally, these are designed to be a type of incoming audit based on the results of the supplier qualification process, along with the supplier's performance records. The inspection frequencies and sample sizes may be changed as needed to raise confidence intervals to reflect the supplier's performance. The scope of the incoming material audit should be directly related to the criticality of the supplied material, the possible failure mechanisms, and the severity ratings of those failure modes. For example, if the installation of a defective component would scrap an entire assembly, depending on the potential scrap cost, the inspection sampling should be raised high enough to significantly increase the confidence interval. The best way to identify nonconforming material is internally! Thus, the incoming material inspections should be designed in accordance with the severity ratings established in the FMEA.

It is crucial that nonconforming material be identified in house rather than through customer notification. Customer notification of a problem indicates an escape of nonconforming material and is a critical failure of the quality management system. Once the nonconforming product leaves the confines of your processing or manufacturing facility, liabilities can increase dramatically. The severity of a nonconformance can range from minimal to severe, as illustrated in the following examples:

- *Minimal problem.* The red in the logo printed on the carton holding 12 cans of soda is darker than the red specified in the art mock-up requirements.
- *Very minor inconvenience.* The carton holding 12 cans of soda has a required perforation that allows consumers to set the case up as a refrigerator roll-out dispenser. The perforation is shallower than the specification, which results in the carton not tearing cleanly and easily along that perforation. The consumer may not even notice the minor inconvenience.
- *Product ease-of-use characteristic.* The perforation referenced above is missing from the carton.
- *Product not meeting standard of acceptability as advertised to the consumer.* The pop-top on the soda can cannot be opened. While the consumer will notice the problem, the beverage can still be opened and consumed. The consumer is very likely to notify you of this type of defect.
- *Product not meeting the consumer's perceived level of quality.* The flavoring in the soda in the can is incorrect. The consumer will notice, and the brand's image may suffer.
- *Customer health and safety risk.* The raw materials in the soda are contaminated with a toxic cleaner, creating an immediate health hazard if consumed.

NONCONFORMING MATERIAL SEGREGATION, ISOLATION, AND NOTIFICATION PRACTICES

Segregation, isolation, and notification processes involve many internal activities, and again, depending on the staffing in your quality department, many of these activities may be performed by the same quality professionals. Once the outside supplier's materials or services have been determined to be nonconforming, stop using them. While this seems like a very simple principle, production departments often forget that "no production" is better than "hold production." Minimize the effects of nonconforming material on your process and isolate all suspect product. This can result in two separate non-conforming lots, one of raw material non-conforming, and a second lot of final product produced with the non-conforming raw material. A well-segregated nonconforming finished product lot will have good product on both ends of the lot, and all nonconforming production will be within the boundaries of the hold lot. If your incoming material traceability is effective, the non-conforming raw material should be isolated to a single lot. The SQP is critical to the proper segregation & confinement, often referred to as containment. Containment may all be localized, or it may be in multiple locations. If the production line with the defect was direct loading onto outbound trucks, the likelihood that all suspect production is contained within the plant will be low. Identification, isolation, and marking will ensure containment. Once identification & containment are complete, the next step is to notify the supplier. The SQP's role now shifts to effectively assist the nonconforming material supplier in root cause analysis and preventive action implementation. This starts with initial data. Gather the best data you can for the initial notification. If you contact your supplier with vague complaints (e.g., "It's junk," "This is a terrible lot of material," "We can't run it"), you will get many more questions than answers. Be very specific and use data. Provide lot numbers, production dates, and job numbers. Let the supplier know roughly when the material was delivered to your facility. Is the nonconforming material direct from a production line or was it pulled from floor stock? Let them know the defect rate that you may have determined. Describe in detail what led up to the initial investigation of a possible nonconformance. Remember, as the SQP you know what type of information is needed to start a root cause investigation and the date you need it by. As the receiver of that information, your material supplier becomes your customer, so give your customer what they need to perform successfully.

DISPOSITION OF NONCONFORMING PRODUCT MATERIAL REVIEW BOARD (MRB)

Now that the nonconforming material has been identified and all affected production has been held, the next step is to determine what to do with it. This is the function of a material review board (MRB). The MRB should be made up of participants from quality and engineering at a minimum and should seek rework input from operations. The MRB's duty is to determine how nonconforming material will be dispositioned. While the operations department has less input if the product will be reworked, they can help determine whether your facility and people have the ability to perform any rework that may be suggested. Dispositioning nonconforming material is sometimes simplified by the final end-use customer.

Some producers of aerospace, medical device, and other high-end technical products simply cannot allow any nonconforming material to be involved with their final assembly or product. In these cases, such requirements are normally stated very clearly during initial product development stages. These situations should be well defined in design FMEA projections. Now, what about those products or customers that have other disposition options besides scrap? Some customers allow MRB authority to be enacted at the site of the identification of nonconforming material. The MRB will follow a process approved by your organization. It evaluates the nonconformance compared with the severity as determined during FMEA analysis and renders a disposition, such as:

- Use as is (either a design change or a one-time use as is)
- Rework
- Repair (again, either a design change or a one-time variance)
- Regrade for alternative use
- Reject or scrap

This is a critical point, as each nonconforming lot (the non-conforming raw material and finished goods produced with the nonconforming raw material) must be dispositioned through the MRB process. Each disposition will be independent of the other lots.

PREVENTIVE MEASURES

The most critical role of the MRB is facilitating the implementation of preventive actions. As an SQP it is your duty to evaluate the preventive actions that your supplier suggests.

A robust preventive action plan starts with a risk assessment. Since you now have nonconforming material, you have some data (facts) to work with. Use the nonconforming material incident report and the data gathered during that investigation as a starting point for a comprehensive risk assessment. The risk assessment has three basic components: risk identification, risk assessment, and risk management.

Risk Identification

Now that nonconforming material has arrived at your facility, you must work with the supplier of that material to identify any other possible risks involved with its product, process, or service. If this nonconforming material was due to a packaging issue with the product, the root cause being an untrained operator, could the same (or a similar) nonconformance be caused by a different root cause, such as a procedural breakdown? Likewise, could the root cause of an untrained operator result in a different type of nonconforming material? In this way, the risk identification being run for one nonconformance could help strengthen many aspects of the supplier processes.

Risk Assessment

Now that the risks have been identified, the impact to the system needs to be evaluated using two specific measures: the *impact* the risk will have on the processes and the *likelihood* of the risk actually occurring. Simple scales (high = 3, medium = 2, low = 1) will do, or you can work with a 1–10 scale similar to FMEA applications. The critical part is to specifically define the high, medium, and low impact and likelihood factors. Once the risks have been assessed in terms of impact and likelihood, the risk management phase can be formulated.

Risk Management

Each identified risk should be prioritized according to the results of the evaluation of the impact and likelihood of occurrence. Each risk should have specific management tasks, and action plans for those tasks should be established. Appendix B provides a sample risk assessment template.

The task list for each identified risk will be similar:

- ***What are we already doing about it?*** What is the current reality of the situation? What current identification and mitigation steps are already in place? What failed in the current system? All systems should be investigated—the system of the supplier's that generated the nonconforming material and the system of yours that missed the opportunity to capture the incoming nonconforming material at initial inspection.
- ***What more can we do about it?*** What is the improvement opportunity? What needs to be changed to prevent escapes from the supplier, and what needs to be improved in our incoming evaluation process?
- ***When will it be done?*** This question can only be answered with a calendar date or a number of days. "ASAP" (as soon as possible) should be considered a non-answer. Set a realistic deadline.
- ***Who will do it?*** This is also to be answered with specificity. "The maintenance department" is not a "who." "The second shift maintenance supervisor" is a "who"; "John Smith" is a "who." Be specific. As soon as we start assigning improvement opportunities to teams, no one is assigned to them.
- ***How will we review progress?*** Again, answer this question with specifics. How much of an improvement is expected? How will the improvement be measured? Increased output? Lower defects per million opportunities (DPMO)? Lower direct labor costs? Finally, quantitatively what is the expected improvement? Is it 5%, 25%, zero defects? You must be specific in the review process.
- ***Review level of risk.*** Once the improvements have been put in place and the results have been qualitatively and quantitatively evaluated and accepted, the risk assessment for that particular risk needs to be reevaluated. If, say, a risk of high possible impact and a high likelihood

of occurring is reevaluated as still having a high impact but now having a very low likelihood of occurring, that risk should be reclassified, and the next-highest assessed risk should become the targeted improvement opportunity.

Chapter 13
Supplier Corrective and Preventive Action (CAPA)

It is inevitable that, on occasion, one of a company's suppliers will provide nonconforming product or services. When this happens, the customer will want to ensure that its supplier performs robust root cause analysis and implements effective corrective actions to identify the true underlying cause of the problem and prevent its recurrence.

Hoerl and Snee (2012) outline a general process for problem solving:

- Define and document the problem
- Identify potential root causes
- Identify potential solutions
- Identify, implement, and test the best solutions
- Measure results and determine effectiveness of the solutions
- Standardize effective solutions

When working with suppliers, this process is generally guided through use of a corrective action request issued by the customer to the supplier. The customer defines the problem that was encountered and enters this information on the formal corrective action request that is sent to the supplier. The customer and supplier should also jointly define the scope of the corrective action request in order to determine if any other processes or products at the supplier may potentially be impacted by the nonconformance.

As part of the problem definition, the supplier should review data provided through tools such as check sheets, Pareto charts, and control charts to understand the extent of the nonconformance and to determine a direction for investigation.

When identifying potential root causes, the supplier should use tools such as 5 Whys analysis, Is-Is Not analysis, cause and effect diagrams, brainstorming, and box plots.

To aid in identifying potential solutions, the supplier should use tools such as the aforementioned brainstorming as well as affinity diagrams and multivoting. When a "best" potential solution is identified, the supplier should implement and test it to verify its efficacy; DOE can be used to test different settings of a potential solution to determine the optimal solution.

Once a solution has been implemented and optimized, the supplier should verify its effectiveness after a reasonable length of time, which is determined based on the cycle time of the process, the volume of product that is produced, and the frequency and speed of feedback from the process and from customers. Pareto charts and control charts should be updated as new data are made available to aid in the analysis and verification of effectiveness.

Finally, once a solution has been verified as effective both internally at the supplier and through the use of customer data and feedback, it should be standardized by updating relevant process documentation, such as flowcharts, FMEAs, HACCPs, control plans, and work instructions. The supplier should also close the loop with the customer by updating the corrective action request documentation and returning it to the customer within the customer's required time frame.

If the solution involved a change to the design of the product, then the supplier should work with the customer to ensure that the customer's change control requirements are fulfilled. At a minimum, the supplier should submit a first article to the customer for review and validation.

ROOT CAUSE ANALYSIS TOOLS AND METHODS

> Evaluate the root cause analysis of a problem using tools such as cause and effect diagrams, Pareto analysis, 5 Whys, fault tree analysis, design of experiments (DOE), brainstorming, check sheets, measurement system analysis (MSA), production records, and review of process flow. (Evaluate)
>
> **Body of Knowledge IV.C.1**

Check Sheets

Check sheets are an invaluable tool for systematically recording data for further analysis. Check sheets can take many forms but they should always include basic identifying information, such as the name of the process and/or product being measured, the location, and the time frame for data collection. Data collectors must also be able to agree on what constitutes a recordable event, which may require the development of a set of documented criteria. Once a check sheet has been developed and data collectors agree on what should be marked as an event, each event is simply entered into the appropriate space in the check sheet as it occurs, usually using a tick mark or an "X."

A check sheet can be as simple as a matrix with reasons/events listed in a column and day, date, or item number across the top (see Figure 13.1). This type of check sheet is useful when attempting to stratify a problem. For example, the team may be investigating several defect types and wish to determine which presents the biggest share of issues, or the investigation may be focused on tracking any cycles or patterns in the data over time.

Product drawings can also be used as check sheets by marking the location of a defect directly on the drawing. In this way, one can quickly determine if defects

Defect	Monday	Tuesday	Wednesday	Thursday	Friday
Program	III	II	𝍸	III	I
Miswire	I		II		
Short		I	I		I
Break/open			I	I	

Figure 13.1 Defect check sheet.

are concentrated in any particular location on the product, which may suggest a direction for further investigation. Figure 13.2 shows an example of such a check sheet. In this example, defects appear to be concentrated on the corners of the product. This may lead the team to further investigate fabrication, material handling, or packaging methods.

Data Collection Sheet
Part number: 50004
Dates: May 2–6, 2016
Line: 2

Record cosmetic defect data for all pieces fabricated this week on this check sheet. For each piece fabricated during the week, mark an X at the location of each cosmetic defect found.

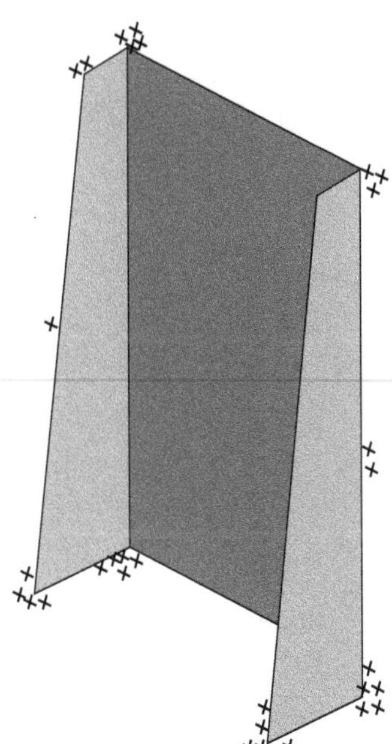

Figure 13.2 Defect data collection sheet.

Pareto Charts

A Pareto chart is a specialized version of a bar chart in which the bars are arranged in descending order. Categories are listed on the horizontal axis, while frequencies are charted on the vertical axis. The ordering of the bars in the Pareto chart provides prioritization of the issues or defects that are being charted. A problem-solving team will want to focus on examining root causes and developing corrective actions for those issues represented by the larger bars on the left of the Pareto chart, as these issues constitute the largest proportion of overall issues.

Figure 13.3 shows an example of a Pareto chart used to track defects in a manufacturing line. In this particular example, collection of a month's worth of data has revealed a clear signal—the team should focus on investigating the "not programmed" defect, as it represents 63.6% of the total defects experienced on that manufacturing line during the month.

Occasionally, a Pareto chart will not reveal a clear signal; there will not be an indication of the "important few" defects that should be investigated further. This usually occurs for one of two reasons:

1. There is not yet enough data. In this case, the data collection time frame should be extended until a signal is revealed.
2. The categories are not correctly identified. The team should reexamine the categories to determine if they need to be adjusted or if a new set of categories (and associated definitions) should be defined. It is also possible that multiple categories are actually very similar in practice and should be combined.

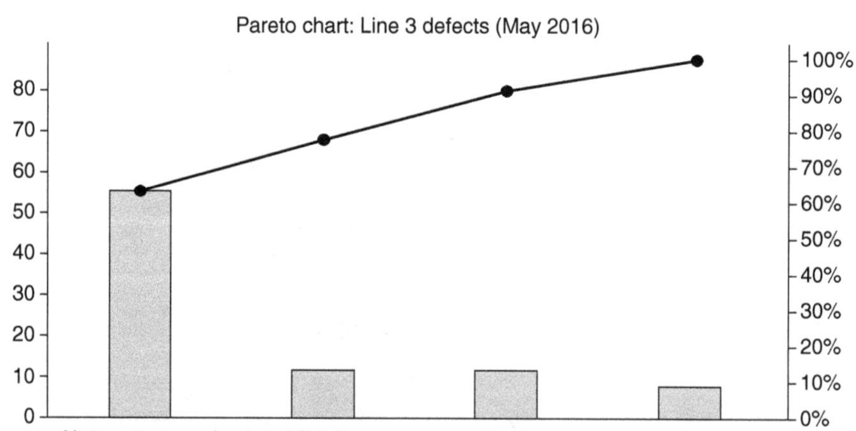

	Not programmed	Miswire	Short	Break/open
Total	56	12	12	8
%	63.6%	13.6%	13.6%	9.1%
Cumulative %	63.6%	77.3%	90.9%	100.0%

Figure 13.3 Pareto chart.

5 Whys

The 5 Whys method is a straightforward yet powerful method of drilling down into an issue to determine its underlying root cause. To use this method, start with the particular issue or defect being examined and ask why this issue occurred. Continue to ask "why" at each step until the answer refers to a process that has failed, is ineffective, or does not exist. While the method is referred to as "5 Whys," it may take the team fewer or more than five steps to reach the true root cause.

There may be instances where the team has more than one answer to the "why" question. In this case, the team should gather and examine data to attempt to confirm which potential answer is correct in this circumstance. Check sheets, Pareto charts, box plots, and DOE may be useful tools in this regard.

Consider the classic example of the use of the 5 Whys method described by Taiichi Ohno at Toyota in the 1950s. His example was that of a robot that stopped operating in the middle of its process:

- Why did the robot stop? Because the circuit overloaded, which caused a fuse to blow.
- Why did the circuit overload? Because there was insufficient lubrication on the bearings, which caused them to freeze up.
- Why was there insufficient lubrication on the bearings? Because the oil pump was not circulating enough oil.
- Why was the oil pump not circulating enough oil? Because the pump was clogged with metal shavings.
- Why were there metal shavings in the pump? Because there is no filter installed on the pump. (Toyota 2006)

Is-Is Not Analysis

Is-Is Not analysis is a problem-solving tool that helps identify and narrow the scope of a problem that works by helping the team think about the problem—specifically what is the problem and what is not the problem. This tool is often used to determine the root cause of a problem when using the CAPA or 8-D problem solving methodology.

Problem	
What is the problem?	What is not the problem?
Where is the problem?	Where is not the problem?
When is the problem?	When is not the problem?
Who is the problem?	Who is not the problem?
How much is affected by the problem?	How much is not affected by the problem?

Fault Tree Analysis

FTA evaluates system (or subsystem) failures one at a time, represented pictorially in the form of a branching tree of fault modes. At each level in the tree, combinations of fault modes are described with logical operators (e.g., AND, OR). FTA relies on experts' process understanding to find causes. A potential area of use for FTA is to establish the pathway to the root cause of the failure. FTA can be used to investigate complaints or deviations to fully understand their root cause and to ensure that intended improvements will resolve the issue and not lead to other issues. FTA is an effective tool for evaluating how multiple factors affect a given issue. The output of an FTA includes a visual representation of failure modes. It is useful both for risk assessment and in developing monitoring programs. Figure 13.4 depicts a basic FTA.

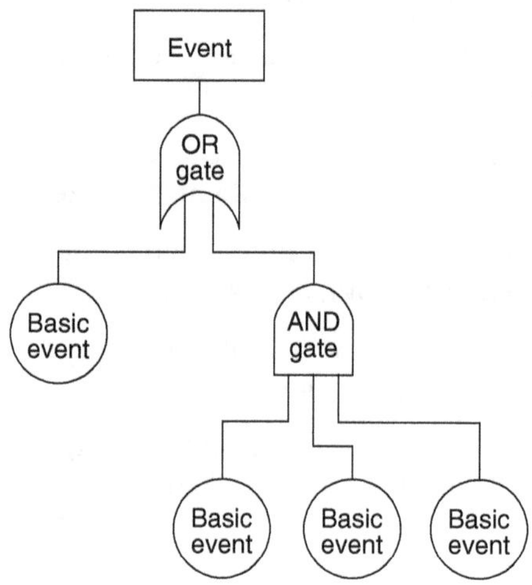

Figure 13.4 Fault tree analysis.

Box Plots

A box plot is a graphical representation of the location and variability of data from a process. Box plots are used for attribute or discrete inputs and continuous outputs. The inputs are listed across the horizontal axis and the values of the output are displayed on the vertical axis.

The box itself provides information about the location of the quartiles of the data. The top line of the box is located at the third quartile (75th percentile), the middle line at the median (second quartile or 50th percentile), and the lower line at the first quartile (25th percentile) of the data. The interquartile range is the difference between the values for the third quartile and the first quartile. Lines extend above and below each box on the chart; the upper line is drawn from the box to the largest value that is less than the third quartile—(1.5 × the interquartile range), while the lower line is drawn from the box to the smallest value that is greater than the first quartile—(1.5 × the interquartile range) (Juran and De Feo 2010, 549). Any value that falls outside of the boxes or lines is charted with a symbol to indicate an

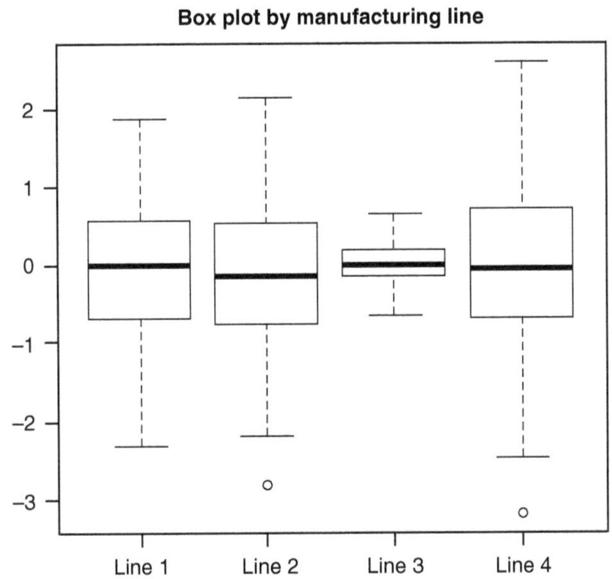

Figure 13.5 Box plot: Line 3 has less variation.

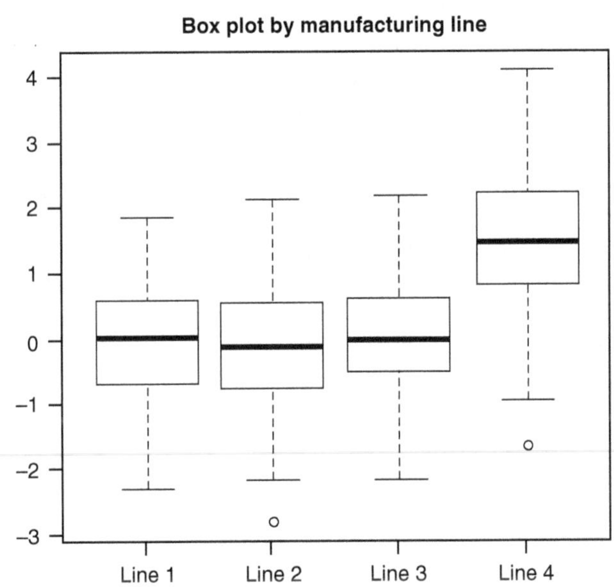

Figure 13.6 Box plot: Line 4 is centered differently.

outlier; the specific symbol used depends on the program that is used to generate the box plot.

If all the boxes on a chart are of similar size and location, then there is not a lot of variation between the groups. Differences in the size of the boxes indicate a difference in the amount of variation between the groups, while differences in the location of the boxes indicate a difference in the values produced by each group. Figures 13.5 and 13.6 provide examples of this. In Figure 13.5, line 3 has less variation than the other three lines. In Figure 13.6, line 4 produces values that are centered differently. In either case, the team should investigate the line that is different to determine the reason for those differences.

Brainstorming

The purpose of brainstorming is to develop a large number of ideas or thoughts in a short time. If a problem-solving team is having difficulty identifying potential root causes of a problem or developing potential corrective action solutions, then the use of the brainstorming technique may be beneficial to jump-start the generation of ideas.

To start brainstorming, the team agrees on the central statement to brainstorm, which is written down and displayed for reference by the team during the activity. Team members contribute ideas related to the central statement, which is generally done using one of two methods:

- Team members methodically take turns giving an idea out loud, going around the room. Each idea is recorded on a flip chart or board at the front of the room. Multiple rounds are conducted, until the team runs out of ideas or a specified time limit is reached.
- Team members write ideas on sticky notes, which are posted so that others may read them. Team members continue to contribute ideas until activity slows down or a specified time limit is reached.

It is essential that all ideas be accepted in the brainstorming activity without comment or judgment, since the goal is to generate a large number of ideas. This also allows team members to piggy-back off others' ideas, which should be encouraged. Brainstorming activity should be kept to a short time frame, perhaps 5–10 minutes. Only after the end of the time allotted should the team review and process the list of ideas generated, and then only to ask clarification questions or to identify ideas that are identical.

Affinity Diagrams

Once brainstorming has been completed, the team may find that an affinity diagram is useful to organize the ideas into categories or groupings. To use an affinity diagram on brainstormed ideas, each idea should be recorded on a sticky note, which is posted on a wall or whiteboard. The team then works silently to move the ideas into natural groupings (see Figure 13.7).

The team then creates a header for each grouping. To do this, the team analyzes each idea within a grouping to determine what the common theme is among all of the ideas. The team should also verify that all of the ideas within a grouping actually belong together and determine if any need to be relocated to different groupings.

At this point, the team may also group several similar groupings into a larger grouping or may subdivide one large grouping into multiple subgroupings. Each grouping and subgrouping should have its own header.

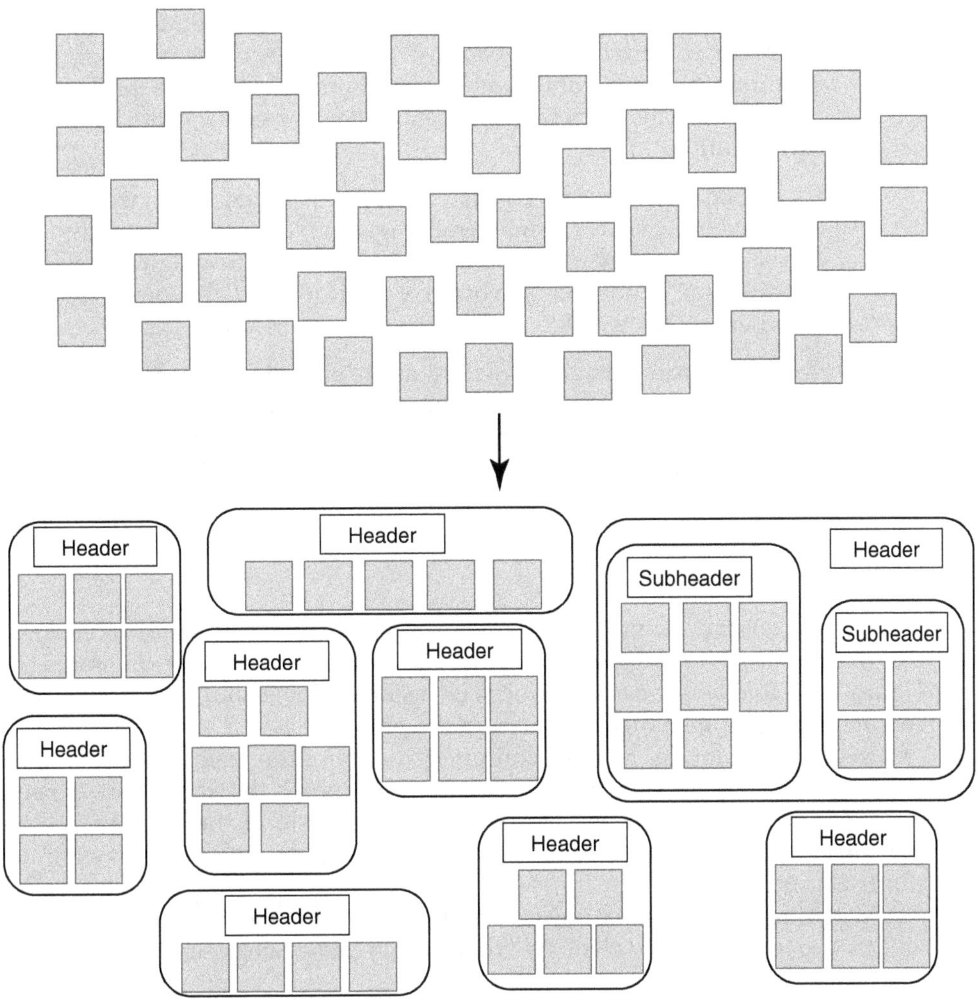

Figure 13.7 List of brainstormed items organized into an affinity diagram.

Multivoting

Multivoting is a tool that can be used to determine which ideas have support from the team and what the team feels the highest priorities should be. Multivoting is useful during the development of potential solutions to aid the team in identifying those solutions that the team feels should be implemented first.

Once the team decides to use the multivoting method, each potential solution should be listed on a flip chart or board in front of the team. At this point, the team may use any of several voting methods:

- Each team member prioritizes all of the items on the list. Team members take turns revealing their rankings, and a score is assigned to each item, usually using a "reverse order" ranking (assigning a value of 1 to the lowest ranked item, 2 to the next-lowest ranked item, and so on).

- Each team member votes for a subset of the solutions that is most important to him/her. The team specifies the size of the subset of solutions that each team member may choose (such as one third or one quarter), and the team adds up the number of members who chose each solution.
- Each team member is given a fixed number of votes, which may be allocated however the team member wishes. A team member may choose to assign one vote to multiple solutions, or to assign all votes to one solution, or anywhere in between. This method allows each team member to add "weight" to the voting for any solution.

Scores or votes for each item are added up at the end, and the team reviews the results to see if consensus has been reached on any items. Once consensus has been reached, then the team can begin to plan for implementation of the consensus solutions. It may take multiple rounds of multivoting for the team to reach consensus on a reasonable subset of potential solutions.

Cause and Effect Diagrams

A cause and effect diagram, also referred to as a fishbone diagram or an Ishikawa diagram, is used to organize and display potential causes of an issue (see Figure 13.8). It allows the team to focus on a stated problem and develop ideas or consensus around the potential causes of the problem.

To start developing a cause and effect diagram, the team must first agree on the problem statement, which should clearly state the effect that is being experienced. This should be written on the right side of the diagram so that the effect under study is clearly communicated to anyone reviewing the charted results.

Next, the major cause categories should be added to the diagram. Most teams will choose to use most or all of the "traditional" cause categories:

- People
- Method (includes processes and procedures)
- Material (includes raw materials and components)
- Machinery (includes equipment and technology)
- Measurement (includes calibration and data collection)
- Environment (includes facilities as well as items such as climate, weather, and operating conditions)

Some of these categories may need to be adjusted depending on the type of operating environment. For example, "material" in a service environment may include information and records, and as such the team may choose to change the category name.

Once the major categories have been identified and placed on the diagram, the team identifies potential causes and adds them to the diagram within the appropriate major category. The team may choose to employ brainstorming to develop

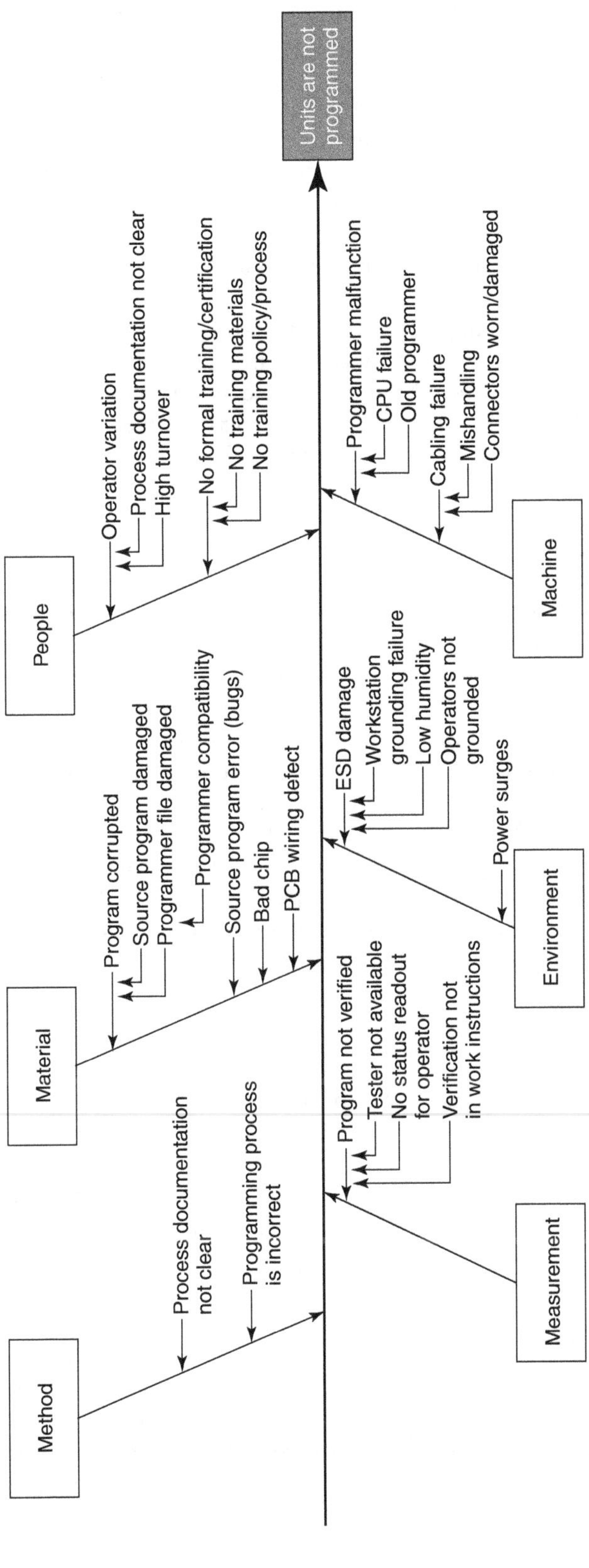

Figure 13.8 Partial cause and effect diagram.

a list of potential causes. Brainstormed ideas can be placed in a category as soon as each idea is developed, or the team may choose to generate the list of ideas first and then categorize and post them after brainstorming has concluded.

As potential causes are added to the diagram, the team should continue to drill down for more detail by using the 5 Whys method. The team may only have enough knowledge to drill down an additional level or two, but this activity may point to areas of future investigation or places where the team needs to collect data to verify the current state. The team should also look across the entire diagram to see if there are any causes that appear repeatedly or that may be related to causes under another major category.

After the team has developed the full cause and effect diagram, it should be reviewed with a larger audience of people that are involved with or familiar with the process. The usefulness of the final cause and effect diagram is limited by the knowledge of those who build it. By reviewing it with a larger audience, the team is able to verify that it represents current knowledge of the process or add information as required.

Design of Experiments

DOE is the simultaneous study of several process variables. By combining several variables in one study instead of creating a separate study for each, the amount of testing required will be drastically reduced, and greater process understanding will result. This is in direct contrast to the typical one-factor-at-a-time (OFAT) approach, which limits understanding and wastes data. Additionally, OFAT studies cannot be ensured of detecting the unique effects of combinations of factors, a condition defined as an interaction.

DOE includes the entire scope of experimentation, including defining the output measure(s) that one desires to improve, the candidate process variables that one will change, procedures for experimentation, actual performance of the experiment, and analysis and interpretation of the results. The objectives of the experimenter in a DOE are to learn how to:

- Maximize the response
- Minimize the response
- Adjust the response to a nominal value
- Reduce process variation
- Make the process robust (i.e., make the response insensitive to uncontrollable changes in the process variables)
- Determine which variables are important to control and which are not

Experimental Resolution

Resolution refers to the ability to separate main effects and low-order interactions from one another. The most significant fractional designs are those of resolutions III, IV, and V. Resolution I and II experiments are not useful, and resolution VI and

VII experiments are inefficient as the additional effort is expended on the estimation of very high-order interactions that seldom occur. The resolution described in Table 13.1 is only used for regular designs. Regular designs have run size with a power of two and full aliasing.

Interactions occur when the effect of factor A differs according to what the level factor of B is (a combined effect that cannot be predicted by the individual main effects). Confounding is the result of defining columns for main effects and/or interactions that are identical or exactly reversed in sign. The resulting confounded effects are aliases of one another. Aliases are effects that are identical except for the sign, that is, confounded effects.

Table 13.1 Experimental resolution.

Resolution	Ability	Example
I	Not useful: an experiment of exactly one run only tests one level of a factor and hence cannot distinguish between the high and low levels of that factor.	2^{1-1} with defining relation I = A
II	Not useful: main effects are confounded with other main effects.	2^{2-1} with defining relation I = AB
III	Estimate main effects, but these may be confounded with two-factor interactions.	2^{3-1} with defining relation I = ABC
IV	Estimate main effects not confounded by two-factor interactions. Estimate two-factor interaction effects, but these may be confounded with other two-factor interactions.	2^{4-1} with defining relation I = ABCD
V	Estimate main effects not confounded by three-factor (or less) interactions. Estimate two-factor interaction effects not confounded by two-factor interactions. Estimate three-factor interaction effects, but these may be confounded with other two-factor interactions.	2^{5-1} with defining relation I = ABCDE
VI	Estimate main effects not confounded by four-factor (or less) interactions. Estimate two-factor interaction effects not confounded by three-factor (or less) interactions. Estimate three-factor interaction effects, but these may be confounded with other three-factor interactions.	2^{6-1} with defining relation I = ABCDEF
VII	Estimates of main factors, two-factor and three-factor effects are not confounded with one another but may be confounded with higher-order interactions. Four-factor and higher interactions may be confounded.	2^{7-1} with defining relation I = ABCDEFG

Table 13.2 Various types of experimental designs.

Analysis of variance (ANOVA)	Basic statistical technique for analyzing experimental data. It subdivides the total variation of a data set into meaningful component parts associated with specific sources of variation, including interactions.
Balanced design	Fractional factorial experiment design in which an equal number of trials is performed for each factor, at each combination of factor levels.
Box-Behnken design	An alternative to central composite designs when the optimum response is not located at the extremes of the experimental region or does not use previous results from a fractional design. All design variables must vary continuously.
Central composite	Finds the optimal levels of the design variables by adding a few more experiments to a full factorial design. All design variables must vary continuously.
Evolutionary operation (EVOP)	Adjusts variables in a process in small increments using a planned pattern of changes in the search for a more optimal point on the response scale. Although this approach may be slower than DOE, the changes are small enough to prevent nonconformances but large enough to establish changes. This process is accomplished in a production environment.
Full factorial design	Evaluates the combination of all levels of all factors in an experiment.
Geometric screening designs	Screening designs that are a power of two (e.g., runs of 8, 16, 32, 64). The confounding of effects is complete. All fractional factorials are part of this family, as are Plackett-Burman designs for those run sizes.
Graeco-Latin square design	A special form of fractional factorial design with a single primary factor and three nuisance factors. It can be used as a form of blocking when there are three blocking factors to be used.
Hyper Graeco-Latin square design	A special form of fractional factorial design with a single primary factor and four nuisance factors. It can be used as a form of blocking when there are four blocking factors to be used.
Latin square design	A special form of fractional factorial design with a single primary factor and two nuisance factors. It can be used as a form of blocking when there are two blocking factors to be used.
Mixture design	A special type of experiment for analyzing mixtures of components that must sum to a constant. This type of design is typically used in manufacturing of chemicals and food production. Triangular (ternary) graphs are used to display the mixture proportions.
Nongeometric screening designs	Screening designs that are a multiple of four and not a power of two (e.g., 12, 20, 24, 28, 36). The confounding of effects is partial. These designs are uniquely Plackett-Burman.
One-factor-at-a-time (OFAT) experiment	Each factor is varied individually while the levels of other factors are held constant. For example, an individual tries to fix a problem by making a change then executing a test. Depending on the findings, something else may need to be tried.

Table 13.2 Various types of experimental designs. (continued)

Plackett-Burman designs	A family of screening designs that can address $n - 1$ factors with n trials so long as n is divisible by four. Designs can be geometric (the number of runs is a power of two) or nongeometric.
Screening experiment	First step of a multiple factorial experiment strategy, where the experiment primarily assesses the significance of main effects. Two-factor interactions are normally considered in the experiments that follow a screening experiment.
Simplex axial design	A design that consists mainly of the points positioned inside the simplex.
Simplex centroid design	A design that only includes the centroid points. Components that appear in a run in a simplex centroid design have the same values.
Simplex lattice design	A space-filling design that creates a triangular grid of runs. The design is the set of all combinations where the sum of the factors is 1.
Taguchi design	Experiments that make robust processes by realizing that not all factors that cause variability can be controlled. These uncontrollable factors are called noise factors.

Types of DOE

There are several types and methods of conducting DOE. Table 13.2 provides an overview of several recognized methods.

Effects are plotted and used to determine how the levels of the factors influence the effect. Interaction plots plot two or more factors to determine whether interactions between the factors are present. There are few simple rules that dictate their interpretation. Factors effect plot interpretation (see Figure 13.9):

1. If an effect is horizontal, the effect is not significant.
2. If an effect is not horizontal, the effect has more significance as the slope of the line increases. The direction of the slope of the line does not matter.

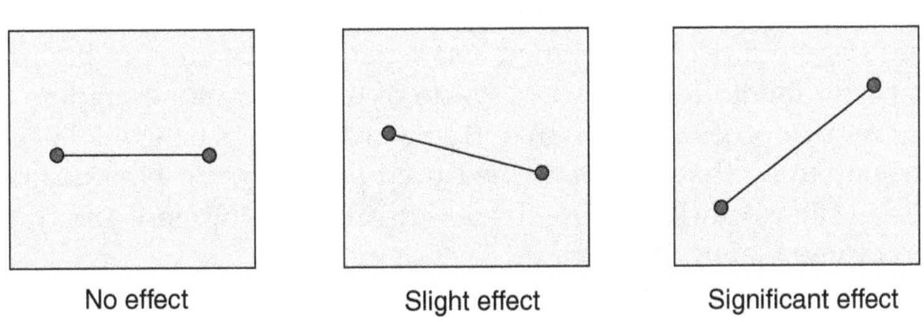

Figure 13.9 Factor effect plots.

Source: *Practical Design of Experiments (DOE)* (Durivage 2016), used with permission.

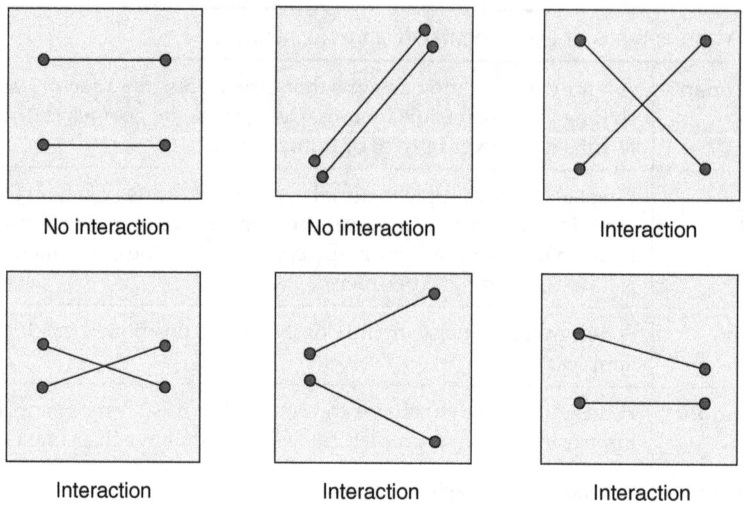

Figure 13.10 Factor interaction plots.
Source: Practical Design of Experiments (DOE) (Durivage 2016), used with permission.

Factor interaction plot interpretation (see Figure 13.10):

1. If the lines are parallel, an interaction is not present, even if the lines are not horizontal. The distance between the lines does not matter.
2. If the lines intersect, or can intersect, an interaction between the factors is present. The slope of the lines and their direction do not matter.
3. A factor with a higher slope is more significant than a factor that is horizontal.

MSA or Gage R&R

MSA or Gage R&R—repeatability and reproducibility—studies analyze the variation of measurements of a gage (repeatability) and the variation of measurements by operators (reproducibility). Gage R&R studies are also referred to as test method validation (TMV). To understand why this type of analysis is so important, recall that the goal of process control is to reduce variation in the process and ultimately the product. To address actual process variability, the variation due to the measurement system must be identified and separated from that of the process. Studies of measurement variation are a waste of time and money unless they lead to action to reduce process variation and improve process control. Since you cannot address something that cannot be measured precisely, the assessment of the gage becomes an early priority during the design and development and transfer phases prior to commercial production.

Before we can continue discussing gage R&R, we have to define "gage." The term "gage" refers to any device used for making measurements. In this book, the terms "gage" and "device" are used interchangeably and refer to any device or equipment used for making a measurement.

Every observation of a process contains both actual process variation and measurement variation (see Figure 13.11). In the case of measurement systems, the sources are:

- *The gage/device.*
  - —*Calibration.* Is the gage accurate?
  - —*Stability.* Does the gage change over time?
- *The operator.* Does the operator have the necessary skill and training?
- *Within-sample variation.* Variation within a sample is a part of process variation that is often mixed with measurement variation.
- *Repeatability.* The variation observed when an operator measures the same sample using the same gage several times.
- *Reproducibility.* The additional variation observed when several operators use the same gage to measure the same sample.
- *Linearity.* Is the gage more accurate at low values than at high values or vice versa?
- *Bias.* Is there a shift of the average measurements from the reference value?
- *Discrimination.* Is the gage sensitive enough to measure the part?

Gage R&R studies assess repeatability (gage variation; see Figure 13.12) and reproducibility (operator variation; see Figure 13.13). Repeatability is the variation observed when an operator measures the same sample using the same gage several times. Reproducibility is the additional variation observed when several operators use the same gage to measure the same sample. The combination of both sources of variation is referred to as gage R&R (see Figure 13.14). Note that gage R&R does not address the total measurement system but is narrowly defined and gage specific.

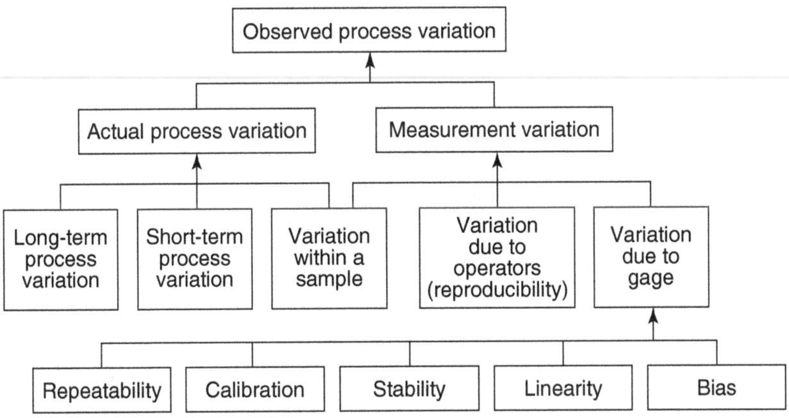

Figure 13.11 Possible sources of process variation.

Source: Practical Attribute and Variable Measurement Systems Analysis (Durivage 2015), used with permission.

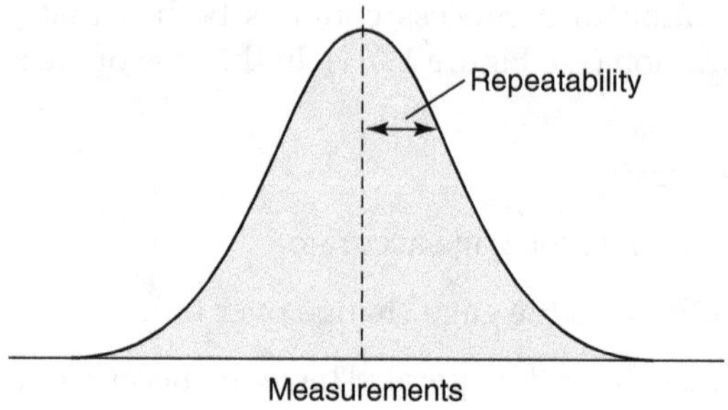

Figure 13.12 Repeatability.

Source: Practical Attribute and Variable Measurement Systems Analysis (Durivage 2015), used with permission.

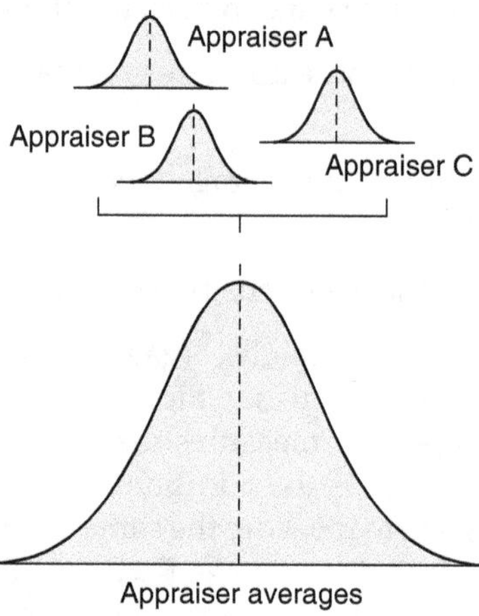

Figure 13.13 Reproducibility.

Source: Practical Attribute and Variable Measurement Systems Analysis (Durivage 2015), used with permission.

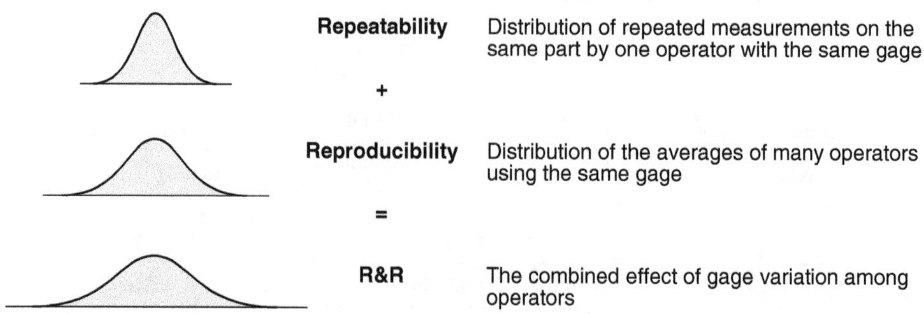

Figure 13.14 Repeatability, reproducibility, and R&R.

Source: Practical Attribute and Variable Measurement Systems Analysis (Durivage 2015), used with permission.

A gage R&R of < 10% is considered acceptable, between 10% - 30% is marginally acceptable, and > 30% is unacceptable. Recall the observed process variation is the combination of the actual process variation and the measurement variation. A gage R&R > 30% means that nearly one third of the observed process variation is due to the measurement system. Reducing the gage R&R is extremely important to ensure products are not being scrapped due to inadequate measurement.

Attribute Measurement System Analysis

An attribute MSA is similar in concept to a gage R&R but is performed when evaluating the effectiveness of an attribute inspection scheme. Attribute inspection classifies a sample as either conforming or nonconforming (or good/bad, pass/fail, go/no go, etc.). In a perfect measurement system, every unit that was inspected would always be classified correctly.

Attribute inspections are often subjective, so it is crucial that inspection criteria be properly and clearly specified and that inspectors be well trained. An attribute MSA provides a way to evaluate the effectiveness of the inspection scheme, which includes the inspection criteria and inspector training.

Earlier we defined repeatability and reproducibility for gage R&R. These definitions can also be used in the context of an attribute MSA. So, repeatability is the ability of an inspector to make the same decision when presented with the same sample for inspection multiple times, while reproducibility is the ability of different inspectors to reach the same decision when presented with the same sample for inspection. Repeatability is calculated by counting the number of times an inspector's decisions agreed for the same sample that was presented multiple times, divided by the total number of unique samples inspected. An average of repeatability values for all inspectors involved in the study gives an indication of the overall repeatability of the measurement system. Reproducibility is calculated by counting the number of times all decisions of all inspectors involved in the study agreed on the same sample, divided by the total number of unique samples inspected. For an attribute MSA, higher results for repeatability and reproducibility are desirable.

One concept that is unique to an attribute MSA is the direct calculation of inspection accuracy. If prior to the study each sample is judged by a master appraiser, such as a customer, senior inspector, or other expert, then this master judgment can serve as the standard for the study. Individual effectiveness, or the accuracy of individual inspectors, can be measured by calculating the percentage of times that an inspector's repeated results agree with each other *and* with the standard. Overall effectiveness can be measured by calculating the percentage of times that all inspectors' results agree with each other *and* with the standard. As with repeatability and reproducibility, higher results are desirable for both individual and overall effectiveness in an attribute MSA. For attribute gage studies, > 90% agreement is considered acceptable, < 90-80% marginally unacceptable, and < 80% unacceptable.

If results are lower than desired, the appropriate action to be taken will depend on where the results fall short. In general, areas to review could include the completeness and clarity of the inspection criteria, the usefulness of inspection aids, the consistency of training provided to inspectors, and the inspectors' interpretation of the criteria.

COLLABORATION WITH SUPPLIER

> Evaluate and implement supplier corrective/preventive action and review its effectiveness and robustness with supplier. Understand the process of updating failure mode and effects analysis (FMEA) and process control plan, and understand statistical process control (SPC), 8D, and product/process design change. (Evaluate)
>
> **Body of Knowledge IV.C.2**

The tools presented in this section help monitor and measure processes and aid in solving and or preventing problems. Additionally, these tools can aid the determining the effectiveness of corrections, corrective, and preventive actions. These tools are foundation to corrective and preventive action (CAPA) and the 8D and A3 problem solving methodologies. As issues are identified and resolved, FMEAs, HACCPs, and control plan documentation are updated to reflect the mitigations and countermeasures that were implemented including, changes to the product, service, or process.

Often, the corrective and preventive actions taken may result in redesign, of products, processes, and systems. Some corrective and preventive actions may be relatively easy and quick to implement, but other times the actions may require timely and complex solutions such as updating production and inspection capabilities. Verification of effectiveness involves assessing if the corrections, corrective, and preventive actions resolved the issue(s) and there was not an adverse effect on the product, process, or service.

Failure Modes and Effects Analysis

Failure modes and effects analysis (FMEA) is extended to investigate the degree of severity of consequences, probability of occurrence, and detectability. FMEA can identify places for additional actions. The corrections and corrective and preventive actions may indicate and require an analysis of the probability of occurrence (O) rating and/or the probability of detection (D) rating requiring an adjustment to the RPN or AP.

Control Plans

Control plans are used to document the process characteristics and variables that must be controlled to ensure that the process remains stable ("in control"). The corrections and corrective and preventive actions may require additional controls. These actions could include an increased inspection activity, by modifying the

AQL to require a larger sampling or by adding a new action for the feature that was identified and associated with the issue being investigated.

Hazard Analysis Critical Control Point

Hazard Analysis Critical Control Point (HACCP) is used to manage risks with physical, chemical, biological, allergenic, and radiologic hazards. The corrections, corrective, and preventive actions may indicate and require an analysis of the prerequisite program, the process flow, and ultimately a review of the control points. The investigation may require additional mitigation(s) including escalating a CP to CCP or enhancing the mitigations for a current CCP to prevent contamination.

Control Charts for Continuous Process Monitoring

Control charts are an essential tool to satisfy continuous process monitoring requirements for a validated process. Control charts are decision-making tools that provide information for timely decisions concerning recently produced products. Control charts are also problem-solving tools that help locate and investigate the causes of poor or marginal quality. The data from control charts can provide useful information when performing revalidation or retrospective validation activities.

Control charts contain a centerline, usually the mathematical average of the samples plotted; upper and lower statistical control limits that define the constraints of common cause variation; and performance data plotted over time.

Control Chart Types and Selection

There are two general classifications of control charts: variables charts and attributes charts (see Table 13.3). Variables are things that can be measured: length, temperature, pressure, weight, and so on. Attributes are things that are counted: dents, scratches, defects, days, cycles, yes/no decisions, and so on.

Control Chart Interpretation

A process is said to be in control when the control chart does not indicate any out-of-control condition and contains only common causes of variation. If the common cause variation is small, then a control chart can be used to monitor the process. See Figure 13.15 for a representation of stable (in-control) and unstable (out-of-control) processes. If the common cause variation is too large, the process will need to be modified.

When a control chart indicates an out-of-control condition (a point outside the control limits or matching one or more of the criteria below), the assignable causes of variation must be identified and eliminated.

Improper control chart interpretation can lead to several problems, including blaming people for problems they cannot control, spending time and money looking for problems that do not exist, spending time and money on new equipment or process adjustments that are not necessary, taking action where no action is warranted, and asking for worker-related improvements where process or equipment improvements need to be made first.

Table 13.3 Variables and attributes control charts selection.

Type	Distribution	Sample	Application
Variables control charts			
$\overline{X}$ and R	Normal	$2 \leq 10$	Measurement subgroups
$\overline{X}$ and s	Normal	> 10	Measurement subgroups
Attributes control charts			
c	Poisson	Constant	Count number of defects per item
u	Poisson	Varies	Count number of defects per item
np	Binomial	Constant	Count of defective items
p	Binomial	Varies	Count of defective items
g	Binomial	Individual	Interval between rare events
Variables or attributes control charts			
X and mR	Normal	1	Individual counts or measurements

Source: Practical Engineering, Process, and Reliability Statistics 2nd ed. (Durivage 2022), used with permission.

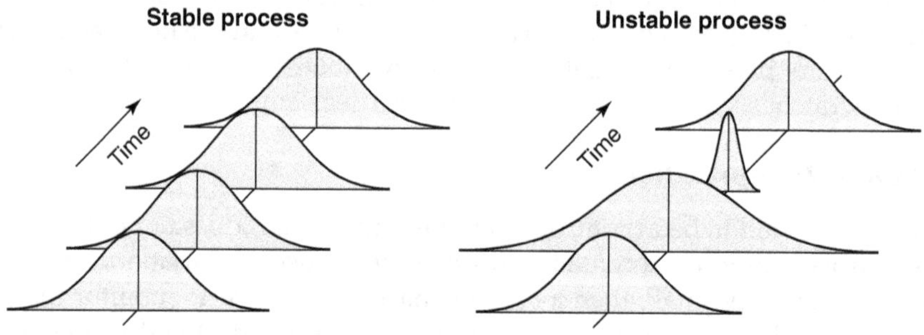

Figure 13.15 Stable and unstable variation.

Source: Adapted from Stephen A. Wise and Douglas C. Fair, *Innovative Control Charting: Practical SPC Solutions for Today's Manufacturing Environment* (Milwaukee, WI: ASQ Quality Press, 1997), Figures 3.3 and 3.5. Used with permission.

The following rules should be used to properly indicate that a process is out of control:

Rule 1—One point beyond the 3σ control limit

Rule 2—Eight or more points on one side of the centerline without crossing

Rule 3—Four out of five points in zone B or beyond

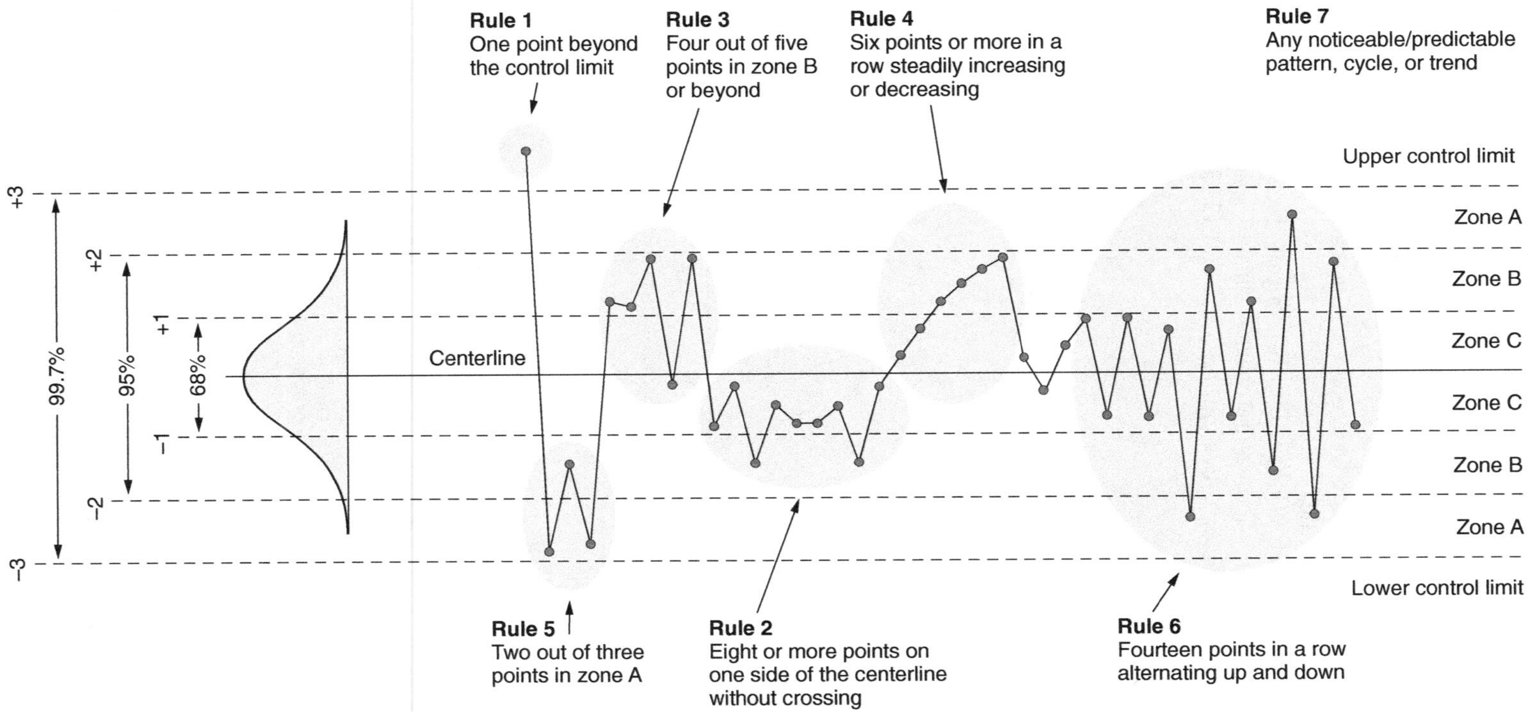

Figure 13.16 Control chart interpretation rules.

Source: Adapted from Stephen A. Wise and Douglas C. Fair, *Innovative Control Charting: Practical SPC Solutions for Today's Manufacturing Environment* (Milwaukee, WI: ASQ Quality Press, 1997), Figures 3.3 and 3.5. Used with permission.

Rule 4—Six points or more in a row steadily increasing or decreasing

Rule 5—Two out of three points in zone A

Rule 6—14 points in a row alternating up and down

Rule 7—Any noticeable/predictable pattern, cycle, or trend

Please note that depending on the source, these rules can vary. See Figure 13.16 for control chart interpretation rules.

The target value (which is hopefully the control chart centerline) is closely related to process accuracy. The range chart is closely associated with process precision (spread or dispersion). (See Figures 13.17 and 13.18.)

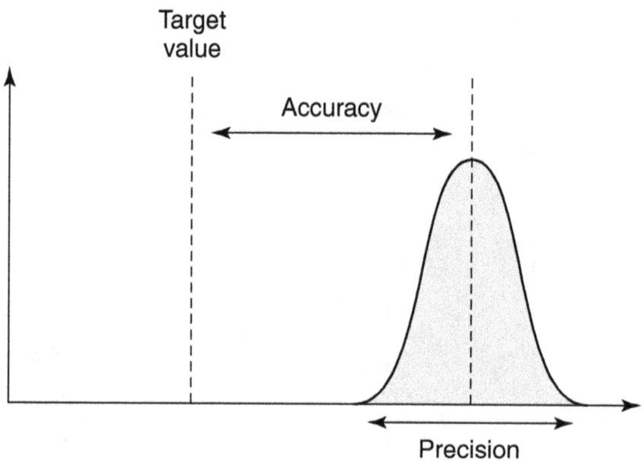

Figure 13.17 Control chart accuracy and precision.

Source: Practical Engineering, Process, and Reliability Statistics 2nd ed. (Durivage 2022), used with permission.

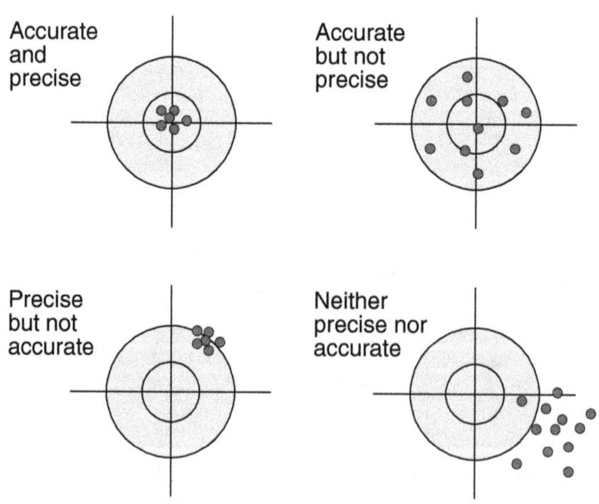

Figure 13.18 Accuracy versus precision.

Source: Practical Engineering, Process, and Reliability Statistics 2nd ed. (Durivage 2022), used with permission.

$\overline{X}$ and R Control Charts

$\overline{X}$ and R control charts assume a normal distribution and are usually used with a subgroup size of less than 10 (typically 3–5). A minimum of 25 subgroups is necessary to construct the chart. (See Figure 13.19.)

$\overline{X}$ and s Control Charts

$\overline{X}$ and s control charts assume a normal distribution and are usually used with a subgroup size of 10 or more. A minimum of 25 subgroups is necessary to construct the chart. (See Figure 13.20.)

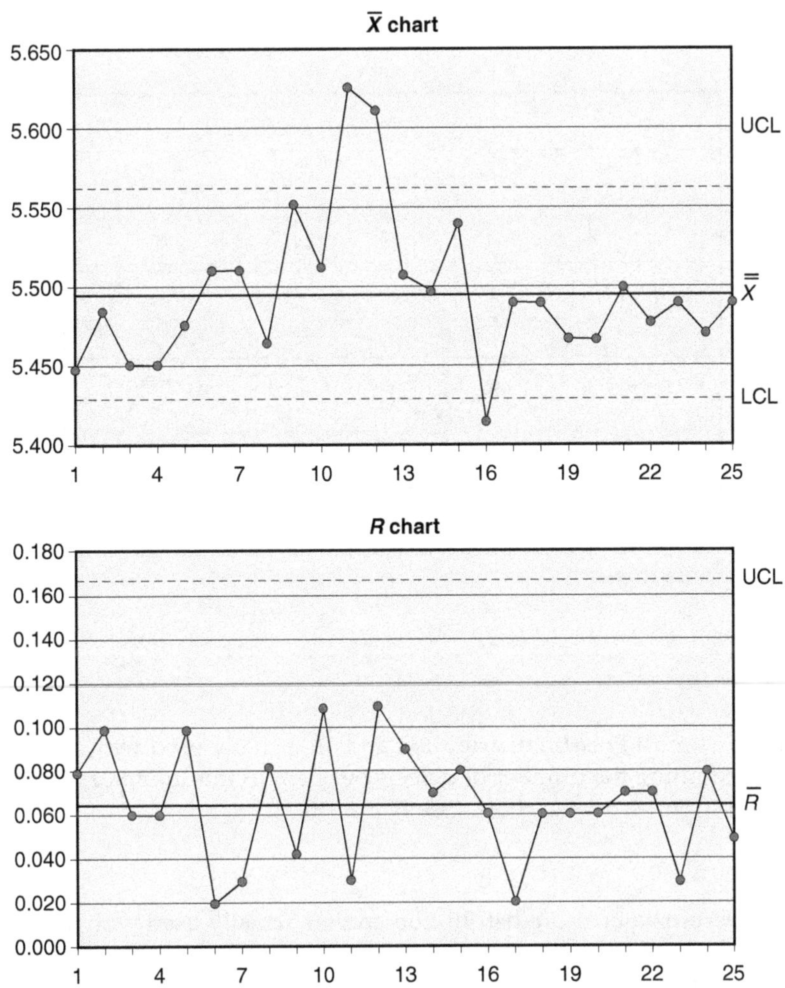

Figure 13.19 $\overline{X}$ and R chart.

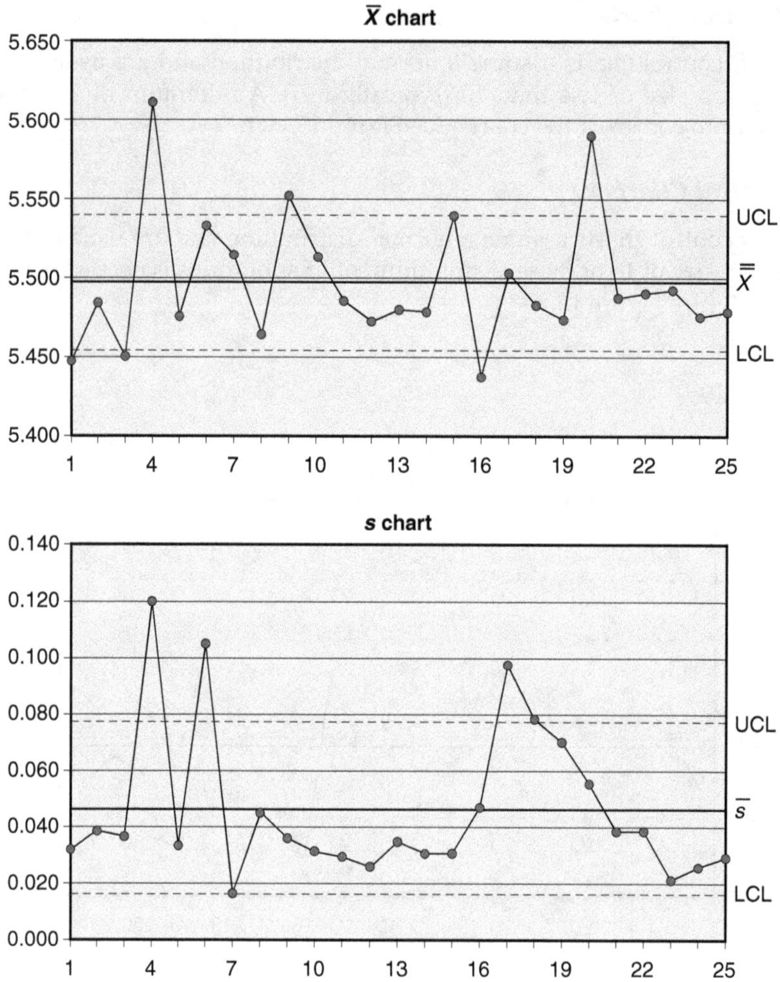

Figure 13.20 $\bar{X}$ and *s* chart.

c-*Charts*

c-charts assume a Poisson distribution and are usually used with a constant sample size, counting the number of defects per item. A minimum of 25 subgroups is necessary to construct the chart. (See Figure 13.21.)

u-*Charts*

u-charts assume a Poisson distribution and are usually used with a variable sample size (although a constant sample size may be used), counting the number of defects per item. A minimum of 25 subgroups is necessary to construct the chart.

If variable sample sizes are used, the control limits will vary for each subgroup because the sample size *n* varies from group to group. (See Figure 13.22.)

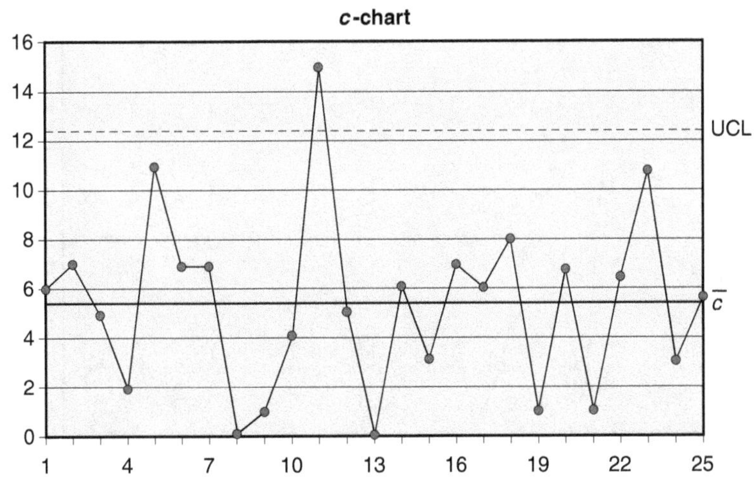

Figure 13.21 *c*-chart.

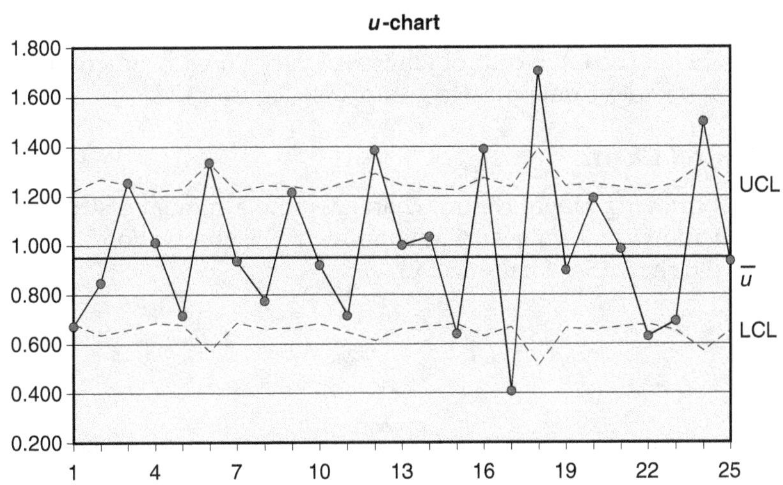

Figure 13.22 *u*-chart.

np-*Charts*

np-charts assume a binomial distribution and are usually used with a constant sample size, counting the defective items. A minimum of 25 subgroups is necessary to construct the chart. (See Figure 13.23.)

p-*Charts*

p-charts assume a binomial distribution and are usually used with a variable sample size (although a constant sample size may be used), counting the defective items. A minimum of 25 subgroups is necessary to construct the chart. If variable

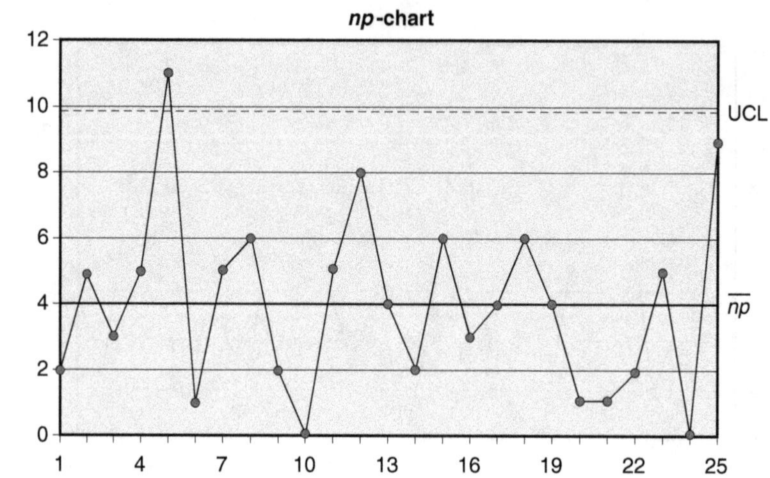

Figure 13.23 *np*-chart.

sample sizes are used, the control limits will vary for each subgroup because the sample size *n* varies from group to group. (See Figure 13.24.)

X and mR Control Charts

X and mR (moving range) control charts assume a normal distribution and are used with individual values. A minimum of 25 observations is necessary to construct the chart. (See Figure 13.25.)

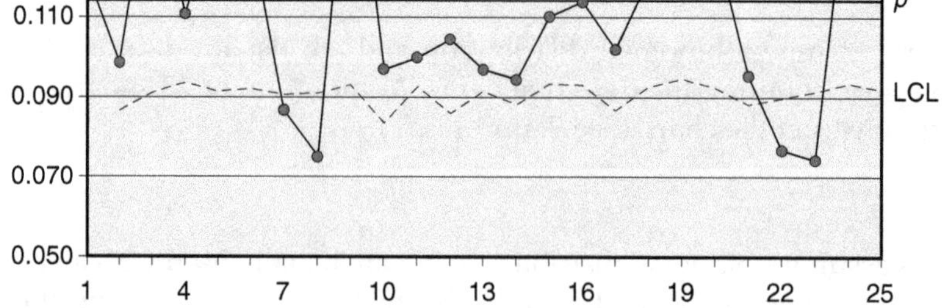

Figure 13.24 *p*-chart.

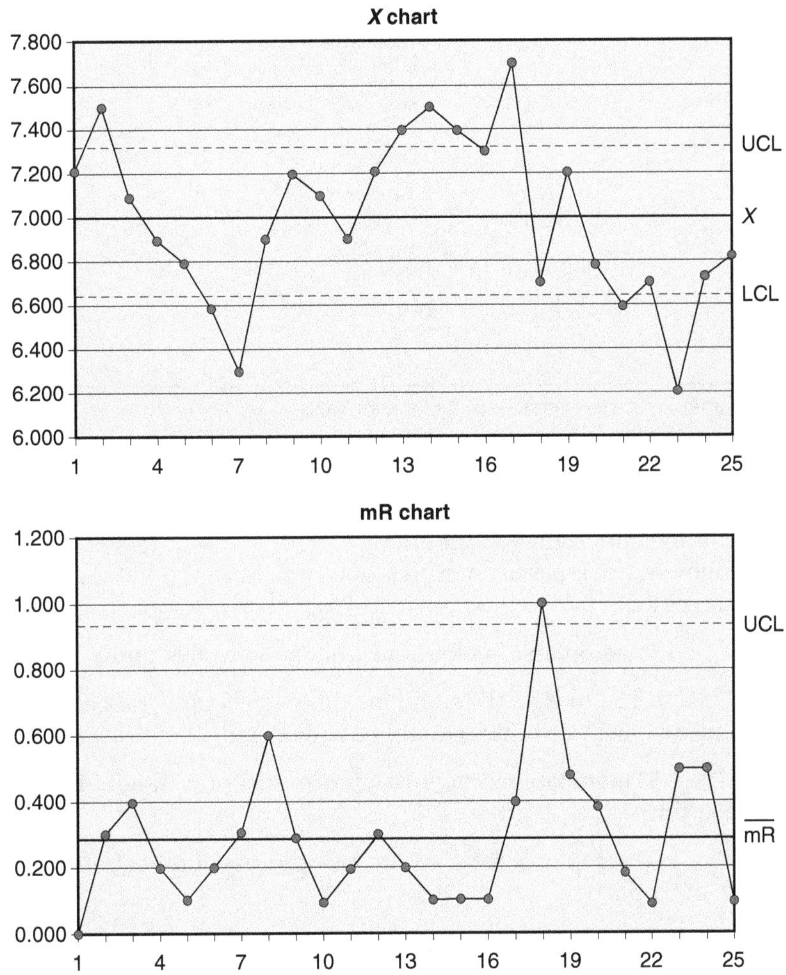

Figure 13.25 *X* and mR chart.

Process Capability

While control charts tell us about the historical performance and stability of a process, these charts do not tell us anything about the ability of the process to meet customer requirements. In order to determine if a process is capable of conforming to specifications, we must compare the process performance, including both location and variability, with the specifications. (See Figures 13.26 and 13.27.)

There are three potential process capability indices for variables data: C_r, C_p, and C_{pk}. Additionally, there are three process capability indices for variables data: P_r, P_p, and P_{pk}. The main difference between the "C" and "P" indices is that C_r, C_p, and C_{pk} are used with sample data to determine if the process is capable of meeting customer specifications (short term), whereas P_r, P_p, and P_{pk} are used with population data to determine if the process is capable of meeting customer

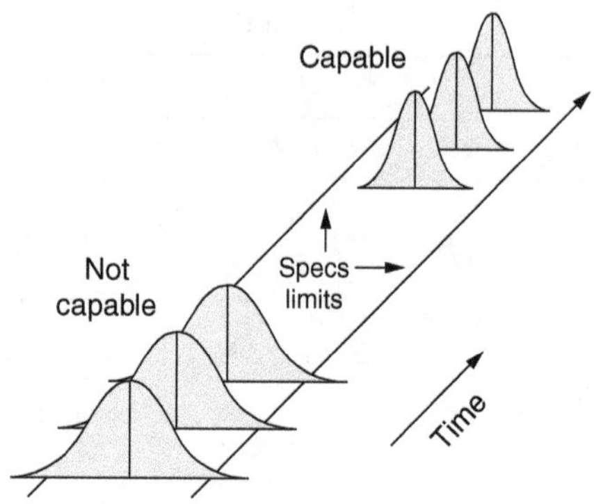

Figure 13.26 Capable process versus not capable process.

Source: Adapted from *Innovative Control Charting* (Wise and Fair 1997). Used with permission.

specifications (long term). These capability indices (short and long term) assume normality of the underlying distribution.

The following risk-based interpretation of C_p and C_{pk} values are commonly used and accepted:

- $C_p < 1$ unacceptable (fallout rate greater than 2699 ppm)
- $1 \leq C_p \leq 1.33$ marginal (fallout rate of 63–2699 ppm, process improvement activities should be considered)
- $C_p \geq 1.33$ acceptable for low-risk process outputs (fallout rate of less than 63 ppm)
- $C_p \geq 1.50$ acceptable for medium-risk process outputs (fallout rate of less than 7 ppm)
- $C_p \geq 1.67$ acceptable for high-risk process outputs (fallout rate of less than 1 ppm)

Short-term capability	Desired ratio
$C_r = \dfrac{6s}{USL - LSL}$	<1
$C_p = \dfrac{USL - LSL}{6s}$	>1
$C_{pk} = \text{Minimum of } \dfrac{\overline{X} - LSL}{3s}, \dfrac{USL - \overline{X}}{3s}$	>1

Long-term capability	Desired ratio
$P_r = \dfrac{6\sigma}{USL - LSL}$	<1
$P_p = \dfrac{USL - LSL}{6\sigma}$	>1
$P_{pk} = \text{Minimum of } \dfrac{\mu - LSL}{3\sigma}, \dfrac{USL - \mu}{3\sigma}$	>1

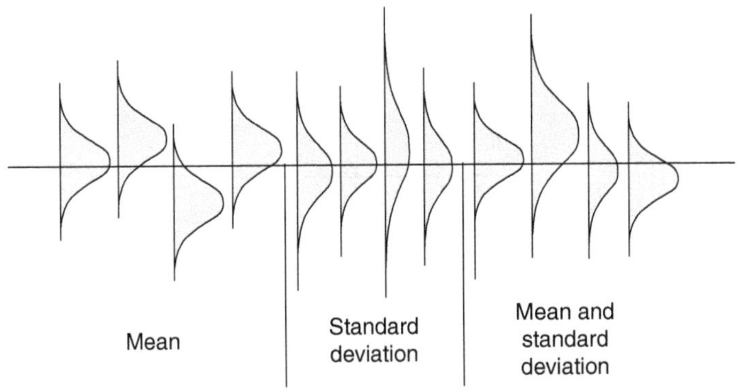

Figure 13.27 Process control issues.
Source: Practical Engineering, Process, and Reliability Statistics 2nd ed. (Durivage 2022), used with permission.

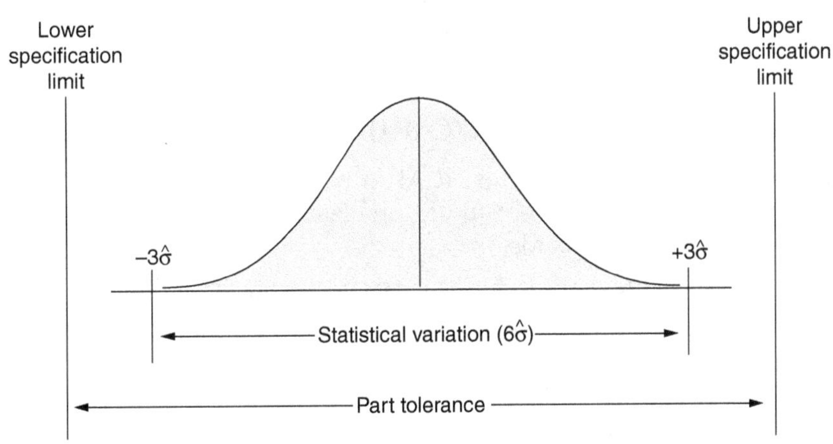

Figure 13.28 Relationship between statistical control limits and product specifications.
Source: Practical Engineering, Process, and Reliability Statistics 2nd ed. (Durivage 2022), used with permission.

These indices compare the process statistical control limits to the product specifications to determine the process capability. Figure 13.28 demonstrates the relationship between statistical control limits and product specifications.

Figure 13.29 is a graphical representation of process capability ratios compared to the statistical control limits and part specification limits.

There are three ways to increase process capability: decrease the process variation, center the process or increase the product specification limits.

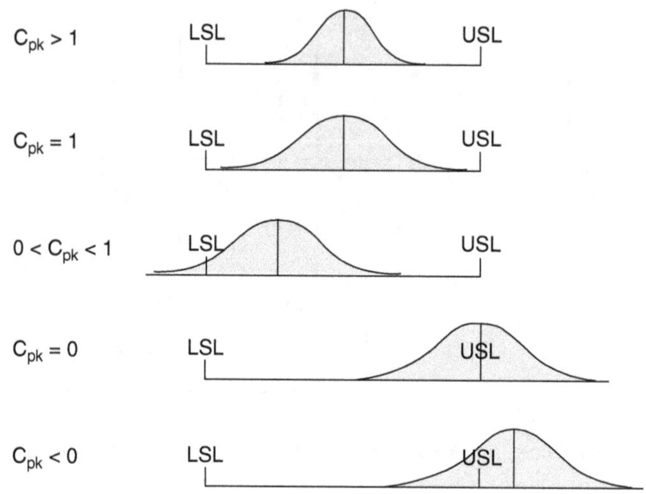

Figure 13.29 Graphical relationship of process capability ratios.
Source: Practical Engineering, Process, and Reliability Statistics 2nd ed. (Durivage 2022), used with permission.

Corrective and Preventive Action (CAPA)

Corrective and Preventive Action (CAPA) is a widely used technique to solve existing problems and prevent future problems. CAPAs are required by many regulations and ISO Standards.

1. **Problem Identification and CAPA Initiation:** The problem identification and CAPA initiation phase requires documenting the issue to begin the CAPA process. The description should be as complete as possible, including who, what, when, where, why, and how many.
2. **Risk Analysis:** A risk analysis should be performed based on patient/user/business/compliance risk. The results of the risk analysis should drive CAPA timelines. It should be obvious that low-risk issues do not have the same sense of urgency as high-risk issues.
3. **Correction/Containment:** Correction and containment should be completed as soon as possible to prevent further production and distribution. The organization should review related processes and products to see if there are wider issues. In the case of product-related issues, a field correction and/or recall may be required. Correction and containment activities are an effort to reduce the impact in the short term.

Correction activities can include sorting, reinspection, reworking, and reprocessing.

4. **Investigation/Root Cause Analysis:** Several of the most common tools/methods for performing investigations to determine the root cause of a problem during CAPA include but are not limited to:

- Brainstorming
- The 5 whys

- Flowcharting
- Cause and effect diagrams
- Affinity diagrams
- Is/Is Not
- Design of Experiments
- Measurement System Analysis

5. **Corrective/Preventive Action(s):** Corrective and preventive actions are considered long-term solutions to resolve or eliminate the cause of a nonconformity or eliminate the cause of a potential nonconformity. A corrective action is an action to eliminate the cause of a nonconformity and to prevent recurrence. A preventive action is an action to eliminate the cause of a potential nonconformity or other potential undesirable situation. A correction is an action to eliminate a detected nonconformity.
6. **Implementation:** The implementation phase is where corrective and preventive actions are initiated and applied to address the root cause or causes of a nonconformity. Examples are procedural updates, training, and process modifications.
7. **Verification of Implementation:** The verification of implementation phase documents that corrective and preventive actions were actually implemented. The verification of implementation (VOI) phase is often confused with the verification of effectiveness (VOE) phase of a CAPA, but these are separate, distinct phases.

VOIs are used to verify that corrections, containment, corrective actions, and preventive actions were indeed implemented as promised and planned.

8. **Verification of Effectiveness Plan (VOEP):** The effectiveness plan phase is used to establish and define predetermined criteria to verify that corrective and/or preventive actions were indeed effective.
9. **Verification of Effectiveness (VOE):** The verification of effectiveness phase includes reviewing and documenting the predetermined criteria established in the verification of effectiveness plan. A successful verification of effectiveness should demonstrate the true root cause of the issue was properly identified and the corrective and/or preventive actions were effective. The FMEA, HACCP, process control plan, and SPC may be updated to assess the product, service, or process change.
10. **Closure:** This is the last phase of the CAPA process. A CAPA should only be closed once the verification of effectiveness has been successfully completed.

The CAPA system is most effective when used for systemic issues. In other words, the feeder systems to the CAPA system (nonconformances, out of tolerances, out of specifications, deviations, and complaints) are used to triage and sometimes resolve the issue without escalation to CAPA. The escalation to CAPA determination should consider the risks to patients, users, workers, products, and processes. A risk-based escalation process should use pre-defined criteria and provide rationale for the escalation decision.

When a customer issues a CAPA to supplier, it is generally referred to as a supplier corrective action request (SCAR). The level and depth of the supplier's response to a CAPA or SCAR should be proportionate to the risk of the problem.

8D Problem Solving Methodology

The 8D Problem Solving Methodology is widely used in the automotive industry, however 8Ds are also used in other manufacturing industries due to structured approach. The 8D Problem Solving Methodology is named due to the eight steps or phases involved. The steps are as follows:

- **D0: Plan** – Plan for solving the problem and determine the prerequisites.
- **D1: Use a team** – Select and establish a team of people with product/process knowledge.
- **D2: Define and describe the problem** – Specify the problem by identifying in quantifiable terms the who, what, where, when, why, how, and how many (5W2H) for the problem.
- **D3: Develop interim containment plan; implement and verify interim actions** – Define and implement containment actions to isolate the problem from any customer.
- **D4: Determine, identify, and verify root causes and escape points** – Identify all applicable causes that could explain why the problem occurred. Also identify why the problem was not noticed at the time it occurred. All causes shall be verified or proved, not determined by fuzzy brainstorming. Several of the most common tools/methods for performing investigations to determine the root cause of a problem during an 8D includes but are not limited to:

  —Brainstorming

  —The 5 whys

  —Flowcharting

  —Cause and effect diagrams

  —Affinity diagrams

  —Is/Is Not

  —Design of Experiments

  —Measurement System Analysis

- **D5: Choose and verify permanent corrections** – Through preproduction programs, quantitatively confirm that the selected correction will resolve the problem for the customer.

- **D6: Implement and validate corrective actions** – Define and implement the best corrective actions (CA). The FMEA, HACCP, process control plan, and SPC may be updated to assess the product, service, or process change. If the corrective actions are not effective, close the 8D and repeat the process.
- **D7: Take preventive measures** – Modify the management systems, operation systems, practices, and procedures to prevent recurrence of this and all similar problems.
- **D8: Congratulate your team** – Recognize the collective efforts of the team. The team needs to be formally thanked by the organization.

Many times, when issues arise, the customer may only require completion of Phases 0–3. However, if the issue is significant, all phases are required.

A3 Problem Solving Methodology

The A3 problem solving methodology process is typically used as part of lean manufacturing. A3 problem solving provides a procedure that guides problem solving by workers. The approach uses a single sheet of ISO A3-size paper to document and convey information used to solve the problem, hence the source of its name.

Step 1: Background: Make the case for selecting the particular problem and clearly state the impact of the problem.

Step 2: Problem Statement: List the specifics of the problem including the magnitude, where and when the problem occurred, and the problem's impact.

Step 3: Goal Statement: Define the goals and set a timeframe for completion.

Step 4: Root Cause Analysis: Perform a root cause analysis to determine the actual cause of the problem. Several of the most common tools/methods for performing investigations to determine the root cause of a problem during an A3 include but are not limited to:

- Brainstorming
- The 5 Whys
- Flowcharting
- Cause and effect diagrams
- Affinity diagrams
- Is/Is Not
- Design of Experiments
- Measurement System Analysis

Step 5: Countermeasures: Determine the needed to solve the problem. Assign responsibilities for the actions.

A3 Problem Solving Template	
Step 1: Background	Step 5: Countermeasures
Step 2: Problem Statement	Step 6: Effect Confirmation
Step 3: Goal Statement	Step 7: Follow Up Action
Step 4: Root Cause Analysis	

Figure 13.30 A3 Problem Solving Template.

Step 6: Effect Confirmation: Assess if the implemented the countermeasures were effective in meeting the goals. The FMEA, HACCP, process control plan, and SPC may be updated to assess the product, service, or process change.

Step 7: Follow Up Action: Determine if the effective countermeasures can be applied elsewhere. If the counter measures are ineffective, evaluate the root cause and look for alternative countermeasures.

Figure 13.30 provides a basic A3 Problem Solving Template.

Part V

Supplier Quality Management

Chapter 14
Supplier Quality Monitoring

SUPPLIER AUDIT

> Apply the stages of a quality audit, including audit planning, conducting the initial audit, and executing periodic reevaluation. Understand and apply the various types of quality audits (e.g., product, process, and management system) and audit methods (e.g., virtual, on-site, and desktop). (Apply)
>
> Body of Knowledge V.A.1

Supplier audits are an essential part of supplier quality monitoring. There are several types of audits that can be used to assess supplier capability and performance, including product, process, and management systems. Figure 14.1 depicts the five stages of an effective planned audit, including who is responsible. A risk-based system should be used to determine which supplier should be audited, when it should be audited, and the frequency of audits.

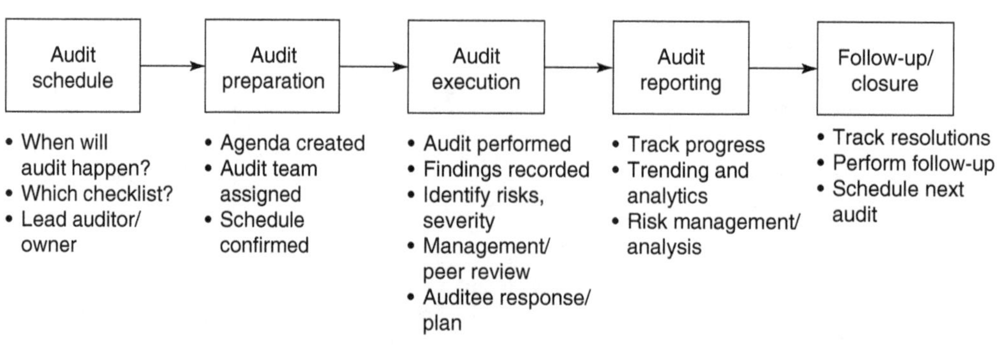

Figure 14.1 The five stages of an audit.

Audit Preparation

Audits are formal and should be well planned and documented. Audits should follow prescribed guidelines and have clear objectives and purpose. The formal audit plan should include:

- Audit objectives
- Audit criteria and reference documents, including the standard that the team is auditing against.
- Audit scope, including the processes to be audited
- Audit dates and locations
- Expected times and durations of on-site activities
- Roles and responsibilities of audit team members, including identification of the lead auditor, auditors, observers, SMEs
- Allocation of resources to areas to be audited
- Identification of the auditee point of contact
- Logistics
- Matters related to confidentiality, such as the need for an NDA
- Any audit follow-up actions

Once approved by the appropriate unit (i.e., company QA), the audit plan should be presented to the auditee for review and any clarification should be provided before finalization of the plan.

The initial full system audit of a supplier requires that the auditee's quality manual be fully reviewed to ensure that it meets the requirements of the QMS to be audited. The manual, policy, procedures, and work instructions are audited and used to determine if the QMS is effectively deployed.

Before conducting the supplier audit, the auditors should review the contract to identify any agreements made between the auditing company and the auditee or the customer and the auditee for third party audits. A review of the contract or purchase order is required to ensure proper planning and preparation of the audit.

Before an audit can be effectively planned and performed, the purpose of the audit must be established. The scope and objective of the audit must be clearly defined and understood by the auditors and auditees. *Scope* refers to the areas that the audit will cover. It helps define the type of audit to be performed and the depth of the audit. This is where the determination must be made as to whether the audit is a product, process, or full system audit. The scope must be documented in a statement, which is communicated to the auditee and the audit team and is used in the opening meeting, closing meeting, and audit report. Establishing a well-defined scope is necessary for audit planning. This is key to developing the audit checklist and determining the depth of the audit. The resources committed to the audit should be sufficient to meet the stated scope and depth.

The lead auditor is responsible for selecting auditors that are qualified to perform the type of audit that is called for. This means balancing the auditor's skills and experience against the audit requirements. This is critical to a successful audit. The number and qualifications of auditors are dependent on the subject, purpose, scope, and depth of the audit.

When the audit schedule has been developed and approved, the lead auditor must formally notify the auditee organization in writing one to three months prior to the audit date. In some companies, the purchasing department is responsible for coordinating the supplier audit. A courtesy telephone call usually takes place prior to the formal notification. The formal notification should include the following:

- Purpose and scope of the audit
- Type of audit
- Dates and duration of the audit
- Auditor names, including which auditor is the point of contact

It is a good practice to provide the auditee with the formal audit checklist and require that they complete a self-assessment and return it to the lead auditor prior to the visit. This greatly enhances the audit and improves preparedness for both the auditor and the auditee. This saves valuable time for both companies.

The lead auditor is responsible for preparing the audit agenda. The audit agenda should include:

- The opening meeting time frame
- Times for the specific audit activities
- Name of the auditor(s) assigned to each activity
- Projected times for beginning and ending the daily audit activities
- Projected times for auditor conferences
- Projected times for daily meetings with the auditee
- Projected times for the closing meeting and requested attendance

As part of the audit planning, audit logistics must be addressed, including:

- Travel and accommodations
- Maps and security passes, including any training needed prior to entering the auditee location, such as safety
- Internal audit physical layout, including process flowcharts
- Escorts and meeting areas (a knowledgeable escort is critical to a successful audit)
- Requirements for personal protective equipment (provided by the auditee) based on the scope of the audit
- NDAs
- Security considerations in some foreign and domestic environments

Audit Execution

Audit performance or execution consists of several major areas, including:

- On-site audit management, including time management
- Opening meeting
- Audit data collection and analysis
- Establishment of objective evidence
- Organization of objective evidence
- Closing or exit meeting

The lead auditor is responsible for all phases of the audit, from selection of the audit team members to reporting audit results. According to ISO 19011, the lead auditor's responsibilities include:

- Selection of audit team members
- Preparing the audit plan
- Representing the audit team with the auditee
- Maintaining the ethics of the audit team
- Directing the activities of the auditors
- Defining the requirements of each audit assignment
- Complying with audit requirements
- Preparing working documents and briefing auditors
- Reviewing documentation
- Reporting critical nonconformities and conclusions
- Reporting obstacles in performing the audit
- Reporting audit results

Performance of the audit is broken down into the following activities:

- Opening meeting
- Field work/interviews
- Briefings/caucuses
- Closing meeting and draft report

Opening Meeting

The opening meeting includes:

- Introduction of the audit team, including names, titles, and qualifications

- Audit purpose, scope, and objectives
- Summary of methods and procedures of the audit
- Question-and-answer session
- Audit schedule
- Any necessary agenda changes
- Applicable standards, such as ISO 9001
- Required logistics (escorts, transportation, rooms, wi-fi)
- Tentative time and date for the closing meeting
- Communication channels between the auditor and auditee

Audit Data Collection and Analysis

Information is collected during the audit through:

- Questioning and interviewing
- Field investigation and verification

Closing Meeting or Post-Audit Meeting

At the end of the audit and prior to completion of the audit report, a meeting is held to present audit evidence to auditee senior management to ensure that the audit results are clearly understood. These are three key characteristics of a closing meeting:

- Attendance includes the lead auditor, auditor(s), and appropriate auditee personnel (top management but not middle-level management)
- Findings and objective evidence are presented; a presentation is made, but copies are not provided until the end of the meeting
- The audit team provides a rough draft of the final report, or provides information according to the sponsoring company policies, which includes:
  - —Name of the audited organization
  - —Date(s) audit performed
  - —Audit purpose and scope
  - —Findings and audit evidence, in order of importance (worst first)
  - —Date and signature of lead auditor
  - —Element scores/final scores if scoring checklist is used, or classification
- The lead auditor stresses the importance of correcting the critical deficiencies first and identifies the areas that may require a follow-up visit

- The audit team should not make specific corrective action recommendations in the audit report or verbally to the auditee

Types of Audits

Various types of audits may be performed based on the scope and requirements. These include:

- Management system audits
- Process audits
- Product audits

There are also different methods of auditing. These include:

- On-site
- Virtual or remote
- Hybrid
- Desktop

The method of auditing may be dependent on, but not limited to:

- Criticality of auditee, risk assessment
- Location of auditee, such as another country
- Cost to have auditors on–site
- Location of auditors/subject matter experts
- Urgent need
- Access restrictions, e.g., Homeland Security access

An **On-site** audit is the preferred method of auditing. This is when all auditors are present at the auditee and the entire audit is carried out at that location. On-site auditing should be the first consideration when selecting an audit method.

A **Virtual or Remote** audit occurs when the auditor(s) is not at the auditee location and at their office(s); the audit is conducted using computers and streaming services, such as Webex, Teams, Skype, or Zoom. This method works well for interviewing and document review (document can be shared electronically). However, tours of the facilities and processes may be limited and difficult. Tours would involve the auditee using handheld cameras/devices to stream to the auditor(s). This may be limited by Wi-Fi access and policies regarding cameras in certain areas.

The **Hybrid** method is a combination of on-site and virtual. One way to use this method is to have one auditor on-site at the auditee and another auditor on-line streaming in. Another way this method is used is when the auditor(s) are on-site for part of the audit (e.g., tours and the closing meeting) and online streaming for another part of the audit (document review and interviews).

The **Desktop Audit or Questionnaire** is the last method for auditing. This is when the auditor or sponsoring company sends a questionnaire to the auditee to complete and return. Desktop audits should be the last method considered or used for low-risk suppliers. This is due to many aspects that cannot be controlled, such as a timely response from the auditee, who at the auditee completed the questionnaire, and inability to verify the information provided.

Management System Audits

Management system audits, the largest and most extensive type of audit, are conducted to determine, through objective evidence, whether the QMS and organizational plans are adequately carried out and effectively meet identified requirements. A full system audit includes and exceeds the scope of the process and product audit. A significant part of the system audit can be devoted to the process audit with one or more processes audited during a full system audit.

A system audit should include all phases of the supplier's system and cover six major categories:

1. Drawings and specifications
2. Purchased material
3. Measuring and testing equipment
4. Process control and product acceptance
5. Storage, packing, receiving/shipping, and record retention
6. Quality program management

Process Audits

Process audits should be conducted periodically at the supplier's facility to ensure adherence to documentation. Process audits are usually conducted quarterly to semiannually. Some examples of process audit frequencies are:

- Critical products: quarterly
- Complex, key, or high-volume products: quarterly
- Minor or supplier-controlled products: semiannually or annually

The customer or customer's representative who performs process audits should establish an auditing plan that includes all phases of the supplier's process for manufacturing a particular product or process type. The following is a list of major types of process audits:

- Machine supplier
- Injection molding supplier
- Casting supplier
- Sheet metal or stamping supplier
- Special process audit

The process audit is the most frequently performed and the most convenient audit, often yielding faster results than the system audit. Process audits are most useful in verifying conformance to standards, methods, procedures, or other requirements. Process audits:

- Are less extensive than system audits
- Usually concentrate only on one specific process element
- Require less planning than system audits
- Can be very helpful in improving the process element in question
- May be performed in as little as under one hour to as much as a couple of days depending on the scope of the audit
- Often require fewer auditors than system audits, depending on the process

Product Audits

A product audit is a review of a part to determine whether it duplicates the criteria of the first article inspection. A supplier's inspection documents also may be audited during a product audit. Product audits may be conducted at the supplier's facility by either the supplier or customer representative, or the audit may take place at the customer site and be performed by the customer representative. When the audit occurs at the supplier site and is performed by the supplier representative, the customer representative should witness or verify the process and review the results. If the audit occurs at the customer site, the auditor should use the customer's inspection areas, equipment, and laboratories as needed. The supplier representative may be allowed to witness this process. If the audit uncovers deficiencies, the supplier representative should be shown the method used. This will ensure that inspection of product is consistent regardless of where and by whom it is performed.

The product audit is an assessment of the final product or service and its fitness for use evaluated against specifications and the purpose of the product or service. Product audits are performed on product that has already passed final inspection. Product audits are customer oriented and may be performed by one auditor. Supplier product audits may be done at the supplier site, at the customer site, or with the final customer.

AUDIT REPORTING AND FOLLOW-UP

> Apply and analyze audit reporting and follow up, including verification of the effectiveness of corrective action. (Analyze)
>
> Body of Knowledge V.A.2

The formal audit report is the product of the audit, with the lead auditor being responsible for the report's content and accuracy and for submitting the formal report in a timely manner.

The audit report timing is critical. If a long time passes between the audit and the reporting, auditee management is given the impression that the response is not critical and other priorities are placed above this action. The audit report should be completed as soon as possible and mailed within one week. With technology such as e-mail, the report can be distributed in a short time.

Prior to the distribution of the audit report, it is reviewed for content and accuracy by the audit team and signed by the lead auditor. The use of a standardized format is helpful not only to the auditing organization but to the auditee. The standard format includes a cover sheet and the main audit report, which are composed of the following:

Cover Sheet

- Title of the report, including the name of the auditee
- Date of issue
- Distribution list
- Executive summary (optional, but is beneficial to auditee management)

Main Audit Report

- Name of the auditee, location, audit purpose and scope, dates the audit was performed, audit team members, and lead auditor
- A statement on the tone of the audit—i.e., whether it was cooperative, antagonistic, or supportive. This helps the auditing organization, auditee, and future auditors.
- A list of the findings in order of importance, with observations listed to support each finding
- Closing remarks, including any requests for corrective action and the date the auditee is expected to respond
- The signature of the lead auditor and approval from the audit authority if not delegated to the lead auditor

A request for corrective action is normally presented through a separate standardized form. This form can be a corrective action request form or a similar form customized for audits. The form should have the header information, audit finding, and response due date. The audit findings can be addressed singularly or collectively. The recommended format is to include the finding and the corrective action on the same form. This improves tracking. The corrective action should include unique identification. ISO 19011 indicates that the audit report should provide a complete, accurate, concise, and clear record of the audit and should include the following:

- Audit objective
- Audit scope

- Audit client
- Audit team leader and team members
- Dates and places of audit activities
- Audit criteria
- Audit findings
- Audit conclusions
- Noteworthy accomplishments. These are exceptional practices observed during the audit and should include a description and the impact to the organization.

Audit reports are considered confidential and should not include the following:

- Proprietary information
- Subjective opinions. Only facts are admissible.
- Any recommendations unless in the contract or approved by the audit authority
- Names of employees who were the subject of findings. Position and title may be included for traceability to a finding location.
- Nit-picking. When possible, several observations under one element can be treated as observations to support the finding. For example, if a calibration system does not meet requirements, a list of different observations can support the system nonconformance.
- Argumentative statements
- Any items not covered in the closing meeting. If items are identified after the closing meeting and located in audit notes, they can be discussed with the auditee but not included in the audit report.

An effective report will include background information, an executive summary, prioritized results, graphical tools, and a timeline for the auditee to respond and implement corrective actions. Observations, findings, and conclusions should be ranked by importance and arranged in a logical sequence. The use of bullet points is an effective way to present findings. The criticality of the observations is important and is most often categorized as critical, major, or minor. Critical nonconformances are usually associated with safety-related findings.

A Pareto chart can be helpful in organizing the audit report and assisting management with the selection of the most serious problem to address. If the auditor uses nonconformance categories such as major and minor, this will help the auditee to focus on the most serious identified findings.

The final audit report should be approved by the lead auditor or audit authority. The report can be distributed by the lead auditor or by the audit authority. In some cases, the purchasing lead distributes the report. According to ISO 19011, the audit is considered complete when all activities described in the audit plan have been carried out and the approved audit report has been distributed. The completion and effectiveness of corrective actions should be verified. This verification may be part of a subsequent audit.

A unique identification number is assigned to the corrective action to track the activity and to serve as a cross-reference. This also allows the lead auditor to match the response to the corrective action.

The auditee is given a due date for the response associated with each audit corrective action. For a full quality system audit, the response time is usually 30 days. This does not mean the corrective action will be fully implemented. The auditee may request an extension for the permanent corrective action response or implementation. This should be submitted in writing and maintained as part of the formal audit documentation. Reference to the original due date should be maintained and action taken if the deadline is delayed multiple times.

Record Retention

Audit reports and supporting documentation should be retained for the period of time specified by the audit program or audit authority. ISO 19011 states the audit program records should be maintained and should include:

- Records related to the individual audits:
  - Audit plans
  - Audit reports
  - Nonconformity reports
  - Records of corrective/preventive actions
  - Audit follow-up reports, if applicable
- Results of audit program reviews
- Records related to audit personnel covering subjects such as:
  - Auditor competencies and performance evaluation
  - Audit team selection
  - Maintenance and improvement of competence

Audit Follow-Up and Closure

The supplier's upper management is responsible for developing and implementing a corrective action plan based on the corrective actions issued by the lead auditor. Permanent actions are required to address the root causes with the goal being to eliminate these issues. An effective root cause analysis will allow the auditee to identify the real root cause and not just the symptom.

The supplier corrective action requires at least two improvement responses for each finding: the short-term action and the long-term or permanent action. Short-term actions remain in place until the permanent actions are fully implemented. Both actions should reference the implementation dates. The auditee presents the completed corrective actions to the auditing authority, which is usually the lead auditor, for review. Once the corrective actions are reviewed, the lead auditor informs the supplier of the decision to accept or reject the responses. If the supplier's responses do not address the true root cause, the lead auditor will explain the root cause process. The corrective action responses frequently address only the symptom; when this occurs, the lead auditor should recommend that they move upstream in the process in order to locate the true root cause. The auditor can recommend an effective method for root cause analysis, at a minimum the 5 Whys methodology.

Many supplier audits are based on the ISO standards, which have clearly defined requirements. In ISO 9001:2015 standards, what were previously called preventive actions were replaced with risk-based thinking. This refers to a coordinated set of activities and methods that organizations use to manage and control the many risks that affect its ability to achieve objectives. Risk is the effect of uncertainty on an expected result, and an effect is a positive or negative deviation from what is expected.

The methods used for verification of corrective action changes are dependent on the severity and criticality of the noncompliance. There are several methods for verifying implementation of a corrective action. These include:

- Documentation of completion supplied by the supplier
- Documentation resulting from a specific activity audit performed by the supplier quality function and supplied to the lead auditor
- On-site, narrow scope audit performed by the original audit team to evaluate the specific audit finding
- On-site, full system audit performed by the original audit team

When the nonconformance is related to documentation, the verification of closure may be based on copies of the documentation sent by e-mail. However, with this method there is no assurance that personnel have been trained or the changes have been effectively implemented. On-site audits are normally performed to verify corrective actions for all types of nonconformances. If the relationship with the supplier has a high level of credibility through years of interaction, this can influence the depth of the verification.

Follow-up audits to verify the corrective actions and confirm the actions were effective in correcting the issue (effectiveness checks) are important and should be implemented within the time period agreed to by the audit authority, which is usually the lead auditor, and the supplier. For supplier audits, there are two different follow-up audits associated with the corrective action. The first audit is performed by the supplier in a timely manner. The second audit is performed by the client or customer audit team based on the severity of the finding. If the nonconformance is critical in nature, the follow-up audit should be performed based on timing of corrective action completion and availability of personnel. The follow-up audit is an investigation of the findings identified in the audit report. The objective of the follow-up audit is to determine whether the corrective action was implemented and effective. This includes evaluating new or updated procedures, observing revised processes, and ensuring effectiveness. In addition, if the actions taken were not effective, the auditor should verify that the auditee identify why and implement new corrective actions.

Auditor follow-up on corrective actions is one of the weakest links in the audit process. Schedules, commitments, and lack of focus may reduce this critical activity to a low priority. The audit loop is not closed until the corrective action is complete and assessed. The follow-up activity should generate a report that identifies the audit as still open or closed. The follow-up report (or final report, if the audit is closed) is retained with the other formal audit records. The permanent corrective action requires that supplier management and the auditor determine whether:

- The root cause analysis has identified the full extent of the problem
- The corrective action is satisfactory to eliminate or prevent recurrence (effectiveness check)
- The corrective action is maintainable

If the results of the corrective action appear to be ineffective, the auditor should:

- Reaudit the supplier and specifically investigate the corrective action results
- Reissue the corrective action request before or after the repeated audit
- Escalate the situation to the auditing organization's upper management. This may result in disengagement of the supplier.

Audit Closure

After corrective action has been implemented and verified, the corrective action request may be considered for closure. The auditor must make a decision as to whether the corrective action fixed the symptom or corrected the true root cause of the deficiency. The supplier should be reaudited after completion of the corrective action request. Documentation of the closure of the corrective action request is done on the corrective action form or an attachment. Plans to close the audit nonconformances should not have implementation dates that extend beyond one year. It is standard for the implementation dates to be in the range of one to three months maximum, but for more complex issues a one-year period may be required. The root cause and corrective action plan must be approved in a timely manner. The closure documentation includes the closure date, supporting documentation, and the signature of the person closing the corrective action request. The completed corrective action should be retained as a quality record. The lead auditor is responsible for tracking the planned completion dates for the corrective action.

Periodic Reevaluation

Once the audit is closed, it is not done forever. If a company continues to do business with a particular supplier, then the supplier should be reevaluated on a periodic basis to ensure that the supplier's processes are still capable of providing good product and that critical processes have not drifted out of control. Most companies will choose to perform this reevaluation through a repeat audit; the audit interval should be based on the criticality of the product purchased from the supplier and on past results, including results from prior audits as well as results of supplier metrics, any recent history of nonconforming material (discussed in Chapter 12), and/or results of any corrective and preventive action efforts at the supplier (discussed in Chapter 13). Once a company determines that a reaudit of a supplier is required, the new audit is completed using the same process previously discussed here. In particular, the auditor should focus on any process that was found to be nonconforming during the previous audit as well as any process that has recently produced nonconforming product or may be deteriorating in performance as demonstrated through metrics.

SUPPLIER COMMUNICATION

> Evaluate various communication techniques such as periodic reviews, metric and performance indices, change management, notifications, recalls, change requests, and business updates. Maintain active communication with suppliers to assess risk and take appropriate action. (Evaluate)
>
> **Body of Knowledge V.A.3**

There are many ways that a supplier can cause issues with a customer. These can include:

- Poor product quality
- Poor responsiveness
- Poor on-time delivery
- Increased cost
- Lack of communication

Customers can take steps to circumvent these issues. Planning and periodic review meetings, contract meetings, and communication reviewing both requirements and performance can help prevent supplier disappointment.

Joseph Juran states that joint quality planning requires detailed discussion between the customer and supplier covering three major areas:

- Economic
- Technological
- Managerial (Juran and De Feo 2010)

All three of these areas are important and are part of the up-front stated supplier expectations.

Joint economic planning meetings cover the economic aspects of customer–supplier conversations and should concentrate on the following elements:

- Value rather than conformance to specifications
- Optimizing overall quality costs

Joint technological planning meetings as defined by Juran cover issues related to sophisticated product and include agreement on specifications, performance requirements, qualification of reliability and maintainability, preparation of manufacturing process control, defect seriousness classification detail, standardization of test methods and conditions, establishment of acceptable quality levels, system of lot identification, system of lot traceability, and timely response to defects and delays.

The conventional concepts of managerial control include the following:

- Definitions of responsibilities
  - —Within the buyer's organizations
  - —Within the supplier's organizations
  - —Mutual buyer–supplier concerns
- Documented reporting requirements
  - —Ensure compatibility of product language
  - —Include any documented proofs (certifications)
  - —Include audit reporting language
- The formalization of communication channels may need multiple communication channels
- A formal written contract
  - —May be included in purchase orders
  - —Details product specifications
  - —Details other quality provisions, such as notification of changes and non-conformances, including recalls
  - —Details delivery and related requirements
  - —Details payments and penalties
- Buyer/supplier management requirements
  - —Supplier's adherence to buyer's quality policies
  - —Buyer's organizational chart and general plan with respect to quality—this helps the supplier understand expectations
  - —Buyer's procedures to evaluate supplier quality
  - —An explanation of quality specifications and seriousness classifications
  - —Buyer's plans for supplier surveys
  - —Copies of the buyer's required quality report forms
  - —A glossary of key quality terms used by the buyer

- Actions expected of the supplier
  - —Quality plans
  - —Inspection and testing
  - —Sample submissions
  - —Reliability programs
  - —Reports and records
  - —Nonconforming material controls
  - —Engineering change requirements
  - —Record retention
  - —When notification to the buyer/sponsoring company is required, e.g., process/product changes, non-conformances, recalls, and business operations that could impact the product

Communication During the Contract

There is a need for continuing buyer–supplier cooperation and communication during the execution phase of a contract. The nature and complexity of this communication vary depending on the complexity and criticality of the product. For standard commodities, most communications are channeled through the purchasing department and the supplier marketing/sales department.

For complex engineered products, there may be multiple communication channels (e.g., purchasing, management, quality control [QC], sales, supplier management, design, and test lab). There can be any combination to support this complexity.

Common communication activities considered for emergency or near emergency items may be required. In such cases, typically one or two parties at both the buyer and supplier locations are notified of the issue. The issue is detailed in writing to prevent confusion and should consider:

- Deviations
- Product unfit for use
- Design and change information
- A need for corrective action

Other Communication Channels

Lines of communication should consider the following:

- Emergency communications involving the highest level of management
- Routine communications that include
  - —A quarterly status review
  - —A periodic product or process audit

—An annual systems audit

—A monthly or quarterly supplier rating report

In addition to emergency and routine communications, there is a growing need for supplier communications in the following areas:

- *Positive communications.* Buyers are quick to communicate negative information to suppliers, such as nonconformances, late deliveries, and costs. Positive communications are also needed in the form of letters of praise, awards, and recognition. When appropriate, an increase in supplier business is also a positive communication acknowledging a job well done.
- *Supplier training.* Supplier training can be a cost-effective way to improve quality in purchased goods and services. It is a highly desirable communication link. Benefits of training include:

  —Improved awareness of the organization's needs

  —Improved supplier quality systems, reducing costs and preventing problems

  —Face-to-face communications promote trust and the realization that both companies can work together for mutual benefits

  —Feedback for improvements in design and manufacturability

  —A demonstration that the company is committed to supporting a "partnership" relationship with the supplier

Supplier Performance Assessment

Supplier assessment and feedback are essential for both the supplier and the customer. Both objective and subjective data must be collected and analyzed to determine if corrective action is necessary.

The requirements for a supplier rating system are highly dependent on the relative complexities and criticality of the products involved. Industrial groups, governmental bureaus, and regulatory agencies often define requirements for the control of purchased material. Most routine supplier feedback is reported through some type of computerized document. This can be a supplier report card provided electronically or posted on a supplier portal with firewalls and password protection.

Supplier Assessment Metrics

A wide assortment of measurements are used to provide supplier feedback, including:

- Quality metrics
- Timeliness metrics
- Delivery metrics

- Cost metrics
- Compliance metrics

Quality Metrics

Quality performance is often expressed as:

- Percent defective lots
- Parts per million (PPM)
- DPMO
- Lots accepted vs. lots rejected
- Percent nonconforming
- Special characteristic measurements (mean time before failure [MTBF], etc.)

Timeliness Metrics

Timeliness is commonly reported as:

- Percentage of on-time deliveries
- Percentage of late deliveries

Delivery Metrics

Delivery-related measures, considering the growing emphasis on JIT operations, include:

- Percentage of over-order quantities
- Percentage of under-order quantities
- Percentage of early deliveries

Cost Metrics

A number of cost items are also considered in supplier ratings:

- Dollars rejected divided by dollars purchased (or vice versa)
- Dollar value of nonconforming product (scrap, returns/warranties, rework)
- Dollar value of purchases (per reporting time segment)

Compliance Metrics

Compliance performance is often difficult to define but can include:

- Percentage of reported quality information (of that required)
- Percentage of supplied certifications (of that required)

Supplier Feedback Reports

Supplier feedback is critical to a strong customer–supplier relationship. This feedback can be provided in many formats. When providing feedback to upper management, use less detail and focus on the associated costs. The following items are examples of this level of feedback; they are most effective when shown in a trend chart format:

- Total dollar value purchased
- Percent defective dollar to dollar value purchased
- Percent defective dollar recovered for each supplier
- Percentage of lots or project rejected
- Corrective action activity that includes critical issues
- Composite supplier rating score when available

Supplier Improvement

TCO considers the total cost (inspection, rework, etc.). The new trend is toward interdependence between the supplier and customer. Suppliers are selected based on quality, delivery, technology, lifecycle cost, and management philosophy. A supplier identified as single source offers several advantages, but there is also risk that must be mitigated and included in the customer's risk assessment process. The advantages include less overall product variation, fewer communication links, and potential price discounts based on volume. The risks can be assessed in many ways. Suppliers can be evaluated through surveys or audits. The formation and completion of timely corrective action plans are crucial in supplier management.

A supplier may be evaluated both before and during the execution of a contract. A formal audit consists of an audit team that visits the supplier's facility to determine if the supplier has the ability to provide adequate product or services. The utilization of international standards such as ISO 9001, AS9100, ISO 13485, IATF 16949, and/or ISO 17025 as the basis for a supplier audit ensures that the key characteristics of a high-performing supplier are included. These are management commitment, resource management, infrastructure, communication (both external and internal), manufacturing (including product and service realization), design control, quality planning, purchasing, QC, quality results (including key performance metrics), and continuous improvement.

Supplier surveillance is a key part of maintaining active communication with suppliers to assess risk and take appropriate action when needed. The two major surveillance approaches are auditing and process surveillance, which consists of monitoring the manufacturing process of the supplier. Process surveillance involves the following steps:

- Witnessing key events
    - Operations
    - Inspection
    - Tests

- Inspection of critical characteristics
  - — By witnessing
  - — By performing
- Joint troubleshooting of mutual quality-related issues

Strong partnerships with suppliers improve the communication process. The benefits to both supplier and customer include better quality, faster response time, flexible manufacturing, lower inventories, and shorter cycle times. The supplier can have larger orders, better forecasts, and shared savings. Both customer and supplier benefit.

Ship-to-stock (STS) and JIT procurement are also benefits of a quality communication process. JIT consists of two principal elements: procurement and inventory. JIT procurement focuses on forecasting, scheduling, inventory cost control, and freight and transportation expense control.

SUPPLIER DEVELOPMENT AND REMEDIATION

> Identify and analyze present and future training needs and gaps, using quality methods and tools such as kaizen and benchmarking. Use process improvement tools such as DMAIC, cycle time reduction, defect rate, and cost reduction. Evaluate supplier remediation to develop and manage improvement plans. (Evaluate)
>
> **Body of Knowledge V.A.4**

Supplier development is defined as any effort of a customer to increase the performance and/or capabilities of its supplier and meet its needs. This development offers benefits to both suppliers and customers. It begins with the identification of present and future training needs through analysis. The analysis is done using quality methods and tools. One of these tools is the kaizen approach. *Kaizen* is Japanese for "continuous improvement" and is referred to as incremental improvement involving everyone. Kaizen refers to:

- Productivity
- Total quality control
- Zero defects
- JIT
- Suggestion system

The kaizen strategy involves considerations for maintenance and improvement of operating standards by management, improvement of processes, use of plan-do-check-act (PDCA), quality as the highest priority, and problems solved using data in the form of numbers or graphs as opposed to qualitative information.

Another method for determining future training needs and gaps is benchmarking. There are best practices and various types of benchmarking, including process, performance, project, and strategic benchmarking. A supplier performing effective benchmarking can identify areas where gaps exist.

As part of a supplier audit, a seasoned auditor might be tempted to communicate best practices that have been identified at different suppliers, but the auditor must be aware that this information cannot be provided due to potential confidentiality concerns.

Suppliers have varying levels of maturity, which influences best practice benchmarking efforts. Some key strategies are:

1. Identify key performance factors
2. Understand your own process and the processes of others
3. Select performance criteria based on needs and priorities
4. Measure the performance with the organization in the criteria areas
5. Determine the leader in the criteria areas
6. Collect information and data on the performance of the benchmark leader
7. Evaluate and compare current practices with the benchmark
8. Drive significant improvement changes to advance performance levels
9. Utilize the new awareness to extend performance through breakthroughs
10. Incorporate the new information in business analysis and decision making
11. Seek alternate benchmark leaders or process areas for further improvement

Another improvement sequence for best practices/benchmarking is:

1. Determine current practice
2. Identify best practices
3. Analyze best practices
4. Model best practices
5. Repeat the cycle

In situations where there is no known or related process available, a supplier may have to resort to either reengineering an existing process or developing a new process to improve the operation, project, product, or activity of concern. SWOT analysis is a good tool for competitive benchmarking. Benchmarking compares the current project, methods, or processes with the best practices known and uses this information to drive improvement of overall company performance. The standard for comparison may be competitors within the industry but is often found in unrelated business segments.

There are several varieties of benchmarking, including the following:

- *Process benchmarking.* Focuses on discrete work processes and operating systems, such as the strategic planning process. This form of benchmarking seeks to identify the most effective operating practices among those that perform similar work functions.
- *Performance benchmarking.* Enables management to assess the organization's competitive position through product and service comparisons. This form of benchmarking usually focuses on elements of price, quality, product/ service, speed, reliability, and other performance characteristics.
- *Project benchmarking.* Project management is easier to benchmark than many business processes because of the opportunities for selection beyond direct competitors. The projects share the same constraint factors of time, costs, resources, and performance. Benchmarking prior to the start of a new project is useful in selecting new techniques for planning, scheduling, and controlling the project.
- *Strategic benchmarking.* Examines how companies compete. It is seldom industry focused and moves across industries seeking to identify the winning strategies that have enabled high-performing companies to be successful in their marketplace.

In most cases, the primary improvement strategy that a company should employ depends on its current performance.

When areas of process improvement are identified through kaizen activity or benchmarking, several key improvement tools such as define, measure, analyze, improve, control (DMAIC); cycle time reduction; and defect rate and cost reduction can be used to realize these improvements.

Define, Measure, Analyze, Improve, Control

DMAIC, a Six Sigma process, is used to ensure the best possible project results (see Figure 14.2). The following process steps make up this process improvement technique:

- *Define* the customer, their CTQ issues, and the core business process involved
- *Measure* the performance of the core business process involved
- *Analyze* the data collected and the process map to identify root causes of defects and opportunities for improvement
- *Improve* the target process by designing creative solutions to fix and prevent problems
- *Control* the improvements to keep the process on the new course

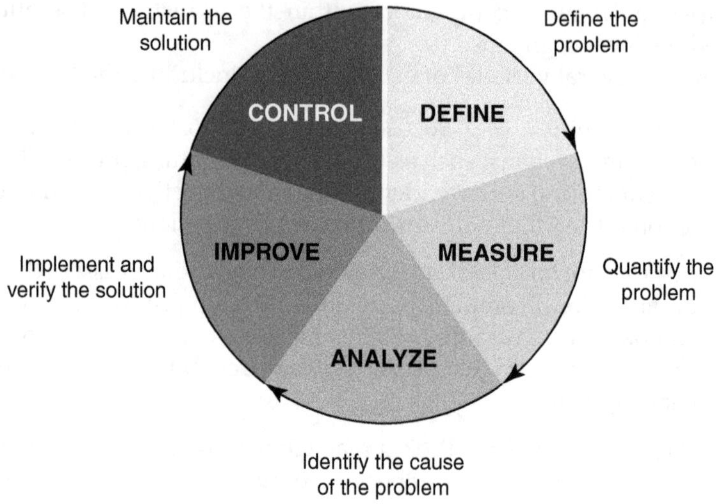

Figure 14.2 Six Sigma DMAIC cycle.

Cycle Time Reduction

Cycle time reduction is a critical concept for the operation of a lean enterprise. Cycle time is defined as the amount of time needed to complete a single task and move it forward in the process. Efforts to reduce cycle time are customarily undertaken to please a customer, to reduce waste, to increase capacity, or to reduce product damage. Training concepts and principles associated with cycle time reduction include:

Introduction to the total systems concept, fixing the system

- Problem-solving tools such as 5 Whys
- "Next process as the customer"
- Nonjudgmental approach to problem solving
- Identification of value- and nonvalue-added work
- Identification of muda—the seven wastes:
  - —Overproduction
  - —Inventory
  - —Repairs/rejects
  - —Motion
  - —Processing
  - —Waiting
  - —Transport
- Principles of motion study
- Workflow patterns (simple flow, U-shaped)

- Standard operations
- 5S workplace organization
- Visual management principles
- JIT—producing only the necessary quantities at the necessary time using the necessary resources
- Poka-yoke principles/error proofing
- Team dynamics

Defect Rate Reduction and Cost Reduction

Defect rate reduction and cost reduction can be achieved through the proven techniques listed in the previous section. Quality costs consist of those costs associated with company efforts devoted to planning the quality system, those associated with efforts to verify that quality is being obtained, and those associated with failures resulting from inadequate systems. Quality cost categories are prevention costs, appraisal costs, and failure costs. The definitions are:

- *Prevention costs.* The costs of activities specifically designed to prevent poor quality in products or services
- *Appraisal costs.* The costs associated with measuring, evaluating, or auditing products or services to ensure conformance to quality standards and performance requirements
- *Failure costs.* The costs resulting from products or services not conforming to requirements or customer/user needs. The costs resulting from poor quality. Failure costs are made up of internal and external components:
  - —*Internal failure costs.* Occur prior to delivery or shipment of the product or the furnishing of a service to the customer
  - —*External failure costs.* Occur after delivery or shipment of the product or during or after furnishing of a service to a customer. This is the most costly type of cost

The advantages of a quality cost system include:

- Provides a manageable entity and a single overview of quality
- Aligns quality and company goals
- Provides a problem prioritization system and a means of measuring change
- Provides a way to distribute controllable quality costs for maximum profit
- Improves the effective use of resources
- Provides emphasis on doing the job right every time
- Helps to establish new product processes

Quality cost systems do have limitations and pitfalls. They do not solve quality problems, nor do they suggest specific actions to improve. In addition, important costs may be omitted from the reports and inappropriate costs may be included; these costs are susceptible to measurement error.

An effective cost of quality process can be a measure of the overall effectiveness of the process improvements identified through the use of tools listed in this section and result in bottom-line impact to the financial statements.

The process improvement tools listed are key to a successful remediation process. This includes preparation and execution of the plan. When issues are identified that require remediation, these can be effectively managed. The plan can be driven by the results of an audit, the receipt of defective product, late deliveries, or higher-than-projected costs. Risk assessments with associated contingency plans can also be part of the remediation plan.

PROJECT MANAGEMENT BASICS

> Understand and apply various types of project reviews, such as phase-end, management, and retrospectives or post-project reviews to assess project performance and status, to review issues and risks, and discover and capture lessons learned from the project. Apply forecasts, resources, schedules, and task and cost estimates to develop and monitor project plans. (Apply)
>
> **Body of Knowledge V.A.5**

Project management review is an exercise undertaken at the end of each project phase to identify the current status of the project. The review identifies the deliverables that have been produced to date and determines whether the project has met the objectives set. The outcome of the project management review is documented on a project phase review form, which states whether the project is currently on track, on schedule, and under budget.

Once the project phase review form has been completed, the form is presented to the project owner or sponsor, who decides whether to proceed to the next project phase.

Project reviews are conducted at the end of the initiation, planning, and execution phases of a project. A good project phase review form helps document the results of the project review and is a visual document that states:

- Project is currently on schedule
- Budget allocated is sufficient at this point
- Deliverables have been produced and approved

- Risks have been controlled and mitigated
- Issues have been identified and resolved
- Changes have been properly managed
- Project is on track

The project phase review form achieves the following:

- Documents the results of the project management review
- Clearly communicates the progress of the project to the sponsor
- Identifies any risks or issues that have impacted the project
- Shows the sponsor the deliverables produced to date
- Facilitates approval to proceed to the next phase

By implementing project reviews, the project owner puts in place the necessary checkpoints to monitor and control the project to ensure success.

Phases of Project Management

Project management is the application of knowledge, skills, tools, and techniques to a broad range of activities in order to meet the requirements of a particular project. The five phases of project management are depicted in Figure 14.3. If the lifecycle provides a high-level view of the project, the phases are the road map for accomplishing it.

Phase 1: Project Initiation

The project begins with a business case, and the goal at the start of the project is to define the project at a broad level. During the initiation phase, the project owner researches whether the project is feasible and whether it should proceed. Any feasibility testing needed is performed during this stage. The project owner will engage with important stakeholders to determine whether the project is a go. If a determination is made to proceed, the project owner will create a project charter or project initiation document that outlines the purpose and requirements of the project. The project charter should include:

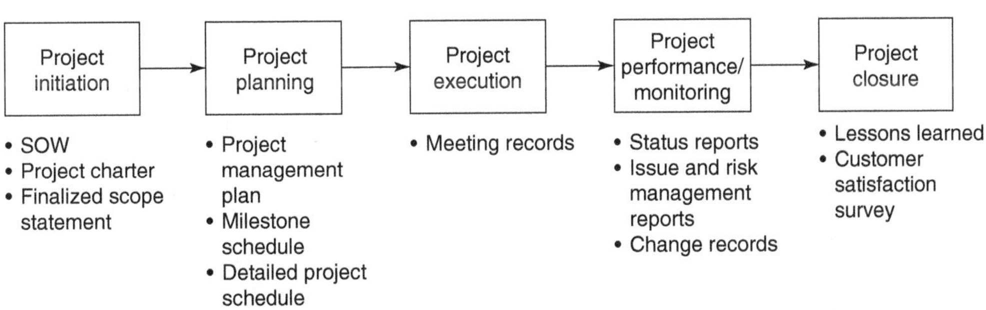

Figure 14.3 The five stages of project management.

- Business needs
- Stakeholders
- Business case

Phase 2: Project Planning

This is a key phase of successful project planning and focuses on developing a road map that everyone will follow. Goals are normally set in the planning stage. Two of the most popular methods for setting goals are SMART and CLEAR:

- SMART Goals—This method helps ensure that the goals have been fully investigated and provides a way to clearly understand the implications of the goal-setting process. Figure 14.4 depicts the SMART Goal methodology.
  - —Specific—Set specific goals by answering the following questions: who, what, where, when, which, and why
  - —Measurable—Create criteria that you can use to measure the success of a goal
  - —Attainable—Identify the most important goals and what it will take to achieve them
  - —Realistic—You should be willing and able to work toward a particular goal
  - —Timely—Create a timeline to achieve the goal
- CLEAR Goals—A newer method for setting goals that takes into consideration the environment of today's fast-paced business. Figure 14.5 depicts the CLEAR Goal methodology.
  - —Collaborative—The goal should encourage the project team to work together.
  - —Limited—The goal should be limited in scope and time to keep it manageable and achievable.

S	Specific
M	Measurable
A	Attainable
R	Realistic
T	Timely

Figure 14.4 SMART Goals.

C	Collaborative
L	Limited
E	Emotional
A	Appreciable
R	Refinable

Figure 14.5 CLEAR Goals.

— Emotional—Goals should tap into the passion of employees and be something they can form an emotional connection to. This can optimize the quality of work.

— Appreciable—Larger goals should be broken down into smaller tasks that can be quickly achieved.

— Refinable—As new situations arise, be flexible and refine goals as needed.

During the project planning phase, the scope of the project is defined and a project management plan is developed. This plan involves identifying costs, quality, available resources, and a realistic timetable. This also includes establishing performance measures. The project measures are generated using the scope, schedule, and cost of a project. The use of a baseline is important to determine whether the project is on track. During this phase, roles and responsibilities are clearly defined to ensure that the team understands how they will be measured. The following are documents a project manager will create during the project planning phase:

- Scope statement—This document clearly defines the business need, benefits of the project, objective, deliverables, and key milestones. The scope statement may change during the project, but this will require the approval of the project manager and the project sponsor.
- Work breakdown schedule (WBS)—The WBS is a visual representation that breaks down the project scope into manageable sections for the team.
- Milestones—The milestones identify high-level goals that need to be met throughout the project. These are included in the Gantt chart.
- Gantt chart—A visual time line that can be used to plan tasks and visualize the project time line.
- Communication plan—A communication plan is important if the project involves outside stakeholders. Communication around the project and creation of a schedule of when to communicate with team members should be based on deliverables and milestones.
- Risk management plan—A risk management plan is needed to identify all potential risks. These include unrealistic time and cost estimates, customer review cycle, budget cuts, changing requirements, and lack of committed resources.

Phase 3: Project Execution

This is the phase where deliverables are developed and completed. Project execution is where many activities are occurring. These include status reports, meetings, development updates, and performance reports. A kick-off meeting usually marks the start of the project execution phase. This is where the teams involved are informed of their responsibilities. Some of the tasks completed during the execution phase include:

- Development of the team
- Assignment of resources

- Execution of project plans
- Procurement management if needed
- Project manager directs and manages project execution
- Tracking systems are set up
- Task assignments are executed
- Status meetings are held
- Project schedules are updated
- Project plans are modified as needed

Phase 4: Project Performance/Monitoring

The project performance/monitoring phase often occurs simultaneously with the project execution phase and is focused on measuring project progression and ensuring that all activities align with the project management plan. Project managers use key performance indicators (KPIs) to determine whether the project is on track. A project manager normally picks two to five KPIs to measure project performance. These include:

- Project objectives—Measure whether the project is on schedule and within budget. This provides visibility as to whether the project will meet stakeholders' objectives.
- Quality deliverables—Determine whether specific task deliverables are being met.
- Effort and cost tracking—Project managers will account for effort and cost of resources to see whether the budget is on track. This type of tracking informs whether a project will meet its completion date based on current performance.
- Project performance—Monitors changes in the project and takes into consideration the amount and type of issues that arise and how quickly they are addressed. These changes can occur from unforeseen hurdles and scope changes. Project managers may adjust schedules and resources to ensure that the project is on track.

Phase 5: Project Closure

As the name indicates, this phase represents the completed project. Key team members are recognized. Special events may be organized to thank personnel for their efforts. Any contractors hired to work specifically on the project are terminated at this time. When the project is completed, the project manager may hold a postmortem meeting to evaluate what went well in the project and identify project failures. These lessons learned are helpful to improve the results of future projects. Once the project is complete, project managers have additional tasks to complete. The project managers create a project punch list of things that didn't get accomplished during the project and work with team members to complete the open actions. The project manager will collect all project documents and deliverables and store them in a designated location.

Lessons Learned

The Project Management Body of Knowledge (PMBoK) from the Project Management Institute (PMI) defines lessons learned as the learning gained from the process of performing the project. Lessons-learned sessions are traditionally held during project closeout near the end of the project. Lessons learned may be identified and documented at any point during the project lifecycle. The reason for documenting lessons learned is to share knowledge gained from experience to promote the recurrence of desirable outcomes and preclude the recurrence of undesirable outcomes. This includes the process for identification, documentation, validation, and dissemination of lessons learned. Incorporation of those processes includes identification of applicable lessons learned, documentation of lessons learned, archiving lessons learned, distribution to appropriate personnel, identification of actions that will be taken as a result of lessons learned, and follow-up to ensure the appropriate actions were taken.

Chapter 15
Teams and Team Processes

A team is a group of people who share responsibility for accomplishing a specific purpose and maintaining specific performance results. The supplier management process utilizes various types of teams and team members, including:

- The customer team, which seeks out a supplier team
- Members of the customer team
- Members of the supplier team
- The key quality team members from the customer and supplier teams who are responsible for final supplier selection, supplier maintenance, and quality documentation
- Purchasing to ensure procurement of quality products at minimum cost
- Quality auditor team members representing the supplier and the customer

Wherever successful communication is needed, teams and team members play a key role. It is critical that team members playing any role in ensuring the maintenance of the supply chain for any industry be good team players who excel in professional communication and who have background knowledge on the materials, goods, and services being bought and sold. Irrespective of whether the team members represent the customer or the supplier, a can-do attitude and participation skills are crucial.

Understanding the types of team members and whom they represent (e.g., technical, quality, regulatory, sales, marketing, R&D, engineering) is critical to understanding what their expertise is, which allows the team to tap into their special knowledge in order to create a successful supplier quality management process throughout the product lifecycle and beyond for similar products.

TEAM DEVELOPMENT

> Identify and describe the various types of teams and the classic stages of team development: forming, storming, norming, performing, and adjourning. (Apply)
>
> **Body of Knowledge V.B.1**

Teams can be either be formal or informal. Teams are classified according to the roles or tasks performed and may include:

- Natural Teams or Work Group – a permanent part of an organization which perform ongoing business activities together usually led by a manager or supervisor
- Process Improvement Team – a team that is brought together for a specific project with a focus on improving a specific process. Process Improvement Teams can be associated with a kaizen blitz (rapid improvement)
- Cross-Functional Team – a team representing different functional areas addressing an organizational-wide issue
- Virtual Teams – a type of team where the individuals meet using digital communication technology such as video teleconferencing due to travel constraints or the location of the team members
- Cellular Teams – a type of team usually associated with manufacturing cells. Cellular Teams are generally lead by the lead operator. The team is charged with cross-training of the team members for improving the effectiveness and efficiency of the cell
- Self-Managed Teams – a type of natural team in which the decision making authority and responsibility residing with the team as opposed to a manager or supervisor. Self-Managed Teams are usually authorized to make decisions on a wide range of issues within the scope of their particular process or department
- Temporary or Ad Hoc Teams – a temporary team which may be identified due to a specific problem or situation

Teams are said to progress through a development cycle. The development stages are identified as forming, storming, norming, performing, and adjourning (see Figures 15.1 and 15.2). These stages were developed for cross-functional teams with a specific purpose, long-term projects.

The typical progression of a team development cycle follows:

1. *Forming.* The team is organized and members are welcomed. There is a certain amount of distrust, and team members push their own ideas.

Forming	Storming	Norming	Performing	Adjourning
• Individual • Own ideas • Cautious • Organizing	• Draw on own experience • Start interacting • Test leadership • Test objective	• Focus on goals • Open to ideas • Team coalesces • Cooperation	• Cohesive • Focus on process • Resolve issues • Achieve goals	• Issues resolved • Goals achieved • Team disbanded

Figure 15.1 Classic team development stages.

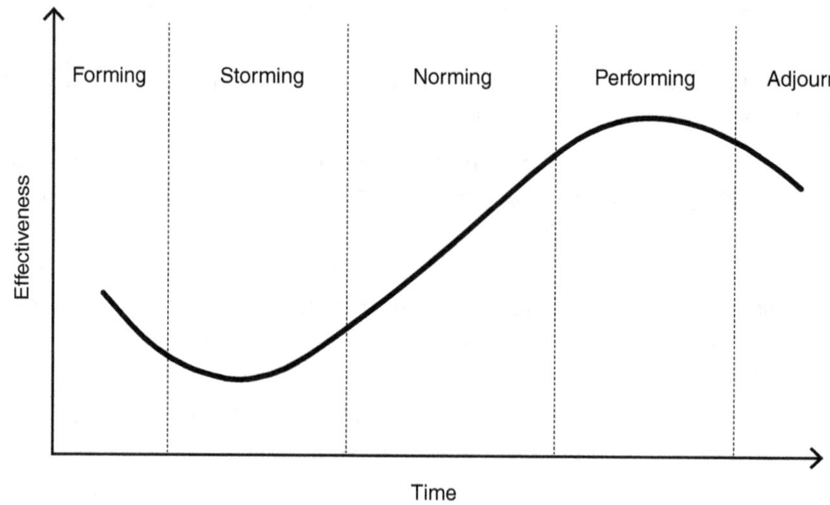

Figure 15.2 Team development cycle.

This is when the team gets organized (defining purpose, team member roles, objectives, etc.).

2. *Storming.* Storming suggests rushing around, attacking, and behaving impetuously. Individual team members attempt to win others over and test the authority or legitimacy of the team. As team members start to interact and the responsibility they have been given sinks in, interpersonal conflicts may arise, or they may become testy to see if management supports the team and its objectives.

3. *Norming.* Members start accepting the team rules and guidelines and start cooperating.

4. *Performing.* In this stage, the team is cohesive and has most likely worked out its differences. The members trust each other and build on team member strengths. The team has matured, has worked through its individual and turf issues, and is able to achieve objectives. Team objectives are described well and understood cohesively by the team members, who are ready to deliver.

5. *Adjourning.* Once delivery of team objectives is complete, the team adjourns and moves on to the next project.

If the basis for the team (goals, support, and purpose) changes, the team dynamics can slide back to forming or storming.

Before adjourning, teams should conduct a deep-dive review session on what went well, identifying lessons learned to carry over as best practices for the team's next project.

TEAM ROLES

> Define and describe various team roles and responsibilities for leader, facilitator, coach, and individual member. (Understand)
>
> **Body of Knowledge V.B.2**

There are several types of team member roles with various levels of responsibility and authority including:

- Sponsor – Responsible for team's effort and supports team plans, activities, and outcomes. Allocates resources
- Champion – Initiates idea or concept. Provides advocacy and periodic reviews. Usually a business leader or executive
- Leader – Directs team efforts from start to finish, shares progress with all stakeholders, and creates successful working environment
- Member – Commits to helping team achieve goals, provides subject matter expertise, and contributes as needed
- Facilitator – Acts as advisor, teacher, and moderator. Deals with process, not content or effort. Coach or gatekeeper
- Scribe – Helps team form thoughts, captures group's thinking
- Recorder – Captures meeting minutes
- Timekeeper – Tells the facilitator and team when time is running out on an agenda item and suggests to place the item in the "parking lot"

Team Facilitation and Bringing Out the Best in Individual Members

Facilitation techniques needed for the cross-functional team and the long-term project team normally include the following:

- Ensuring that necessary team-analyzing resources are available, such as flip charts, a data projector, etc.
- Recognizing and defusing clashes between team members
- Ensuring that all team members have a voice and that the team is not controlled by domineering personality types

- Using diverse views to strengthen team deliverables
- Keeping the team focused on the audit objectives, evidence-based decision making, and the schedule
- Using organizational tools (e.g., outlines, plans, flowcharts, sorted information) as necessary to facilitate decisions
- Cultivating a professional, impartial, and objective environment (e.g., testing outputs to ensure that no bias could be perceived by the auditee)
- Identifying problems or potential problems that influence team effectiveness (e.g., disgruntled auditee, known hostile or difficult situations, potential conflicts of interest)
- Establishing team rules or agreements as necessary (e.g., members must be on time to all scheduled meetings, members must be willing to listen to other members' views)
- Keeping communication lines open between team members and the auditee
- Keeping every individual member informed of team progress
- Presenting team views or issues for discussion and resolution
- Ensuring that everyone is on the same page

The need for facilitators and facilitation techniques decreases with team member training, experience, and successful team history. The better qualified the team members and the more they have worked together on other teams, the less active facilitation/coaching will be necessary.

Audit Teams

Audit teams are ad hoc teams formed to carry out an audit plan and achieve the supplier audit objectives. Most audit teams are informal (i.e., there are no team meeting agendas). Audit teams may or may not be cross-functional. The leader of the team is normally selected by the client or the audit program manager. If specialty areas such as sterilization, design controls, or chemical testing are involved, representatives from each area are included on the team, if possible.

There are seven primary elements of task-driven audit process teams:

- Following the audit plan and considering modifications as necessary
- Monitoring team progress and the audit schedule
- Completing assigned tasks
- Recording the results of the audit
- Ongoing verification of timely achievement of audit objectives
- Making evidence-based decisions relative to the audit findings and audit progress and documenting them for audit records and audit report
- Identifying new action items, responsibilities, and timing issues

There is always a lead auditor. If only one auditor is conducting the audit, that auditor is the lead auditor. For audit teams, the lead auditor should apply the following team management techniques:

- *Team member roles.* The lead auditor ensures that everyone understands his or her assignment before the audit starts.
- *Team membership.* The lead auditor ensures that all team members are aware of the other members, their assignments, and their contact information. Audit team members should be given time to get acquainted, such as at a preaudit meeting.
- *Methods and deliverables.* The lead auditor should ensure that audit team members understand their deliverables (e.g., completed checklist, records of evidence, written nonconformities, opportunities for improvement) and procedures they are to follow.
- *Resources.* The lead auditor should provide the necessary resources for the team members, such as forms, examples, and working papers, and if necessary instruct team members on the use of working papers.
- *Leadership.* The lead auditor should lead by example, show enthusiasm for other members, and support and congratulate members for a good job. The lead auditor should ensure that members know where to go when they need help. He or she should also make sure that assignments are equitable (i.e., treat everyone the same).
- *Work environment and personal needs.* The lead auditor should ensure that the audit team members' work environment needs are met. If a team member has a personal emergency, the lead auditor must take action to support the team member. (Note: The lead auditor should have access to emergency contact information for each auditor.)
- *Good time management.* The lead auditor should ensure that the audit starts and ends on time, that team members are on time for meetings, and that team member requests are responded to.

Audit Team Member Roles

Supply chain management involves several members both at the receiving end and at the shipping end. In addition, all supplier audits involve three key types of participants who interact throughout the audit starting at the audit planning stage. Audit teams have a leader (the lead auditor) and auditors (audit team members). For overseas global auditing, it may be appropriate to include individuals who are familiar with the culture and language.

Lead Auditor

The lead auditor is responsible for calling the meetings, making meeting arrangements, running the meetings, and reporting progress to the client and the auditee. The lead auditor should take ownership of the audit process. The lead auditor normally controls the team output or deliverables. Lead auditors must be able to guide the team and make decisions as needed to ensure that the team is effective. The lead auditor should be a qualified (competent) auditor.

Audit Team Member

Team members may include qualified auditors, auditors in training, and technical experts. Team members accept assignments and report progress to the lead auditor or the team as a whole. Team members may be given responsibilities such as keeping records or taking meeting minutes.

PERFORMANCE AND EVALUATION

> Describe various techniques to evaluate training, including evaluation planning, feedback surveys, pre-training testing, and post-training testing. (Understand)
>
> Body of Knowledge V.B.3

Team member training is paramount to achieve and sustain a high-function team. There are several methods for evaluating the effectiveness of training including surveys, pre-training testing, and post-training testing.

Surveys solicit input from the trainee about the training course including the quality of the materials, preparation of the instructor, the technology utilized to conduct the training, and if applicable the room/environment where the training was conducted using a simple Likert Scale. A Likert Scale is simply a rating scale that is used to rate or grade each response to a survey question. Generally, a score of 1 is considered undesirable and a score of 5 is considered outstanding.

Additionally, surveys will generally solicit what you liked the most about the course, what you liked the least, if there a topic that was not fully covered, etc. The information should then be compiled and analyzed and used as feedback to make improvements to the course and/or overall training program.

Pre-training testing and post-training testing is one of the most effective methods to evaluate the student's knowledge of the course content before taking the course and after completing the course. A pre-training test is also a valuable tool for the instructor to gage the level of knowledge of the students and what areas or content may need additional emphasis. The training delta, the difference between the pre-test score and the post-test score is a great way to demonstrate competency and assess the overall effectiveness of the training.

$$\frac{\text{Post-Test}}{\text{Pre-Test}} = \text{Training Delta}$$

Example: A pre-test was administered and scored 75 before the training. After the training, a post-test was administered and scored 87. What is the training delta for this student?

$$\frac{87}{75} = 1.16$$

The training delta for this student was 1.16. The training delta is a relative score and is used to gage the effectiveness of the training. Therefore, the higher the training delta, a general assumption can be made about the effectiveness the training.

Direct observation is another way to gage the effectiveness of the training. Using this method, the trainees are taught the subject, shown how to perform the task, perform the task under supervision, and then observed performing the task unsupervised to demonstrate competency. The amount of effort involved should be directly proportional to the risk(s) and complexity involved with tasks being performed.

There is no one best technique to evaluate training effectiveness. However, the method(s) chosen should be documented and planned as part of a comprehensive training program.

CHAPTER SUMMARY

Managing teams involves working with different personalities and ensuring that the team is an effective, cohesive unit. The duration of teams can be a short-term or long-term issue. However, there are still some basic techniques for ensuring that the team works together in the most effective and efficient manner for overall team performance.

The supplier audit team process is normally task driven and established for only one audit. Task-driven process teams stay focused on team objectives such as the audit scope and objectives. On occasion, audit team activities may include actions to preserve the team spirit and intrateam relationships. For overseas global auditing, it may be appropriate to include individuals who are familiar with the culture and language.

Project management skills, effective collaboration, communication, relationship-building skills, and the ability to plan well ahead are essential for identifying and onboarding a key supplier and managing a team. Experience in performing risk assessment and planning for a backup supplier for key processes such as sterilization is important for effective supply chain management teams.

Supplier CAPAs at times require unique technical skills and the ability to collaborate with the key players at the customer and supplier facilities. Pretraining and post training testing of the key players (after onboarding of new auditors and managers of the supply chain) requires critical communication and training evaluation skills. Supply chain management often includes initiation and management of supplier feedback surveys and follow-up with survey respondents.

Supply chain strategies and planning invariably start with the customer, as customers form the foundation of the business. Services and products provided by the supplier require customers who need the product or service and want to pay for it. In order to win customers effectively, the supply chain team must link its supply chain strategy to its business strategy. Performance evaluation of the teams in this arena is closely linked to the levels of planning and decision-making tools and techniques used by the teams involved.

In summary, the teams involved in a firm's supply chain can perform effectively only if they understand the value of the different networks of both internal and external relationships. The important point in performance evaluation is to view the supply chain team performance as a system—not as a group of individual entities doing their own thing.

Chapter 16
Compliance with Requirement and Supplier Categorization

> Understand and evaluate compliance with regulations and industry standards (e.g., RoHS, Governmental regulatory authorities, and ISO), specifications, contracts, agreements, and certification authority. Evaluate and categorize suppliers based on risk and performance. (Evaluate)
>
> Body of Knowledge V.C

Compliance with requirements and supplier categorization are two important elements of supplier quality management (SQM).

1. Compliance with requirements: In order for a supplier to be considered for selection, the supplier must be able to meet the company's quality, delivery, and cost requirements. This involves verifying that the supplier has the necessary capabilities, processes, and systems in place to consistently deliver products or services that meet the company's specifications.
2. Supplier categorization: Once a supplier has been evaluated and found to be capable of meeting the company's requirements, the supplier should be categorized based on the level of risk it poses to the company. For example, a supplier that provides critical components or services may be considered a high-risk supplier and subject to stricter monitoring and management, while a supplier that provides low-risk goods may receive less attention.

The categorization of suppliers helps companies prioritize their supplier management activities, allocate resources, and minimize risks. For example, a company may choose to focus its SQM efforts on high-risk suppliers, while relying on the supplier's own quality management systems to manage the quality of low-risk suppliers such as those supplying commercial off-the-shelf goods (COTS).

By monitoring supplier compliance with requirements and categorizing suppliers based on risk, companies can ensure that they are receiving high-quality goods and services from their suppliers and that their own operations are not disrupted by supplier-related problems.

The categorization of suppliers can be based on a variety of factors, such as:

1. Strategic importance: The importance of the supplier to the company's overall strategy and success
2. Business impact: The potential impact of the supplier's performance on the company's operations, including delivery times, product quality, and cost
3. Risk level: The level of risk posed by the supplier, including the risk of product defects, delivery disruptions, and other problems
4. Performance history: The supplier's past performance, including delivery times, product quality, and responsiveness to customer needs.
5. Compliance with regulations and standards: The supplier's compliance with relevant regulations, standards, and industry best practices

Based on these factors, suppliers can be categorized into different groups, such as high-risk, medium-risk, and low-risk suppliers, or critical and non-critical. This allows companies to focus their SQM efforts.

By using supplier categorization, companies can ensure that they are managing their suppliers in a consistent and effective manner, and that their own operations are not disrupted by supplier-related problems.

RESTRICTION OF HAZARDOUS SUBSTANCES DIRECTIVE

The Restriction of Hazardous Substances (RoHS) Directive 2002/95/EC for supplied material has increased the proof of compliance that suppliers must provide to their customers. RoHS restricts the types of hazardous materials that can be used in electrical and electronic products. All products in the EU market after July 1, 2006, must comply with RoHS.

Under RoHS, the following substances are prohibited from entering the market for electrical and electronic products: lead (Pb), mercury (Hg), cadmium (Cd), hexavalent chromium (Cr VI), polybrominated biphenyls (PBB), polybrominated diphenyl ethers (PBDE), and four different phthalates (DEHP, BBP, BBP, and DIBP). Manufacturers, authorized representatives, importers, and distributors need to understand their obligations to ensure compliance.

The recast RoHS directive 2011/65/EU was published on July 1, 2011, and entered into law on July 21, 2011. It replaced RoHS directive 2002/95/EC on January 3, 2013. The RoHS T directive 2012/65/EU has been amended by a number of delegated regulations, which are likely to be updated every few weeks or months, such as when new highly specified exemptions are included.

Companies and their supply chains should evaluate their process to identify the areas that could be a potential risk and consider the following steps when evaluating the risk against the legislation:

1. Assess the risk

   —Identify any weak links in the process chain

   —Analyze each stage of your operation

2. Establish what you are going to do about it
   — Put reasonable safeguards in place
   — Either eliminate the chance of anything going wrong or control the risks so that errors will be detected early
3. Document your processes and act on your procedures
   — If you cannot prove what has been done, it will be difficult to claim a defense
   — Inform your employees of your actions and provide training
4. Review your system
   — Ensure that it remains effective
   — Correct failures as soon as possible

To use the defense of due diligence, a business or person must prove they have taken all reasonable (positive) steps to comply with the legislation.

POTENTIAL RISKS REGARDING HAZARDOUS SUBSTANCES DIRECTIVE (ROHS)

While the Restriction of Hazardous Substances Directive (RoHS) has been successful in reducing the environmental impact of electronic products, it also presents some potential risks and challenges. Some of these include:

1. Compliance costs: Complying with RoHS can be a costly process for manufacturers, especially those who have not incorporated environmentally friendly practices into their operations. The cost of removing restricted substances from products, testing products for compliance, and updating production processes can be significant.
2. Product substitution: Some manufacturers may substitute restricted substances with other hazardous substances that are not restricted by RoHS. This could result in unintended environmental and health consequences.
3. Lack of harmonization: RoHS is only enforceable within the EU, and different countries may have different regulations and standards for hazardous substances in electronic products. This can create confusion for manufacturers who want to sell their products globally.
4. Limited scope: RoHS only applies to certain types of electronic and electrical equipment and does not cover all products that may contain hazardous substances. This can result in a patchwork of regulations that can be difficult to navigate.
5. Technological limitations: Some technologies may require the use of restricted substances to function effectively. This can limit the development of new technologies and limit the effectiveness of RoHS in reducing the environmental impact of electronic products.

REGISTRATION, EVALUATION, AUTHORIZATION AND RESTRICTION OF CHEMICALS

Registration to meet the EU requirements has been adopted to protect human health and the environment and to prevent risk posed by chemicals. Registration, evaluation, authorization, and restriction of chemicals (REACH) offers an alternative way to assess substances to reduce the number of tests on animals.

REACH has established procedures to evaluate the hazards of substances. Companies must register their substances to comply with the requirements.

The European Chemicals Agency (ECHA) helps companies comply with the legislation for registering their substances. The EU member states evaluate selected substances to comply with requirements regarding human health, the environment, or both. The authorities can ban hazardous substances or make their use subject to prior authorization.

REACH regulations apply to manufacturers, importers, and users. The responsibility for fulfilling the requirements of REACH, such as preregistration and registration, lies with importers established in the EU.

CONFLICT MINERALS

Conflict Minerals requirements became effective in 2018. These requirements were developed by members of the Electronic Industry Citizenship Coalition and the Global e-Sustainability Initiative. The Conflict-Free Sourcing Initiative has grown into one of the most utilized and respected resources for companies from a range of industries addressing conflict minerals issues in their supply chains.

Conflict resources are natural resources extracted in a conflict zone and sold to perpetuate the fighting. The most prominent contemporary examples are the Democratic Republic of Congo (DRC) or an adjoining country, presently Angola, Burundi, Central African Republic, the Republic of the Congo, Rwanda, South Sudan, Tanzania, Uganda or Zambia where various armies, rebel groups, and outside actors have profited from mining while contributing to violence and exploitation during wars in the region.

The most commonly mined conflict minerals are cassiterite (for tin), wolframite (for tungsten), coltan (for tantalum), and gold ore, which are extracted and passed through a variety of intermediaries before being purchased by multinational electronics companies.

Manufacturers must periodically review the policies and procedures of suppliers to ensure continued compliance to Conflict Minerals requirements.

INTERNATIONAL TRAFFIC IN ARMS REGULATIONS

ITAR controls the export and import of defense-related articles and services on the United States Munitions List (USML). These regulations are described in Title 22 (Foreign Relations) of the Code of Federal Regulations.

US personnel must have authorization or an exemption to access information and materials pertaining to defense and military-related technologies. There can be heavy fines for unauthorized personnel who access or provide access to

ITAR-protected defense articles, services, or technical data. ITAR does not apply to general scientific, mathematical, or engineering principles that are taught in schools or are in the public domain.

SPECIFICATIONS

A specification is a grouping of specific parameters that are required to ensure the ability of a product to perform as designed. This is sometimes a difficult set of expectations. Many times the capability of the manufacturing process is the critical factor that results in deviations from specifications. In fact, industry recognizes that changes are necessary in the development and implementation of specification change and deviation procedures. The design/development community must develop specifications that are meaningful and within the manufacturing process capability. The five basic types of specifications are:

1. *Product specifications.* This is the most common type of specification. It defines what is required for a product to perform as expected by the consumer.
2. *Process specifications.* This type of specification defines the parameters of the manufacturing process that must be controlled in order to produce a product.
3. *Analytical specifications.* Where applicable, this type of specification defines the analytical methodologies to measure a required level of accuracy.
4. *Raw material specifications.* This type of specification defines what is acceptable as raw material entering a manufacturing process.
5. *Quality management specifications.* This type of specification defines the management practices under which you wish to have products produced.

These specifications follow a hierarchy. Generally, the product specifications are determined first. They are the driver for the process, analytical, and raw material specifications. These specifications enable the design and manufacturing people to communicate in common terms, which are the requirements at every level of the manufacturing process. Specifications provide checkpoints in manufacturing that ensure quality of the product.

There are components common to all specifications. First, there is a description of the product. It can be a narrative or, in the case of a manufactured part, a profile. After the description, list the required characteristics. These characteristics include the upper and lower tolerance limits. Whenever a tolerance is established, a measurement should be referenced with the units of measurement and, in the case of chemical, physical, and microbiological tests, the test method procedure. The test method describes the instrument, the exact procedure for sampling, the sample preparation, and the limit of detection as well as desired precision and accuracy.

The third component is the handling and packaging requirements. These identify how the product is to be packaged and handled. Different industries attach various addenda to a specification, such as test method descriptions, test method and sampling procedures, environmental testing requirements, certificate of analysis information, product identification elements, codes, expiration date, process control requirements, safety requirements, and various governmental regulatory requirements. In addition to these common elements, product specifications should contain the following:

- Reliability, serviceability, shelf life, and maintainability requirements
- Permissible tolerances and comparisons with process capabilities
- Product accept/reject criteria
- User instructions, such as instability, ease of assembly, storage needs, shelf life, and disposability
- Benign failure and fail-safe characteristics
- Aesthetic and other qualitative specifications and acceptance criteria
- FMEA and FTA
- Ability to diagnose and correct problems
- Labeling, warning, identification, traceability, risk management, and recall requirements
- Review and use of standard parts

Product specifications with these components will contain the information necessary to minimize liability issues.

Process and analytical specifications also have some unique components, which follow:

- Manufacturability of the design, including special process/analytical needs, mechanization, automation, assembly, and installation of components
- Capability to inspect and test the design/prototype, including special inspection and test requirements
- Specification of materials, components, and subassemblies (including approved supplies and suppliers), as well as availability
- Packaging, handling, storage, and shelf life requirements, especially safety factors relating to incoming and outgoing items (ANSI/ASQC 1987, 9)

The completeness and clarity of the specifications are particularly important in the procurement process. There appears to be no limit to how nonconformities can occur or to the questions concerning the characteristics of purchased material. Certain areas, such as visual requirements and workmanship, are subjective. These can be troublesome unless some discussion between the customer and the supplier results in agreement on objective standards to clearly define requirements.

The pertinent specifications and drawings should clearly define the characteristics of the material so that the supplier fully understands what is required, measured, and reinforced. Specifications should be free of ambiguities that could lead to differences in interpretation by the supplier and the customer.

Material specifications should specify the end requirements that must be met. In some instances, the customer may specify raw materials and methods of manufacture, but the customer must then assume responsibility for the resultant product. This manner of specification excludes the supplier's technical knowledge and should be avoided whenever possible.

There are many ways to develop specifications. Everyone has an opinion on how specifications should be developed. Some people work on hunches, some simply trust their instincts, and others use statistics.

The specification team consists of development people, the material supplier, the procurement person, and the manufacturer. This team examines the data that the design people have utilized, the data on the manufacturing capability, and the data on the supplier's capability, and develops a specification over a period of time. The team can utilize other people as appropriate, such as an analytical representative if there is lab testing involved. The team works to develop specifications that are realistic and meet the customer's needs.

Quality Management Specifications

There has been an explosion of consensus standards that define management practices under which we may desire to have our products or services produced. They range from the ISO 9000 series of standards to advanced total quality management models such as the Malcolm Baldrige National Quality Award (MBNQA) criteria.

The ISO 9000 series standards require that you specify the title number and issue of the quality system standard under which you wish to have your product or service produced. If you are working within a registered system, the issue of passing on quality management expectations is moot. You must then set quality management expectations for your suppliers. You do not have to require registration or certification of your suppliers; simply list the standard they are to use.

The growth of competitive recognition systems, such as the MBNQA, provides another set of criteria. These can be used to set in place expectations of both the quality management practices and the cultural style of management you expect to see in place at the supplier.

Government Quality Specifications and Standards

In today's world of consumer affairs, with product liability litigation and an endless list of government and industry regulations, businesses must develop a quality control system that satisfies customer requirements and remains cost effective. To this end, the business manager must rely heavily on customer requirements. Initial contact with potential customers should aim at understanding these requirements and regulations.

Most industries are aware of the need to provide safe and effective products to consumers. Adverse publicity about product safety or a company's negligence in complying with government safety regulations can result in large monetary losses, not to mention product liability lawsuits.

Government quality documents normally begin with a specification of quality program requirements. This requires a documented program for all activities necessary to attain continuous control of product quality during manufacturing. Usually, this program will be reviewed and accepted by the customer after it is satisfied that the necessary activities do exist and are being performed effectively. The quality program specification or the contract typically will refer to other government documents to which compliance is required. These secondary documents may refer to still other applicable documents that establish the requirements of additional program and product details. The second level of quality documents may specify equipment calibration systems, sampling procedures, product detail requirements, and analysis of all raw materials.

Government documents are obtainable through the procuring activity, or the government contracting officer. Those documents that pertain to other governments must be obtained from the appropriate governmental agencies.

The local government contract administrating offices will aid suppliers in setting up procedures that comply with government requirements. In addition to their responsibility for surveying/auditing facilities to determine quality program compliance, they advise on how to comply. Specialists in a wide variety of technical subjects are available to aid in interpreting details and in establishing procedures that will provide the control required. Government quality requirements currently are more detailed than most commercial requirements. The extensive details included in government documents have evolved from past quality experience. It is mandatory to the success of aerospace projects and military endeavors and to the preservation of human life that a deficient level of conformance be eliminated. Therefore, government documents have become concerned with both program concepts and product details, so that quality is thoroughly defined. Product specifications and standards also are necessary to provide interchangeability of products from more than one source.

Continuous source inspection, surveys, and audits are frequently used. In order to ensure that end items have incorporated the requirements specified at all levels of procurement, government contracts rely on source inspections, quality program audits, and surveys. Resident government representatives are assigned when the frequency of shipments, the complexity of the product, or the criticality of performance requires constant surveillance at the manufacturing facility. Their primary responsibility is to ensure the product's conformance when shipped. In addition to outgoing inspection, they also perform continuous audits of control procedures to ensure that quality programs are maintained.

The supplier should become acquainted with all requirements included in the specifications and standards. The total performance and documentation requirements are not apparent from a single or a few specifications listed in a contract. The supplier should also thoroughly review and be informed of all program requirements, product specifications, and documentation details that apply in the referenced documents before committing to a government contract. If suppliers do not become familiar with the specifications imposed, they may learn too late that costs are higher than expected or that the product cannot be shipped because of standards that were not met.

Modifications to government documents may be included in contract conditions. Changes to the pertinent specifications and standards are not possible for individual procurement. Hence, justified changes must be made as additions or exceptions in the contractual agreement. Identification and clarification of questionable details must be completed before the contract is signed. This ensures performance on schedule and avoids delays that may be costly to both supplier and customer.

The supplier and customer should be aware of added cost factors resulting from the requirements for conformance to government documents. Some may not be apparent unless all specifications are reviewed in detail. The additional work required can affect costs significantly and should be considered in pricing.

Some of the cost factors are certifications from subcontractors, inspections to approve quality programs, certification of specific processes, additional documentation required to provide objective evidence of compliance to details, changeovers to military-type materials and finishes, additional environmental testing, and time required for program reviews.

A critical aspect of the customer–supplier relationship is acquisition of realistic specifications. The customer is responsible for clearly defining the requirements; the supplier is responsible for accurately determining whether it has the capability to meet those specifications. Some up-front planning is necessary to develop plans to meet the agreed-upon specifications. Failure to do so will only result in unwanted cost to both customer and supplier. Realistic specifications meet the needs of both sides.

Registration or certification of the supplier's QMS by an outside company is becoming a viable option in many industries. The true third-party registration system offered by many under the ISO standards is rapidly becoming the most popular. In industries such as the automotive industry, acceptable quality system audits are becoming acceptable to competitors as a cost and consistency factor. Registration or acceptance of a QMS by some other party is just that. A recognition of the system will not replace detailed discussions about product, process, materials, and analysis. You still need to gain agreement on those specifications and work at the product and service delivery level in order to continue to satisfy the ultimate customer of the product or service.

Contracts

Contrary to common practice with many materials, there is no such thing as a standard specification for a service. Also, there is no common applicable legislation in this regard. Consequently, before you approach the market, you must have a clear, well-documented definition of what that service means to your organization. In other words, you must draft a contract. A contract is, simply put, the blueprint of the business you intend to carry out; it must encompass operational, legal, financial, and all other relevant aspects of the service.

The model contract must be ready before any quotes are requested for the following reasons:

- The drafting process will force your organization to reach a consensus on the service it expects to receive.
- The model contract will communicate clearly to all potential suppliers what your organization wants, and it will ensure that all the quoted prices correspond to identical service levels.

- Once a winning bid is selected, the contract sets down the framework for contract management and evaluation of supplier performance.

There are no general formulas for contract drafting. Every industry and service category has different expectations and requirements, and creativity is a must when addressing each situation. Nevertheless, some best practices can be applied.

Regardless of what you intend to contract, chances are someone has tried doing it before. Consequently, your first task is to gather any available background information. Generally, this takes one of two forms:

- *Contracts used by other companies for the same service.* Ideally, besides getting a copy of the contract, information should be sought on whether results met expectations and what problems, if any, were encountered. This avoids "tripping twice over the same stone." Don't forget to track down experiences within your own company.
- *Checklists.* There are many published checklists describing the key points that a contract should cover. No one checklist will be perfect for you, but reading a few of them will get your mind working on what aspects you should cover. (DeYong and Case, 1996; Pohlig, 2002a, 2002b; Garrison, Khalil, and O'Reilly, 2001)

Once all background information has been studied, the contracting organization should gather all internal stakeholders and conduct brainstorming sessions with the aim of creating a business model for the service under study. In other words, your company should put into words a thorough description of the service in all its aspects. This process generally requires a few iterations. Once this is done, the end result is given to the lawyers, who draft the agreement. The final document is distributed to all stakeholders, and, if necessary, a final round of talks is held to iron out remaining issues.

During the contract drafting process, the purchasing and contracting department must play the roles of facilitator, coordinator, and umpire of the brainstorming sessions. It is common for contradictory positions to emerge, and it takes both tact and knowledge from the moderator to get all the stakeholders to articulate their ideas and then achieve a balanced end result.

Statement of Work

The Statement of Work (SOW) identifies the part of the contract where all the operational aspects of a service are described. This document contains—expressed in terms your organization understands—the definitions developed in the brainstorming sessions of what a good service is. Drafting a good SOW is critical to success.

In some cases, the SOW is a separate exhibit; in other cases, it is embedded in the body of the contract. Whichever form is used, always bear in mind that a contract is a single entity, and due attention must be given to the balance and harmony of all its parts. Blind cut-and-paste exercises, where a SOW is inserted into some boilerplate clauses, are the shortest path to disaster. Nevertheless, for our discussion, it helps to think of the SOW as a separate document that must have certain traits in order to be successful.

It is normal for a supplier to offer its own version of the SOW. Though this document should be considered as background information, it is not advisable to sign it without close scrutiny. A detailed study on how to draft SOWs is beyond the scope of this chapter; however, there are some key points the SOW must cover. Not all of them are relevant in every situation, but it is always a good idea to go through this checklist:

- *Service definition and scope.* There must be a clear definition of the service you are contracting. All relevant contractor obligations must be listed. Emphasis must be placed on the end results to be accomplished, giving the contractor as much leeway as practical to decide how to organize its work.
- *Deliverables.* Define clearly what service deliverables your company expects to receive. The contractor must be made responsible for these deliverables, and its compensation (payments) must be tied as much as possible to their timely production. Also, acceptance criteria must be set forth for these end products.
- *Company-provided inputs.* In some cases, your company will provide goods or services necessary to perform the work. Examples might be office space, computers, travel expenses, and electricity. These must be clearly defined.
- *Contractor-provided inputs.* Conversely, it might be the contractor's responsibility to provide certain materials. For example, in outsourced plant maintenance services, the deliverables might include provision of spare parts and lubricants. Once again, be careful to list these.
- *Performance evaluation—bonus/penalty clauses.* Performance review mechanisms must be defined and applicable penalties or bonuses stated. A correct definition of performance indicators is vital to close the feedback loops essential for a successful outcome.

QUALITY AGREEMENTS

A quality agreement is a written agreement that defines and establishes the obligations and responsibilities of each party involved in a contractual relationship.

Generally, quality agreements are not formal business agreements and should be separate from commercial supply and distribution contracts. Quality agreements should include all requirements which must be flowed down to sub-tier suppliers as applicable. Quality agreements usually contain notification of change language requiring the supplier to provide written notification prior to making changes to the raw materials and methods of manufacture or inspection so the changes can be evaluated for potential impact on the product including regulatory consequences.

Elements of a Quality Agreement

A well-drafted quality agreement uses clear language to define the roles and responsibilities of each party. It should specify which products or services are involved and who will provide final approval for the various activities. A quality agreement may contain the following sections:

- Purpose and scope
- Terms such as effective date and termination clause
- Dispute resolution
- Responsibilities, including communication mechanisms and contacts
- Change control and revisions

CERTIFICATION AUTHORITY

Certification plays an important role in meeting customer expectations. Certification can demonstrate that a company's product is certified to an applicable industry standard such as one written or administered by ANSI, Underwriters Laboratories (UL), CSA Group, NSF International, etc.

Recent years have seen a growing trend toward environmentally friendly products driven by demand from consumers and retailers. The CSA sustainability mark is one way to show that a product meets environmental performance standards. CSA criteria and attributes include:

- Materials of concern, such as toxic or hazardous material
- Material efficiency, such as product recycled content, packaging recycled content, and efficient use of raw material resources
- Energy consumption during product use
- Product performance, including functionality, reliability, and reparability
- End of life cycle management, such as recyclability, design for recycling, and landfill diversion

Additional considerations include the European Union Decision No. 768/2008/EC for conformity assessment (CE) marking. CE marking affirms the manufacturer's or importer's good conforms with European health, safety, and environmental protection standards.

STANDARDS

More than 24,600 international standards are in use globally. The ISO standards are well-recognized and used in more than 167 countries. The ISO 9000 series of standards provides the foundation for the requirements and a vocabulary for quality management systems. Being certified to the ISO standard demonstrates that a firm's QMS meets the requirements, but it does not guarantee that a manufacturer produces a quality product.

One of the key benefits of obtaining ISO 9001 certification is that it can be used as a marketing tool to convince prospective buyers of a firm's ability to comply with stringent quality requirements. Other benefits of ISO certification include, but are not limited to:

- Allows companies to be consistent competitors in the marketplace
- Helps meet customer needs
- Broadens business opportunities by demonstrating compliance
- Improves operational performance, which reduces errors and impacts the bottom line
- Improves communication and collaboration globally

Part VI

Relationship Management

Chapter 17
Supplier Onboarding

> Understand and apply processes for orientation of suppliers such as providing overview of company, vision, mission, guiding principles, overall requirements, expectations, and criticality of product, service, and delivery requirements. (Apply)
>
> **Body of Knowledge VI.A**

Every journey starts with small steps. Small incremental steps create a foundation and provide a road map toward the final destination. Supplier onboarding is a key element of the supplier management system. During the initial phase of building the relationship, onboarding provides an opportunity to introduce suppliers to the fundamentals of the work requirements and involved processes. Supplier orientation benefits suppliers by providing a good, basic understanding of work requirements; it also provides an opportunity for buyers to work with suppliers in a collaborative environment. Key players in this process include the functional teams responsible for implementing the work and supervising the activities from the supplier and customer organization.

Supplier management effectiveness requires collaboration within the supply chain. Supplier management collaboration yields mutual benefits to companies within the supply chain. Supply chain partners find innovative ways to make collaboration work to everyone's advantage. In today's competitive business environment, many companies recognize the importance of having strong relationships with their suppliers.

SUPPLIER ORIENTATION PROCESS

The supplier orientation process is a framework of activities and plans that are carried out with suppliers to provide them with an understanding of the work requirements and methods. This framework is extremely important as it helps the customer define expectations through communication with the supplier.

During the project start-up phase, it is critical that the selected suppliers have a good understanding of the work requirements, roles and responsibilities, and expectations as perceived by the customer. To facilitate this understanding, the

Table 17.1 Supplier orientation process phases and activities.

Phase	Key activities
Planning	• Identify key resources • Identify requirements, processes, work instructions, and methods to be shared with suppliers • Set up communication with supplier contacts • Schedule orientation session
Development	• Gather work requirements and methods • Prepare work requirements for suppliers per scope of work • Establish expectations for suppliers
Execution	• Formally communicate the orientation package, including requirements and expectations, means and methods, schedules, and deliverables per scope of work

customer should provide a comprehensive supplier orientation (see Table 17.1). The phases of supplier orientation are planning, development, and execution.

Supplier Orientation Planning

During the planning phase, the customer establishes key contacts with the supplier, identifies key processes to be followed for the work, determines requirements, and sets up communication with the supplier. This phase may include but is not limited to people in engineering, production, legal, regulatory affairs, sales, marketing, quality, and purchasing.

Supplier Orientation Development

During the development phase, the customer gathers and prepares required processes, including plans and procedures, work requirements, company management objectives, and supplier expectations. These processes and expectations should be developed based on the scope of work and should be formally communicated to the supplier.

Supplier Orientation Execution

During the execution phase, the supplier orientation session is established and executed. The orientation session provides the supplier with the work requirements, roles and responsibilities, means and methods, deliverables, and expectations set forth by the customer.

ORIENTATION PROCESS DESCRIPTION

This section provides a detailed description of the supplier orientation process. Figure 17.1 illustrates the process flow, whose steps and corresponding activities are described below.

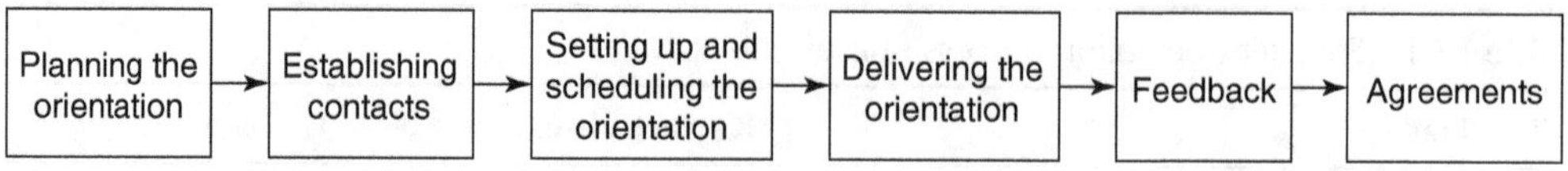

Figure 17.1 Supplier orientation process diagram.

Planning the Orientation

In this process stage, the appropriate person plans for the preparation of orientation material. This may include the company's vision and/or mission, guiding principles related to the scope of work, work requirements, necessary process documentation, criteria and expectations related to the criticality of products and services, and delivery requirements. The customer may prioritize topics according to the scope of the work.

Establishing Contacts

During this phase, the customer identifies supplier contacts. This could mean establishing contacts in the supplier's cross-functional team responsible for supporting the work subcontracted. The customer contact forms a team composed of the supplier representatives.

Setting Up and Scheduling the Orientation Session

The customer schedules and sets up an orientation session with the supplier representatives. This may include but is not limited to people from engineering, production, legal, regulatory affairs, sales, marketing, quality, and purchasing.

Delivering the Orientation

At this stage, the customer delivers and conducts the orientation session by presenting the key topics related to the work. This may involve face-to-face meetings as well as the use of interactive digital technology.

Feedback

During this stage, the customer and supplier contacts discuss and clarify any issues with the delivery requirements, work expectations, reporting schedule, and work requirements.

Agreements

This is the final stage of the process where the elements discussed by the supplier and customer teams are agreed on. This step provides good understanding and agreement regarding the work requirements and expectations. Agreements can cover roles and responsibilities, pricing structure, and quality requirements.

OBJECTIVES OF THE ORIENTATION PROCESS

An effective orientation process ensures the fulfillment of the following objectives:

- Identification of key resources and points of contact at the supplier and customer organizations
- Understanding of the key work processes and requirements
- Identification and clarification of any discrepancies and issues through effective communication
- Identification of any supplier training or development needs and formulation of a plan to address these needs
- Reporting of means and methods, including status tracking activities
- Identification of any issues related to schedule, quality, and deliverables

ROLES AND RESPONSIBILITIES

It is critical to define the roles and responsibilities of the customer and supplier in the orientation process. The customer organization is primarily responsible for conducting these orientations; however, in order for the orientation process to be effective, the supplier must play an active role.

Customer Responsibilities

- Identify key contacts at the supplier organization, including those from engineering, quality, production, procurement, and support, per the scope of the work and contract
- Set up communication sessions including the orientation, follow-up, and feedback as needed. Define the frequency of this communication per work requirements
- Liaison with the supplier organization on any product- and work-related issues
- Develop an understanding of the work processes and requirements and set forth expectations on work requirements

Supplier Responsibilities

- Gain an understanding of the following:
  - Work processes and requirements
  - Work expectations as perceived by the customer
  - Reporting frequency on activities and status
- Clarify any issues, discrepancies, or ambiguities with the customer
- Work with the customer to resolve issues through effective communication

CONCLUSION

Supplier orientation is a process of communicating and creating an understanding of the customer's management objectives, work requirements, and guiding principles. It lends visibility to the expectations and criticality of the scope of work as perceived by the customer.

Developing a collaborative customer–supplier relationship is not just about *what* the parties do together but also what they believe about each other and *how* they interact. Supplier orientation provides an environment where the supplier and the customer can collaborate and gain better understanding and agreement. It is important that this effort involve the key players representing the customer and supplier organizations. The orientation activities contribute to the success of the work project objectives. Effective supplier management requires a strong working relationship between the customer and supplier. The supplier orientation helps to create a better customer–supplier relationship and, if managed well, can create a win–win situation.

Chapter 18
Communication

TECHNIQUES AND MEDIATION

> Identify and apply communication techniques (e.g., oral, written, and presentation) specifically for internal stakeholders and suppliers to resolve issues. Apply different techniques when working in multi-cultural environments. Identify and describe the impact that culture, communication, and Diversity, Equity, and Inclusion (DEI) can have on an organization. (Evaluate)
>
> **Body of Knowledge VI.B.1**

Several important elements must exist in order to manage the relationship with a supplier successfully. The individuals managing the supplier relationship need to understand the overall expectations of the supplier from the customer's viewpoint. The product or service being purchased should have clearly defined specifications. Quality agreements should be established where needed. Systems to address nonconformances, which usually lead to corrective action, should exist. The metrics used to monitor the quality performance of a supplier must be agreed upon, measurable, accurate and easy to report. The efficiency of these elements, and of other important elements directly related to relationship management, is highly dependent on effective communication practices. This chapter will focus on some proven techniques for effective communication and the use of quality tools in the reporting of metrics.

Throughout this chapter, the terms "sender" and "receiver" are used. While anyone engaged in a communication session can be one or both of these, the term "sender" will primarily refer to an internal stakeholder (e.g., a supplier quality engineer, a buyer, an inventory management specialist). "Receiver" will primarily refer to an individual within a supplier's organization (e.g., a regional sales manager, a technical support specialist, a customer service representative).

Effective communication between internal stakeholder customers and suppliers is important to the success of both organizations for many reasons. These reasons typically involve, but are not limited to:

- Establishing initial contact
- Addressing routine and nonroutine issues
- Establishing rapport between parallel groups of internal stakeholders and suppliers (e.g., stakeholder engineering and supplier engineering)
- Working out the details of a quality agreement
- Developing component/widget/ingredient/item specifications
- Handling nonconformances
- Conducting root cause analysis
- Generating corrective action
- Status reporting
- Maintaining ongoing relations

Forms of Communication

The three primary forms of communication that will be discussed here are oral, written, and presentations (a combination of oral and written communication). Each form of communication has advantages and disadvantages for both internal stakeholders and suppliers. One of the best practices of effective communication is to choose the most appropriate form to use in specific situations.

Oral communications are effective for quick and more personal delivery of information between internal stakeholders and suppliers. They can be as simple as making a phone call or setting up a video conference call, which requires more organization and planning. If the oral communication takes place internally, a visit to a colleague's work space is a simple and personal way to exchange information. Oral communications, whether in person or by phone, are interactive and more personal, and they often lead to faster resolution. The parties involved, sender and receiver, should determine if an oral communication session is the best form for the discussion of a specific topic.

Advantages of oral communications include, but are not limited to:

- Minimal planning
- Spontaneous (real-time)
- More personal than written
- Resolutions are often obtained quickly

Disadvantages of oral communications include, but are not limited to:

- Unplanned responses
- Emotions of the sender/receiver are seen (if face to face) or heard (on the phone) by the other party

- Unless someone takes notes, the conversation goes undocumented
- Participants might feel more pressure to respond immediately to a question or request
- Time zone difference challenges for global conversations

When a supply chain person is faced with establishing an internal discussion with a colleague or an external discussion with a supplier, the pros and cons of an oral communication session should be considered.

Written communications are more formal and require some thought. The most common types of written communication are memos, reports, and e-mails. Since written communications are often summaries, notifications, or formal requests, they are usually well thought-out and have a predetermined distribution list. The audience of written communications is a major consideration while the author is drafting the document verbiage. Again, the sender (primarily) and receiver should determine if a written communication session is the best form for the delivery of specific information.

Advantages of written communications include, but are not limited to:

- They are usually well planned
- It is easy to include supporting documentation (e.g., graphics)
- They are more formal than oral communications
- They provide a documented audit trail of what the communication was about

Disadvantages of written communications include, but are not limited to:

- More planning is typically needed
- The initial receivers of the communication are limited to the distribution list
- If responses are needed, they won't occur in real time

Presentations, a combination of oral and written communications, usually involve the formal delivery of some topic to a selected audience that incorporates visual information accompanied by oration and/or dialogue. Presentations are typically used to instruct or inform audiences of various sizes during training classes, conference sessions, company management review meetings, keynote speaker meetings, etc. Once again, the audience of a presentation is a key consideration when the author is developing the content. For example, a presentation on good nutrition habits given to grade school children will have very different verbiage and graphics than a presentation on last month's manufacturing throughput numbers given to a company's senior management staff.

Advantages of presentations include, but are not limited to:

- Written content is enhanced with live oration by the presenter
- The audience has the opportunity to discuss the topics being presented
- Typically, a well-planned, concise, and effective way to communicate specific information to a targeted audience

- They provide maximum flexibility between sender and receiver during the communication process

Disadvantages of presentations include, but are not limited to:

- They usually require much more planning and development work by the sender
- Some receivers may be less apt to ask questions if the presentation is given to a large audience
- Most senders don't feel comfortable making presentations to large groups of people

Whichever form of communication is chosen by the sender (and sometimes the receiver), the following factors should be considered:

- The complexity of the subject matter
- The knowledge and experience of the sender, whether an internal stakeholder or a supplier
- The timing required to deliver the message
- The makeup of the audience (receivers), whether an internal stakeholder or a supplier
- The physical locations of the sender and receiver(s)
- Whether a documented record of the communication is needed
- The formality of the communication environment
- Any cultural differences between the sender and receiver (this is discussed later in the chapter)

Communication Style

The communication style of the communication event (phone call, meeting, memo, report, etc.), usually established by the sender, can also have a significant effect on how the receivers feel after information is shared. If vital and innovative information is given by a sender whose style is not appreciated by the receivers, the importance of the message could be diminished and the receivers might be less motivated to implement the tasks. Thus, wonderfully strategic quality plans may fall by the wayside if the sender's style cannot deliver the message effectively and create motivation among the receivers. These considerations are particularly concerning when internal stakeholders are trying to establish and maintain several good, reliable suppliers of goods and services.

Reece and Brandt (1990) identify four basic communication styles:

1. Emotive
2. Director
3. Reflective
4. Supportive

Research has shown that these four basic communication styles have varying degrees of dominance and sociability, which are two dimensions of human behavior. Dominance involves how controlling and helpful a sender is. A low-dominance person is more collaborative and willing to lend a hand to find solutions. A high-dominance person tends to dictate and attempt to control others. Sociability involves a person's desire for interaction with others. People with high sociability are more outgoing and friendly, and they actively seek out collaboration with others. Those who are lower on the sociability scale tend to keep to themselves and are less apt to seek personal interaction with others to resolve issues.

An *emotive* communicator shows high degrees of dominance and sociability (see Figure 18.1). This person typically displays action-oriented behavior, prefers informality, and is adept at persuasion. An example of an emotive communicator is an easygoing, down-to-earth quality manager who leads by example, has a friendly demeanor, and gets people to follow him or her without difficulty.

A *director* communicator ranks high on the dominance scale and low on the sociability scale (see Figure 18.2). A stern, serious attitude, strong opinions, and

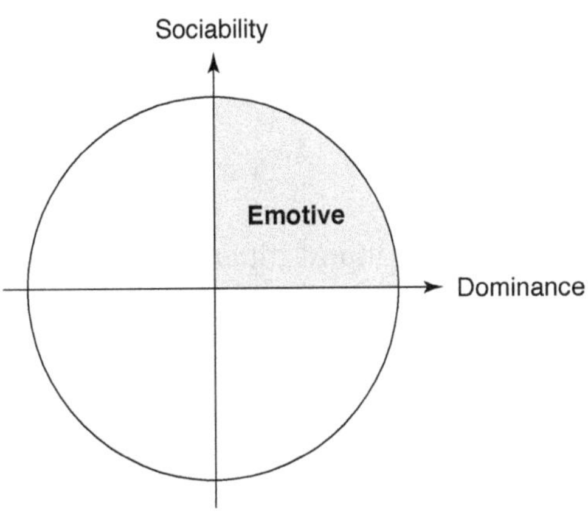

Figure 18.1 Emotive communicator.

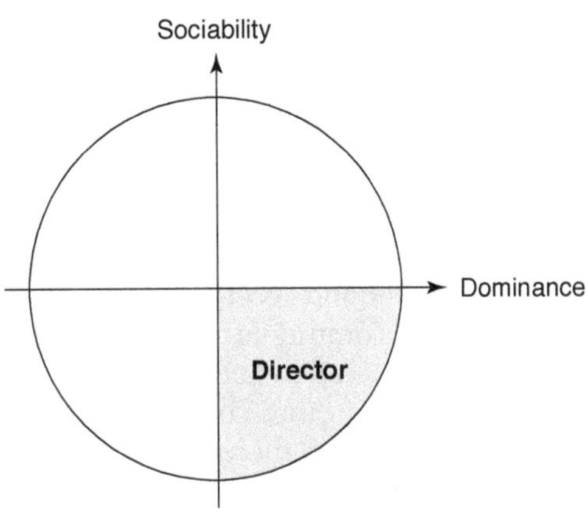

Figure 18.2 Director communicator.

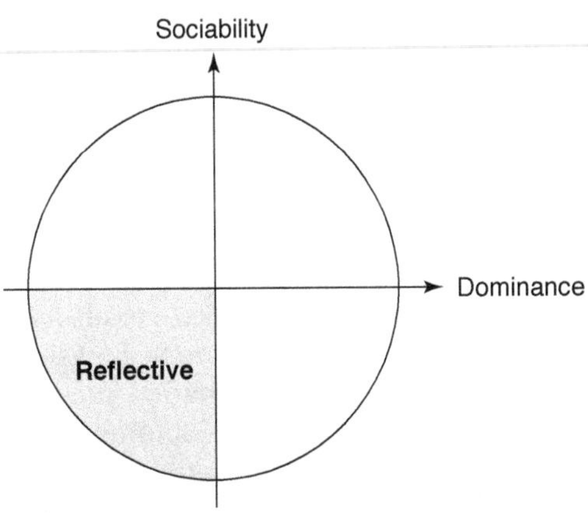

Figure 18.3 Reflective communicator.

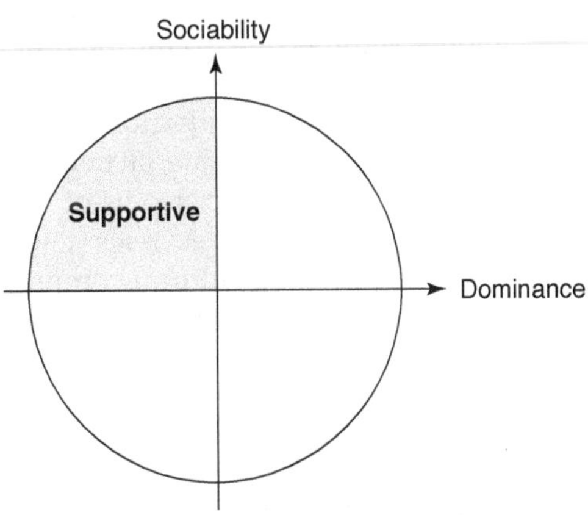

Figure 18.4 Supportive communicator.

indifference are typical characteristics of this type of communicator. An example of a director communicator is a company president who is very businesslike, expresses his or her opinions with conviction, and appears to be inflexible.

Someone who demonstrates low dominance and sociability is a *reflective* communicator (see Figure 18.3). These individuals express themselves in a more deliberate manner, may seem more quiet or introverted, and make decisions more slowly. An example of a reflective communicator is a research scientist known for his or her lab work who prefers the ability to summarize and write white papers without dedicated due dates.

A *supportive* communicator possesses a low degree of dominance and a high degree of sociability (see Figure 18.4). This person is a good listener, avoids displays of power, and uses collaborative methods to find solutions to problems. An example of a supportive communicator is a Six Sigma Black Belt practitioner who leads open, low-stress meetings, encourages participation, and facilitates finding solutions that can be supported by all those involved.

Recognizing one's communication style can help a sender to maximize favorable outcomes when dealing with internal stakeholders and suppliers (receivers). Determination of a communication style does not guarantee a successful outcome of a communication session; however, recognizing the strengths and weaknesses of the senders and receivers involved will certainly improve the probability that an effective event will take place.

Cultural Considerations

With the increase in global business transactions, cultural differences between the sender and receiver can have a significant bearing on the effectiveness of all forms of communication (oral, written, and presentations). Internal stakeholders headquartered in one country very often purchase components and services from suppliers in other countries, oftentimes on the other side of the globe. Cultural differences are more difficult to address during oral communications because the communication event typically takes place in real time (face to face or over the phone). If the sender and receiver already have some knowledge and experience in communicating across cultures, that can go a long way toward achieving a more effective meeting. During an oral communication session, both the sender and receiver have to remain cognizant of the culture of the other party so that misunderstandings due to cultural differences (not the subject matter) can be avoided or minimized. Avoiding cultural misunderstandings during written and presentation forms of communication is easier because the sender usually has much more time to develop the material before the delivery of the information. Other cultural factors can affect the outcome of a communication session as well.

Gorman (2011) discusses several factors that should be considered when dealing with cross-cultural communication events. One factor to consider when the sender and receiver are not from the same country is *high-context* versus *low-context*. If communication participants are from a *high-context* culture (e.g., Asian, Latin American, Central European, African), much of the information may be unspecified, left to the expected understanding of the participants, or communicated through unspoken cues, context, and between-the-lines interpretation. When dealing with people in *high-context* countries, maximizing the amount of detail when communicating about specifications, nonconformances, metrics, etc., will probably help to minimize quality problems in the future. The more details are discussed up-front, the less likely a future misunderstanding will occur that could lead to failures, different goals, quality metrics with different specifications, corrective actions not directly tied to root causes, etc. In communication sessions with *low-context* cultures (e.g., most English- and German-speaking countries), the sender and receiver are expected to communicate with explicit and specific verbiage to get the information delivered and understood.

Another factor to consider when dealing with cross-cultural communications is the *sequential* versus *synchronic* time factor. Communicators in *sequential* cultures (e.g., North American, English, German, Swedish, Dutch) tend to give full attention to one agenda item after the other, hence the term "sequential." In contrast, communicators in *synchronic* time-based cultures (e.g., Asia, southern Europe, South America) view the flow of time as a continuous stream, like tracing the circumference of a circle, and the past, present, and future are intertwined when issues are dealt with. When dealing with participants from synchronic cultures, insisting on clarification of the quality history of an item, specific task target completion dates, calling out responsible individuals by name, etc., will go a long way toward preventing future misunderstandings.

Whether the participants are from an *affective* or a *neutral* culture can greatly influence the outcome of a communication session. In *affective* cultures, people tend to show their emotions and feelings by smiling, laughing, frowning, scowling, etc. They may actually walk out of a room during a live oral communication session if they don't like what is going on. People in *neutral* cultures refrain from showing their emotions and feelings. The United States, France, Italy, and Singapore are examples of cultures that are more tolerant of emotional displays, while Japan, the UK, and the Netherlands are less tolerant of emotional displays. Whether or not communication event participants outwardly display their emotions, specific details of root cause analyses, corrective action development, quality metrics to be monitored, task assignments coupled with target completion dates; and responsible individuals should be pursued with as much information as possible.

Communication Outcomes

Favorable outcomes from communication sessions between internal stakeholders and suppliers may include:

- Meaningful product specifications
- Quality agreements that both parties can fully support
- Improved customer–supplier relationships
- Favorable pricing and deliveries

Less favorable outcomes from communication sessions between internal stakeholders and suppliers may include:

- Compromised specifications
- Strained customer–supplier relationships

- Delayed deliveries
- Additional time needed to resolve issues

To maximize the efficiency of customer–supplier (sender–receiver) communication events, the form of communication (oral, written, presentation), communication style (emotive, director, reflective, supportive) and cultural considerations (high versus low context, sequential versus synchronic, affective versus neutral) should be part of the planning process. A well-planned communication session, taking into account the various factors that directly relate to the effectiveness of the session, can go a long way toward achieving and nurturing successful customer–supplier relationships.

Diversity, Equity, and Inclusion (DEI)

Diversity, equity, and inclusion (DEI) is an increasingly important part of doing business in today's global landscape. By definition, diversity is acknowledging all of the different people (resources) that one must work with, taking into account variables such as gender, ethnicity, age, religious beliefs, socioeconomic status, sexual preferences, etc. Equity is recognizing the different levels of knowledge, availability, access, etc., of different individuals and creating avenues for everyone to have equal opportunities. Inclusion is executing plans, distributing communications, and conducting business that includes everyone in the applicable organization(s). In the past, DEI issues were considered more of an exception, rather than a rule. The opposite is true today. Organizations must include DEI factors as part of their overall strategies, goals, and objectives, in order to remain competitive and continue to grow.

Before attempting to align DEI strategies with an external supplier, an organization must establish effective internal DEI strategies. A company cannot hope to formulate a successful DEI strategy with a supplier unless they have an effective DEI strategy/program themselves. If a customer wants to establish a good, solid relationship with a supplier and include DEI aspects in the working relationship, both the customer and the supplier must have effective DEI programs in place in order to maximize the probability of success of working together.

Customer organizations can be in various stages of internal DEI program development. The details on the development of an organization's internal DEI program are beyond the scope of this book; however, Thompson (2022) recommends organizations use the ADKAR model to promote organizational change and generate an inclusive environment. ADKAR is an acronym that stands for Awareness, Desire, Knowledge, Ability and Reinforcement. An organization's employees need to be *aware* of the need for change. The employees should have the *desire* to support the necessary changes. *Knowledge* of how to change is essential. Once learned, the employees should have the *ability* to demonstrate the important skills and behaviors required for success. Lastly, *reinforcement* is needed to make sure the changes have been implemented correctly and are now part of the fabric of the organization.

Successful and effective DEI programs are developed, nurtured, and supported. DEI programs are not a "one and done" ordeal. An organization can't spend a few resources to conjure up a DEI program, announce it to their employees, check the "done" box and expect everything to run smoothly. DEI programs must be researched, strategically developed, implemented effectively, and monitored continuously to ensure that the organization's goals and objectives are being met. Frost (2022) discusses how unlocking individuals first, teams next, and then organizations is the proper sequence of DEI program development and implementation. Useful DEI principles start with individual efforts, gathers momentum with groups of individuals and hopefully culminates in an organization-wide adoption of the philosophies. As with any other system, methodology, or philosophy of making decisions and doing things, support from executive management is mandatory. The best intentions and strongest enthusiasm probably won't be enough to carry an organization through the journey without the support of senior leadership.

Whether a customer organization wants to utilize a DEI program to improve internal operations or collaborate with a supplier organization to create a mutually beneficial customer-supplier relationship, the DEI program must be treated like any other important system or program in operations management. Thompson (2022) emphasizes the use of the CAPE approach for the development and implementation of DEI programs. The "C" stands for collect. The proper data must be identified and collected. Just gathering data is not good enough. The data must be meaningful, measurable, and accurate. The "A" stands for analyze. Once the proper data are collected, they must be collated and analyzed effectively, to get the most bang for the buck. The "P" stands for plan. Once the data have been analyzed, the correct strategic plans need to be developed, based on what the data are saying and what the organization's goals and objectives are. The "E" stands for execute. After the plans have been developed and approved, they must be implemented properly. Notice the similarity between the CAPE approach for DEI program development and implementation and the Plan-Do-Check-Act (PDCA) cycle quality tool. The CAPE approach and the PDCA cycle both rely on strategic planning, proper implementation, and continuous monitoring for improvement.

When dealing with customer-supplier relationships, organizations must not only work diligently to maximize DEI program success within their own organization but also maintain some level of engagement with a supplier's DEI program as well. It is certainly the supplier's responsibility to develop, implement, and maintain their own internal DEI program. A customer's DEI program should align with a supplier's DEI program for maximum benefit of the customer-supplier relationship. Chapters 2, 3, and 19 of this handbook talk about the importance of collaboration and partnerships in customer-supplier relationships. As part of an organization's supplier management program, greater attention to collaboration and partnerships should be focused on the most critical and/or important suppliers, especially if an organization utilizes hundreds of different suppliers. A customer organization probably won't have enough resources to give equal attention to all of their suppliers. Refer to Chapters 2 and 3 to review ways to properly discriminate between different suppliers based on segmentation and cost analysis criteria. Doing this will help determine which suppliers are more critical and/or important. These are the suppliers that a customer organization need to work more closely with regarding the most effective ways of utilizing the DEI programs between the two organizations.

Distinction Between Diversity, Equity, and Inclusion (DEI) and Supplier Diversity

Diversity, equity, and inclusion (DEI) and supplier diversity are related concepts, but different. According to Porter (2022), "DEI provides the tone and tenor of how the organization views diversity internally. A comprehensive DEI strategy is not complete without a focus on supplier diversity." In other words, incorporating an organizational DEI program provides a framework for the company to recognize differences, provide accessible opportunities, and make sure everyone feels part of the bigger picture. Again, Porter states that "supplier diversity is a proactive business strategy that encourages buying from businesses that identify as belonging to a specific socioeconomic, historically disadvantaged or underutilized demographic as suppliers, vendors and contractors." In other words, supplier diversity is purposely targeting a certain number or percentage of placed purchase orders with suppliers owned/managed by various minority groups (based on gender, religion, socioeconomic status, sexual preferences, etc.).

Davis (2022) discusses several organizational benefits of incorporating a supplier diversity program, where buyers focus on rewarding purchase orders to suppliers that are considered a diverse supplier.

These benefits, summarized by Davis, allow an organization to:

- Expand the talent pool of potential suppliers
- Encourage competition in the supply base
- Create more resilient and agile supply chains
- Generate access to new networks

Impact of DEI

Supplier quality professionals need to realize the positive and negative organizational ramifications of incorporating DEI principles versus not addressing the topic of DEI for both internal operations and relations with suppliers.

Positive organizational outcomes of incorporating DEI principles include (but may not be limited to) the following:

- Demonstrating strategic planning by aligning diversity, equity, and inclusion objectives with the organization's overall strategies, goals, and objectives
- Creating a supportive work environment by showing employees that growth and opportunities will be made available to all, regardless of gender, age, religious or political beliefs, socioeconomic status, sexual preferences, etc.
- Having the ability to adjust to real-time global events, norms, and expectations to maximize organizational efficiencies
- Providing an inviting opportunity to global suppliers that your organization, as a customer, is aware of and actively incorporating measures to address topics that can affect people's perceptions and performance

- Offering assurance to global suppliers that your organization, as a customer, can become a long-term valuable business ally (partner) who can adapt and maintain positive working environments for both customer and supplier

Negative organizational outcomes of not incorporating DEI principles include (but may not be limited to) the following:

- Demonstrating an internal organizational unsureness about how to handle issues that involve diversity, equity, and inclusion
- Lacking the ability to give confidence to employees that growth and opportunities will be given or made available to all employees in a fair and reasonable manner
- Showing a lack of sensitivity and relevancy both internally and externally. Internal employees and external suppliers want to feel like the organization will support them in internal operations as well as in the customer-supplier relationship.
- Creating a potential customer-supplier relationship with a lower level of confidence. If a customer does not have DEI programs in place, it does not necessarily mean they will be a bad customer. Suppliers that care about DEI programs would probably prefer to have their more important customers demonstrate the incorporation of DEI principles in their operations.

Key Takeaways Regarding DEI

Here are some key takeaways regarding diversity, equity, and inclusion:

- An increase in global interactions between organizations has necessitated an increase in diversity, equity, and inclusion (DEI) awareness and implementation.
- A successful DEI program is more than just awareness and training; it is a way of doing business in today's global environment, it is a cultural change.
- The development/implementation of a DEI program is more than a "one and done" or "check the box" activity; it is part of an organization's continuous improvement efforts.
- DEI programs require support from executive management.
- Make use of the ADKAR model to facilitate organizational change and the CAPE approach to continuously improve your organization's DEI program.
- Take advantage of supplier segmentation and cost analysis methods to identify the more critical and/or important suppliers. This will help to maximize the efficiency of customer-supplier DEI efforts.
- Collaborate and form partnerships to a greater degree with critical and/or more important suppliers to incorporate solid, effective DEI

programs in the customer-supplier relationship that will meet the goals and objectives of both organizations.

- You will probably have to develop a semi-customized DEI collaboration program with each of one of your critical and/or more important suppliers, especially if these suppliers are based in different countries.
- Whatever the degree of involvement with specific suppliers and the amount of customization needed with DEI program collaboration, remember that effective communication is one of the keys to success in a collaborative customer-supplier relationship.

REPORTING USING QUALITY TOOLS

> Use appropriate technical and managerial reporting techniques for effective presentation and reporting, including the seven classic quality tools: Pareto charts, cause and effect diagrams, flowcharts, control charts, check sheets, scatter diagrams, and histograms. (Analyze)
>
> Body of Knowledge VI.B.2

Metrics and Reporting

One of the desired outcomes of effective customer–supplier communication sessions is a clear understanding of expectations. To determine whether those expectations are achieved, parameters that are measurable and accurate must be agreed upon. The measured parameters then typically become metrics that periodically appear in reports and summaries. One of the responsibilities of the SQP is making sure the meaningful metrics are summarized and reported in a timely and effective manner. Many forms of quality metric reporting are available to the SQP. The seven classic quality tools, mostly known for their problem-solving capability, are excellent ways to monitor and report the status of specified quality parameters of interest. These tools are very briefly discussed below. Many other resources are available that expand the theories and applications of these tools.

Pareto Charts

A Pareto chart is a graphical depiction of the 80/20 rule in problem solving and reporting. Vilfredo Pareto (an economist) and Joseph Juran (a well-known quality guru) demonstrated that about 80% of issues are caused by about 20% of the factors in a particular scenario.

Figure 18.5 shows a breakdown of plastic widget failures by suppliers. Notice that the majority of the plastic widgets that have been failing at incoming inspection are sourced from supplier B. By collating incoming inspection test results of

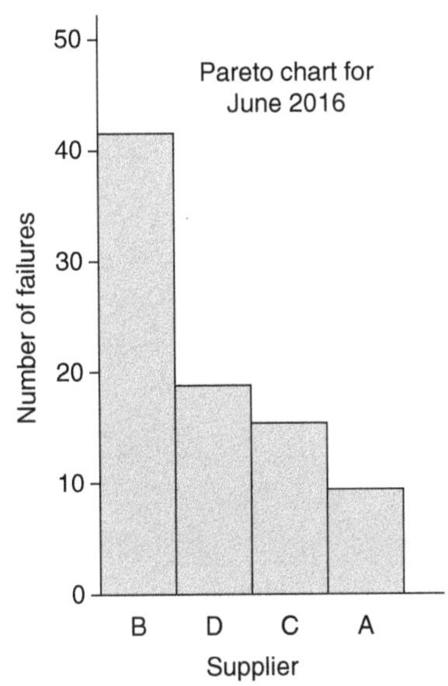

Figure 18.5 Plastic widget failures at incoming inspection by supplier.

the various plastic widgets they receive from numerous suppliers, the customer is able to focus its attention on working with supplier B to drive toward root/probable cause determination and develop associated corrective actions.

Cause and Effect (Fishbone) Diagrams

Cause and effect diagrams, also known as fishbone or Ishikawa diagrams, are useful for determining the root or probable causes of some nonconformance or event. The nonconformance or event (effect) is the head of the diagram, and various potential contributing factors (the causes) lead into the line (backbone) connecting to the event. Organizations very often utilize the 5 M's approach when using fishbone diagrams. The 5 M's are man, machine, material, method, and Mother Nature. A sixth M, measurement, is often used as a component of a 6 M's approach by some organizations. The use of this tool very often results in the identification of a root/probable cause. Figure 18.6 is a cause and effect diagram that displays several factors (causes) that may be contributing to the receipt of wet baseballs (effect). By gathering and collating information using this tool, a customer is able to identify some plausible reasons (causes) why they have been receiving wet baseballs (effect) from one of its suppliers. More efficient communication sessions can then take place between the customer and supplier to resolve the issue.

Flowcharts

A flowchart is a graphical depiction of the various steps in a process. It can serve as an excellent starting point for a team, since it helps to provide a common understanding of the flow of a process. In addition to displaying the basic steps in a process, a flowchart may also contain other relevant information, such as due dates, tools, and responsible individuals.

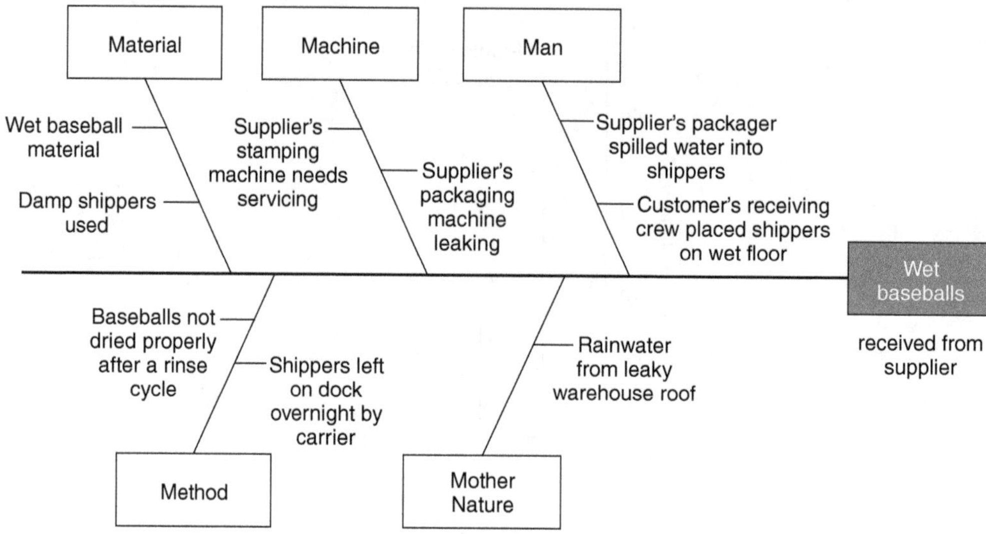

Figure 18.6 Fishbone (Ishikawa) diagram for possible reasons (causes) for wet baseballs (effect).

Figure 18.7 shows a very simplified flowchart of the supplier selection process for a company. The basic sequential steps are indicated, along with various decision points along the way. In this example, the internal stakeholders will have a good idea of whether they will be able to purchase components, ingredients, or services from a particular supplier by following this flowchart.

Control Charts

Control charts are used to identify assignable (also known as special) causes of variation in a process. They are also used to demonstrate how a process is doing over a given period of time. In addition to these two primary functions, control charts can also serve as a means of reporting on parameters of interest.

Selecting the proper control chart to monitor or display a specified parameter is as important as the status of the parameter itself. If an improper control chart is used, valuable data, which could be used by internal stakeholders and suppliers to resolve quality issues, will probably be misinterpreted, leading to erroneous conclusions.

Many excellent books, periodical articles, and white papers have been written on the theories and applications of control charts. Only a few brief monitor/reporting control chart applications for the purpose of improved customer–supplier communications are discussed here.

There are two basic types of control charts, variable and attribute. Variable control charts are used when one wants to monitor parameters that can have continuous values, such as the weight of a person, the length of a baton, or the pH of a reagent, etc. Examples of variable control charts are $\bar{x}$ and R and $\bar{x}$ and s charts. Attribute control charts are used when one wants to monitor parameters that have discrete values, such as the number of red marbles in a bag of colored marbles, how many defects were detected in the last three shipments of component X from

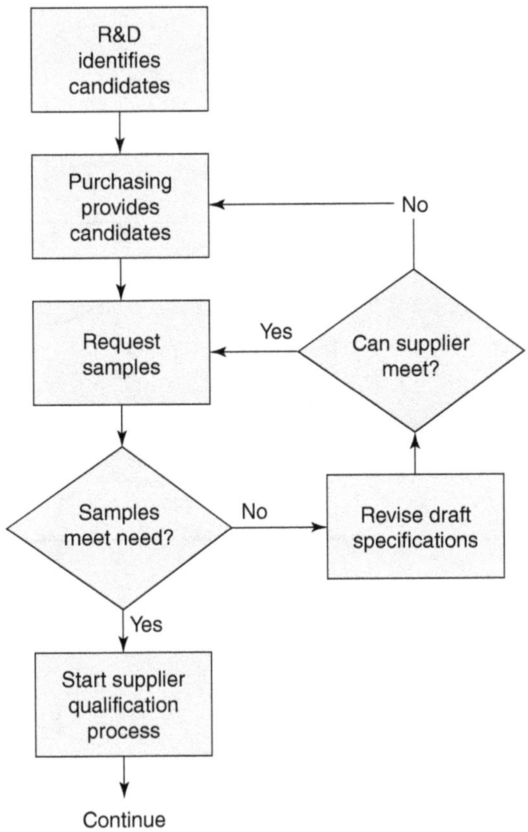

Figure 18.7 Supplier selection process.

supplier C, or the percentage of defective lots received from supplier Y since 2014, etc. Examples of attribute control charts are *c*, *u*, *p*, and *np* charts (see Chapter 13).

Figure 18.8 is a variable $\bar{x}$ chart (the *R* portion of the chart is not shown) that shows the protein content of ingredient lots from a supplier as tested at incoming inspection since the beginning of the year (through June 2016). Notice that the protein percentage in the recently received supplier lots has been creeping below the lower control limit (LCL) and getting dangerously close to the lower specification limit (LSL). Discussing this chart with the supplier may help to prevent problems with future shipments if the supplier can identify why the protein percentage has been decreasing and is able to implement corrective action.

Figure 18.9 is an attribute *c* chart that shows the number of damaged bags (defects) per receipt of recent shipments of an ingredient used to manufacture baked goods. Since the line is moving upward (see the far right side of the chart), supplier quality and purchasing should probably have a discussion with the supplier to determine if a root cause analysis and subsequent corrective action development are needed to lower the number of defects per shipment for future receipts.

Many opportunities exist for customers to communicate information to suppliers through control charts. The information may include topics such as the number of on-time deliveries as a percentage of total deliveries for the past two

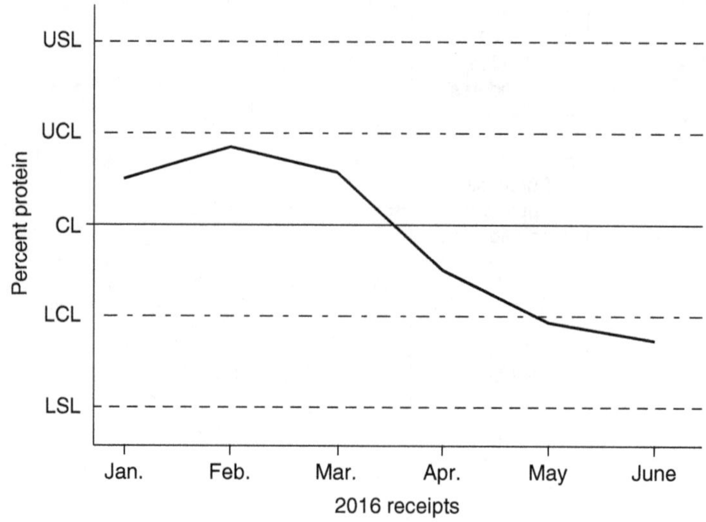

Figure 18.8 $\bar{X}$ chart.

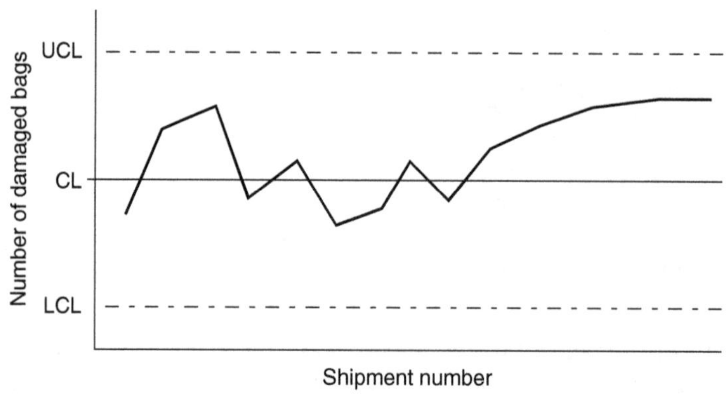

Figure 18.9 *c*-chart.

years, the mean values of specific quality parameters for the last 25 shipments, the number of defects seen in receipts of a specific commodity number over the previous quarter, and so on. The key is to select the most appropriate control chart for monitoring and communicating the parameters of concern to the supplier.

Check Sheets

Check sheets (sometimes referred to as tally sheets) are a very quick and easy way to obtain a graphical depiction of the frequency distribution of a parameter. All it takes is the identification of a parameter to monitor, an approximate range of values to expect, and someone to mark the sheet after each encounter of a data point. The training required to understand and implement this quality tool is very minimal and it has widespread application potential.

Figure 18.10 shows an example of a check sheet. This tally sheet displays the on-time delivery history for shipments of a component from supplier C received during 2015 and 2016 to date. In 2015, the majority of the shipments from supplier C were received either very close to the scheduled date or up to 10 days late. Note that so far in 2016, supplier C has improved its percentage of on-time or early deliveries. This is an example of an opportunity to share some good news with a supplier. Too often, internal stakeholders only communicate problems with their suppliers. Suppliers are businesspeople, too, and they enjoy hearing positive feedback from their customers.

Scatter Diagrams

Scatter diagrams are basically *x-y* plots that give a quick indication of how closely two variables are correlated to each other. Sometimes scatter diagrams are generated by computer software programs, but they can also be drawn by hand to provide a quick indication of correlation between two variables.

Figure 18.11 is a scatter diagram that shows the relationship between the number of out-of-spec widgets (defects) and the size of the received supplier lot. This chart displays a high degree of positive correlation between the received lot size and the number of defects realized in the lot. As the lot size increases, so does the number of defects. The data indicate that the number of defects is about 10% of the lot quantity. Sharing this information with the supplier will be very helpful in moving toward reducing the number of defects per lot.

Histograms

Histograms are frequency distribution graphs showing the number of observations made of a particular parameter. They can be used to monitor parameters such as the number of days it takes for the delivery of a certain commodity from a supplier, the microbiological spore count of samples taken from shipments of

On-time deliveries from supplier C		
	2015	**2016 to date**
8–10 days early	||	|||
5–7 days early	||||	𝍸
2–4 days early	𝍸	𝍸 |||
On scheduled date ± 1 day	𝍸 𝍸 𝍸 |||	𝍸 𝍸
2–4 days late	𝍸 𝍸 ||||	𝍸 ||
5–7 days late	𝍸 𝍸 ||	||||
8–10 days late	𝍸 ||	||

Figure 18.10 Check sheet.

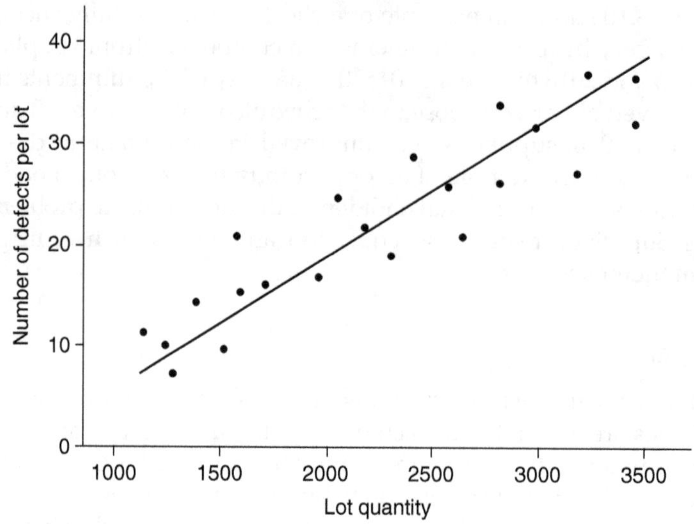

Figure 18.11 Scatter diagram.

a food ingredient, and the weight of 12-volt batteries received by an automobile manufacturer from supplier X, etc.

Figure 18.12 shows a histogram of the lengths of syringe barrel samples taken from lots received by a medical device manufacturer from its most reliable supplier. If the syringe length specification is 5.0 ± 0.1 inches, then all of the samples tested are within specifications. If the specification is 4.95 ± 0.05 inches, then the supplier is manufacturing syringe barrels that are too long, with nearly half of the measured samples having barrel lengths high out of specification. Sharing

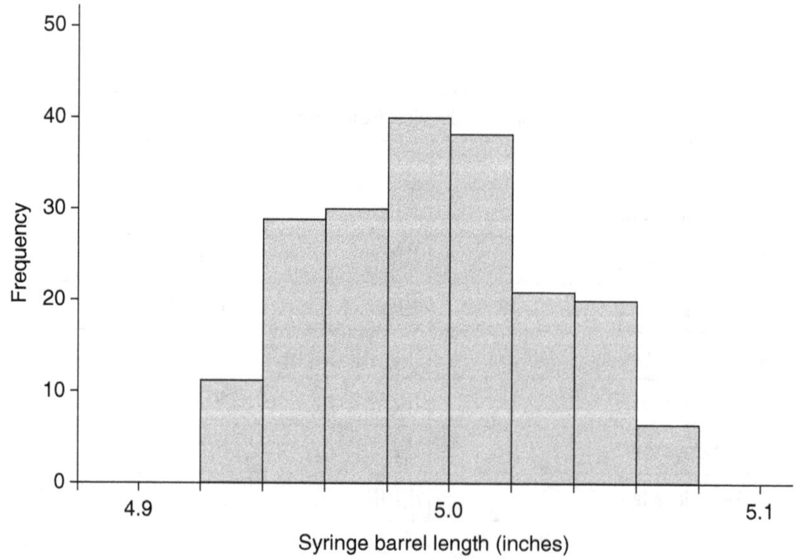

Figure 18.12 Histogram.

this histogram with the supplier will result in more meaningful discussions, as opposed to just telling them that half of their syringe barrels are too long.

As you have seen, the seven quality tools offer a supply chain professional many different ways to present information to a supplier. To get the most out of these tools, the quality practitioner needs to understand how to apply them, which will lead to more effective graphics that can be used to enhance discussions with suppliers.

RECOMMENDATIONS FOR USING THE SEVEN QUALITY TOOLS FOR REPORTING INFORMATION TO SUPPLIERS

When deciding on what and how to communicate important information to suppliers, keep the following points in mind:

- Make sure that a good relationship has already been established between the internal stakeholders (customers) and the supplier
- Determine what specific information would be the most valuable to monitor and report
- Based on the parameter(s) to be monitored and the potential audience for the information received by the suppliers, choose the appropriate quality tool
- Include the use of the quality tool(s) in the contract or quality agreement verbiage, if possible
- Once the proper tool is chosen, make sure to use the tool correctly, e.g., calculate the control limits of variable/attribute charts accurately, allow ideas to flow freely during the development of a cause and effect diagram, show only two variables when using a scatter diagram
- Make sure that once the information is communicated to the supplier, the audience knows how to interpret the chart or graph
- After delivery of the information, follow up accordingly with the supplier to assure a smooth transition from problem reporting to resolution generation
- Continue to hold regular communication sessions with the supplier to maintain a positive ongoing customer–supplier relationship

Try to remember that communication sessions between internal stakeholders and suppliers can often turn into crucial conversations. People have different opinions and viewpoints on things, the stakes are often high, and communication session participants can often get emotional. The supply chain quality professional will need to understand the situation; apply the proper form of communication, taking into consideration the communication style of the sender; acknowledge any cultural differences between the sender and receiver; and choose the most appropriate reporting tool(s). If casual conversations turn into crucial ones, proper planning and choice of communication tools become even more important.

Chapter 19
Leadership and Collaboration

> Understand and apply techniques for coaching suppliers through regular communications, influencing without authority, negotiation techniques, conflict resolution techniques, and establish clear roles and responsibilities of internal stakeholders and suppliers using tools such as a RACI matrix (responsible, accountable, consulted, and informed). (Evaluate)
>
> **Body of Knowledge VI.C**

Take responsibility for making relationships work.

—Ken Blanchard, *Heart of a Leader*

Relationship management is vital to the customer–supplier relationship. In fact, it is vital to the internal supply chain organization as well, and for much the same reason. As much as we would like them to be, suppliers are not part of our company. It is not a "command and control" relationship. There is not a direct reporting structure; there is no definite and formal hierarchy. There is no concrete method of compelling the supplier to do anything. It is a voluntary relationship, mutually entered into by both parties, ostensibly for their respective but mutual benefit. The same can be said for the supply chain organization. In most (but not all) organizations, there is no distinct, coherent organization called "supply chain." Instead, it is a cross-functional, multidisciplinary team performing activities related to selecting suppliers, procuring materials, qualifying materials, and working jointly toward schedule fulfillment. It's more of a matrix structure, similar to the relationship with the external supplier. Given this, relationship management is the most effective skill set to ensure that the supply chain functions smoothly and the "kinks in the links" are minimized.

This chapter outlines the most critical aspects of relationship management for the supply chain professional in general and the SQP in particular. It is impossible to overestimate or overstate the value of leadership in almost any situation and of collaboration in any relationship. In the context of supplier relationship management, the combination is extremely powerful. Leadership from the customer is critical, as the customer–supplier relationship is initiated by the customer, and of course the finished product specifications/requirements are usually (but not exclusively) customer driven. Leadership with the supplier encompasses several areas of impact, including supplier strategy, sourcing strategy, supply chain vision and mission, and supplier development. Good leaders are good relationship managers. Good relationship managers collaborate. Collaboration is mutually beneficial. Indeed, some level of partnership is desirable, depending on the supplier strategy. In order to maximize the value proposition between the parties, it is essential to pursue a win–win outcome on most endeavors and issues. We will discuss how to achieve a win–win outcome, the principles of persuasion, and successful negotiation techniques later in this chapter. As we continue to explore the concepts of leadership and collaboration, we will see how interrelated they truly are in practice. In fact, one could argue that collaboration is a leadership style and a strategy all at once. Because of the voluntary nature of the customer–supplier relationship, it is imperative that collaboration and leadership be appropriately exercised for maximum success.

A WORD ABOUT LEADERSHIP

Leadership is a high calling.

—Ken Blanchard, *Heart of a Leader*

Although the intention here is not to separate leadership from collaboration, it is important to discuss leadership at a high level before drilling down into its more concrete aspects and discussing its relationship to collaboration. Many definitions of leadership have been crafted over the years, but perhaps the most simple, most powerful, and most useful was offered by Peter Drucker: "Management is doing things right; leadership is doing the right thing" (Daskal 2015). Furthermore, it is important to remember that "leadership is not something you do to people. It's something you do with people" (Blanchard 1999, p. 170). This is especially true in the context of relationship management. A servant style of leadership not only allows us to build a meaningful relationship but also makes it easier to persuade, to coach, to influence, and to negotiate a win–win outcome. In turn, the better our relationship is, the more effective our persuasion, negotiations, coaching, teaching, and overall leadership are. We will necessarily be more collaborative. The concepts feed into and out of each other in a continuous and reinforcing feedback loop. But it starts with leadership.

Good leadership develops a realistic, effective, and sustainable vision of the supply chain that allows all stakeholders, internally as well as externally, to understand the mission and the course for the future. Good leadership serves the stakeholders. Good leadership seeks, and inspires stakeholders to seek, the most efficient, productive, and effective courses of action, whether in everyday execution or during crisis. Good leadership insists on true ownership of processes, inputs, and outputs. It provides stewardship: direction, guidance, and oversight that benefits all stakeholders. Good leaders effectively utilize relationship management. In the final analysis, good leadership—good stewardship of the supply base—sets our operations up for success and ultimately serves to benefit our customers, end users, and patients. Combining leadership and collaboration in the context of supplier relationship management can have a powerful and lasting impact.

Leadership and collaboration affect the SQP primarily in the areas of coaching, negotiation, and influencing without authority. Tables 19.1 and 19.2 list very useful principles and tools that can be effectively utilized in all three areas. As we move through all phases of the supplier management lifecycle, these concepts

Table 19.1 Leadership principles.

Principle	Explanation	Examples
Classic leadership traits	Leaders embody values, demonstrate disinterest (willingness to sacrifice, caring), and possess practical management skills.	Nelson Mandela Values: Justice, racial equality, democracy, reconciliation Disinterest: 20+ years in prison Skills: Head of the ANC
Our leaders embody our best selves	We tend to interpret issues and concepts by way of our own values.	Intellectuals want a super-smart leader; veterans want courage; quality professionals want a person who utilizes data.
Identity motivates; we feel best when we live up to our values	People will do almost anything to enforce their best sense of themselves. Listen for a sentence like "I'm really a . . . " The first thing mentioned is the primary identity.	"I'm a Marine. Semper fi!" This person values membership in an elite group and believes in loyalty.
Define issues in plainest possible terms	Find the "pith." Where possible, narrow down the issue to a few words.	Defects are bad. Processes need controls. Shipments must be delivered on time.
Find the values	Find the values by discovering the words and phrases that pop up most frequently.	Quality, compliance, process control, cost, delivery, etc.
Symbolize the values	Take the issue as you have defined it and extract a little piece of the issue to form a powerful symbol.	Defects: "Quality is job 1" (Ford) Delivery: "Our solutions connect people and possibilities" (FedEx) Control: "Welcome to a world where everything runs more smoothly, smartly, simply & safely" (Johnson Controls)

Table 19.2 Leadership communication tools.

Tool	Explanation	Examples
Identity	Get people to describe themselves—most people will do almost anything to live up to that identity.	Appealing to the business sense of a procurement professional Appealing to the need to make data-driven decisions by the quality professional We (the supplier) are a great . . . (injection molder, metal stamper, coater, etc.)
The halo (very useful in coaching!)	Sum up the issue succinctly and connect that brief summary to the expressed values.	During the 2008 presidential race, Joe the Plumber symbolized working-class aspirations to own a business and identify with the Republican party ASQ: Excellence Through Quality "A molder of your caliber would/would not . . . " "Surely a processor as capable as you could . . . " "This is a disappointing lack of control for someone with such a robust quality system . . . "

remain relevant. Supplier selection requires negotiation skills and the leadership to choose the right supplier. Supplier qualification requires negotiation, influencing, and possibly some coaching. Part/material qualification requires all three. Teaching and coaching are integral to supplier development activities. Continuous quality improvement (CQI) also requires all three but relies heavily on influencing, as some CQI efforts might not have immediate (or even short-term) ROI and might take some convincing.

Moving directly to the topics in this section of the BOK, let's discuss coaching through regular communication, influencing without authority (persuasion), and negotiation techniques. These are the soft skills that are the hallmark of the relationship manager. They also make collaboration possible and make good collaboration even better. Remembering our vision and mission, we can put these skills and techniques into action to serve the vision and mission. In this way we can realize our supplier strategy to its fullest.

COMMUNICATION

A good definition of coaching is offering advice and guidance. There are many situations in which the SQP would coach a supplier: for development or CQI activities, in communicating with an internal stakeholder, remediation with a poor-performing supplier, or on regulatory matters. In any of these communications we would employ leadership communication techniques and tools. In the context of supply chain leadership and collaboration, we can learn a great deal about leadership and communication tools and techniques by studying a few of history's greatest, all of whom mastered rhetoric to a degree and therefore relationship management, albeit on a grand scale. Jay Heinrichs's book *Thank You for Arguing* (2013) clearly demonstrates the importance of mastering some of the techniques of classical rhetoric in being able to persuade and influence. Motivating your audience plays a big role in leadership communication, as does setting a personal example. We have already established the link between leadership and relationship management. There is also a direct link between rhetoric and leadership. What is that link? According to Heinrichs, it is connecting with values. Table 19.1 lists some important values-based principles of leadership from Heinrichs's chapter titled "Leading Your Tribe." Table 19.2 lists some communication tools to keep in mind, along with examples of their application.

In summary, as we consider coaching and utilizing leadership communications, especially in the context of persuasion and influence, we need to remember a few basics:

- Audience motivation is critical
- Connect with values
  - —Set the example of "virtue" personally
  - —Walk the walk, don't just talk the talk
- Values are the basis of identity
  - —This is the link to motivation and behavior
  - —People want to be their best selves
  - —Symbolize the values succinctly to create a halo
- Use the halo to motivate desired behaviors

PERSUASION, NEGOTIATION, AND INFLUENCING WITHOUT AUTHORITY

Leadership communication is a great segue into persuasion, negotiation, and influencing without authority, as classical rhetoric is indeed one very important foundation of persuasion. As a general rule, persuade if you can, negotiate if you must. Seek a mutually agreeable outcome by looking past "demands" for the underlying interests of each party. It is here that true collaboration and successful negotiation is most effective. Finally, negotiation is not compromise, nor is it "splitting the difference." Let's consider these concepts in a little more detail.

Persuasion is the optimal strategy in any relationship management opportunity. If we can persuade, then there is no need to negotiate. Luckily the rhetorical skills we discussed above will provide value in this activity as well. Two general rules of negotiation apply in almost all situations:

1. Don't negotiate if you can (and should) persuade
2. Increase the level of need so the other party will be less concerned about the cost of the solution

Persuasion takes considerable skill. A would-be persuader must understand the pain points of the other party and be able to position the need for a solution based on these pain points. As pain increases, the need increases, as does the willingness to accept the solution. In order for this to be effective, however, the need *must* outweigh the cost of the solution, usually severalfold. It is therefore the persuader's task to position the cost–benefit of need in this favorable light. Do not be afraid to redefine the problem. Out-of-the-box thinking is very useful here, as is operating in the gray areas. The persuader should consider what-if and future scenarios as well as the concrete and immediate pain and need when positioning and defining/redefining the problem to the other party. In the end, this is an exercise in creative thinking. Planning, preparation, and research are necessary to make this truly effective.

The classic text on collaborative, win–win negotiation style remains *Getting to Yes: Negotiating Agreement without Giving In* by Roger Fisher and William Ury. The authors refer to the method of *principled negotiation*. The principled stance transcends the traditional hard/soft positional bargaining negotiation style. It opts instead for a "wise outcome, reached amicably and efficiently" (Fisher and Ury 2011, 11). The premise is that participants are problem solvers. This is part of the overall negotiation strategy of the "metagame." In the metagame, each move you make in a negotiation is not just a move that deals with rent, salary, or other substantive questions; it also helps structure the rules of the game you are playing. This is largely unconscious, but the way you negotiate determines whether the negotiations continue and how they continue. In this way the participant is able to focus the negotiation, using these four points as strategies:

1. *People.* Separate the people from the problem
2. *Interests.* Focus on interests, not positions
3. *Options.* Invent multiple options looking for mutual gains before deciding what to do
4. *Criteria.* Insist that the result be based on some objective standard

Let's take a look at each of these in more detail and provide some examples.

People

In short, where the soft approach counsels being soft on people and problems (i.e., there is trust between parties) and the hard approach counsels the opposite (i.e., there is distrust between parties), the principled approach is trust neutral and thus it counsels us to be soft on the people and hard on the problem. Fisher and Ury state, "Figuratively, if not literally, the participants should come to see each other as working side by side, attacking the problem, not each other" (2011, 12). Making this a reality is where the metagame strategy comes into play. Deal with the people issues right away, and then agree to focus on the problem and its solution.

Interests

Interests are *not* positions. Focus on underlying interests rather than a negotiating position. In fact, the stated position often obscures the true interests. To tell one from the other requires questioning, curiosity (discovery), and listening. In fact, one party may actually be able to, or required to, aid the other in understanding their underlying interests. Interests are most closely related to the actual need that sparked the negotiation in the first place.

Options

Because of the difficulty of designing optimal solutions under pressure and the high stakes involved, it is a good idea to set aside some time to create a wide range of options that are mutually beneficial before making decisions. Choose among those options when it comes time to decide on a course of action or a solution.

Criteria

Sometimes a negotiator will dig in and become entrenched. In these situations, counter this tactic (position) with an insistence that this position of immobility is not sufficient and the result must be based on a fair and objective standard rather than will alone. The parties can then move the discussion forward by discussing possible standards (Kelley Blue Book, independent appraisal, fair market value, independent audit, and the like) rather than what they will or won't do. Then agreement can be focused mutually on this standard.

Stages of Negotiation

These four points—people, interests, options, and criteria—are to be used in the three stages of negotiation: analysis, planning, and discussion. Analysis is the stage of discovery and diagnosis. Understand biases and personality issues and discover underlying interests on both sides using the four points above. Planning involves the four points, but now the outcome should be ideas and action items. How do we deal with people issues? Set realistic objectives. Clearly be able to state both parties' interests. Generate additional options and decision criteria. Discussion is the stage where the parties work to understand each other's interests and develop mutually agreeable options. The four points should also form the basis of discussion. This methodology provides a framework for making wise agreements that are beneficial to all parties and result in an amicable outcome. Such agreements have a far greater life span and are much easier to modify if and when circumstances warrant. Lastly, because the agreement was achieved through a mutual process of eventual consensus, the buy-in from both parties will be high, and so, too, will be compliance and even goodwill when living up to the agreement. See Table 19.3.

Conflict Resolution Techniques

Obviously, no relationship is without some degree of conflict from time to time. Professionals, and their organizations, can feel very strongly about certain positions they hold. This is no different for supply chain relationships. Some examples the SQP might encounter would be quality issues (rejections, nonconforming material, returns), differing interpretations of responsibility per the contract (testing, functional performance, design responsibility), or payment terms (a procurement

Table 19.3 Negotiating points summary.

Negotiating point	Techniques	Examples
People: Separate the people from the problem	Defuse strong emotions (blow off steam), test perceptions, walk a mile in the other's shoes, discuss perceptions openly.	Someone is offended by being repeatedly selected for a task, perceiving it as punishment, but the selection is based on that person being reliable, competent, and efficient.
Interests: Focus on interests, not positions	Ask why and why not. Listen to the answers (discovery). Be clear and compelling in stating your interests. Be specific. Put interests before your answer.	Two people argue over an open window, unable to agree on whether it will be half open or three-quarters open. Interests are draft vs. fresh air. Solution: Open the window down the hall, providing air without a draft.
Options: Invent multiple options looking for mutual gains before deciding what to do	Separate inventing from deciding—brainstorm first. Change the scope of agreement—consider broader options. Look for mutual gain, identify shared interests.	Think of this nursery rhyme: "Jack Sprat could eat no fat; his wife could eat no lean. And so betwixt them both, you see, they licked the platter clean."
Criteria: Insist that the result be based on some objective standard	Frame each issue as a joint criteria search. Reason and be open to reason. Never yield to pressure, only to principle.	When facing "Take it or leave it," focus on fairness. Ask for reference to a standard. Find a base and keep working from there, referencing standards along the way; e.g., negotiating a car price using Kelley Blue Book (base), ads for similar cars, insurance information, and other dealers (options).

issue for sure but it can reduce the suppliers' willingness to negotiate, expedite solutions or cooperate in general). Luckily there are many strategies to overcome conflict in business relationships.

Perhaps some of the most effective techniques would include Champlain College's Top Five Conflict Resolution Strategies. Simply put:

1. Don't Ignore Conflict – Conflict won't go away over time. In fact most times it can fester and become toxic so address conflicts early.

2. Clarify What the Issue Is – It's important to uncover the facts as well as the positions of both parties. As we discussed in the section on negotiation, it's also important here to understand the underlying needs of each party and address those. Obviously, good, clear communication is essential here too. Start with independent discussions to thoroughly understand not only position but relevant details.

3. Bring Involved Parties Together to Talk – There is no substitute for talking face to face. Resolving conflict is as much about the relationship as the issue at hand.
4. Identify a Solution – Use quality tools and CAPA principles to find a solution that is both equitable, and actionable, that also solves the issue at hand.
5. Continue to Monitor and Follow Up on the Conflict – Just like the PDCA (plan, do, check, act) cycle, it's critical to follow up with planned solutions and be sure they have solved the problem. If not, course-correct as needed. Check in with the parties and the health of relationship.

A good example of how a SQP could utilize these tools in a typical situation might include a disagreement between the customer's IQC (incoming quality control) and the supplier over measurement results, leading to a nonconforming product situation.

1. As soon as the supplier responds with their findings, disagreeing with the disposition, set up a meeting with the supplier to discuss and acknowledge the issue.
2. In this meeting, make sure to understand details such as GRR, measurement methods, equipment used, and their understanding of the specification. Do the same independently with the IQC department. This is also a good time to take a reading of the state of the relationship—is there a negative history? Are there any trends in issues? How cooperative have the parties been in the past?
3. Bring the parties together in a way that will leverage and enhance the relationship, playing to the strengths and not weaknesses. State the issue and the respective positions **factually**. Help each side understand the others' points of view. Always focus on the facts, but strive for a win-win outcome.
4. Use your quality tools (brainstorming, fault tree analysis, fishbone) and CAPA skills to get to root cause and appropriate corrective action. Collaborate together to strengthen the relationships.
5. Sustain the resolution. Monitor the relationship. Measure the outcomes, and take appropriate action if needed.

COLLABORATION

Collaboration is perhaps the most effective tool for ensuring the supply chain functions smoothly and achieving objectives and improvements. As discussed, negotiation should be based on a win–win strategy and a collaborative approach. Collaboration should take place in multiple areas: R&D, CAPA, CQI projects, supplier development activities, problem solving during issue resolution, and, of course, when a supplier is underperforming.

An example of collaboration is sending a team to a supplier to train them in PPAP or a similar set of quality tools. Or to walk them through a failure investigation, simultaneously teaching problem solving and root cause analysis using tools such as cause and effect diagrams, FTA, 5 Whys, and process mapping. Another example is developing a CAPA rubric to share with the supplier and training them to it, delineating precisely what you expect and how you will judge the robustness of their CAPA or SCAR. Moreover, joint kaizen events can lead to breakthroughs, a reduction in potential errors, and cost savings. Lastly, collaborating on a supplier improvement plan to remediate deficiencies at the supplier can be an investment in both the customer's and supplier's futures. More than one success story has been realized using this practice.

In Chapter 1, we discussed some of the advantages of collaboration and communication:

> When suppliers know what is expected of them (quality agreements) and how they are evaluated (measuring and monitoring), they can focus on improvements such as reductions in variation and cost. The cost reductions resulting from variation reduction will enable suppliers to enter into longer-term contracts while also accomplishing their financial goals. At the same time, suppliers are assured of a longer-term partnership with the customer, a benefit to both parties.

DEFINING THE STAKEHOLDERS AND SETTING CLEAR ROLES AND RESPONSIBILITIES

In order to apply the techniques and tools discussed, it is critical to properly and accurately identify the stakeholders, internal and external, and set clear expectations by defining equally clear roles and responsibilities for each stakeholder. This is important for several reasons. First, we cannot tailor our communications if we have not identified our audience (stakeholders). Second, we must find common ground and understand the needs of our audience (stakeholders) to negotiate collaboratively. Third, internal stakeholders must speak to external stakeholders with one voice to ensure maximum effectiveness and efficiency in the execution and management of supply chain activities. Finally, relationship management is most efficacious when all stakeholders are part of the relationship.

Therefore, the first question to be answered is, who has a stake in this supplier or issue? Look for the answer internally and also externally, with the supplier and the end users—patients, customers, or your manufacturing sites. Also, don't limit your definition of a stakeholder. When it comes to the supply chain, stakeholders can be found anywhere: procurement, QA, QC/incoming quality assurance (IQA), operations, sales and marketing, and R&D. Once you have identified the stakeholders, it's helpful to use Venn diagrams, responsible-accountable-consulted-informed (RACI) charts, or SOPs to define roles and responsibilities. This is critical because a supplier must hear one coherent message from its customer, and this is dependent on properly defined roles and responsibilities. Second, it is important to make sure all tasks are covered thoroughly and nothing is left undone. This will assure proper accountability, follow-through, and closure. Since fundamentally the customer–supplier relationship is a business relationship between two separate entities, ensuring accountability and that no tasks are dropped or left incomplete is essential. Finally, once the roles are defined, communication becomes the critical conduit and keeping the functions "in their lanes" is essential.

Figure 19.1 is an example of a Venn diagram defining roles and responsibilities for two particular supplier quality functions: supplier quality engineer and supplier quality program manager. Notice the areas of independent responsibility and the overlap area of joint responsibility. This division of duties keeps everyone working efficiently, reduces redundancy, and provides a clear delineation and unambiguous accountability for the duties outlined.

The RACI chart in Figure 19.2 goes one step further: It defines both lines of authority *and* lines of communication. RACI stands for responsible, accountable, consulted, and informed. It outlines who is responsible for execution, who is accountable for the activity or duty to be carried out, who might need to be consulted (either for approval or advice), and who only needs to be informed about the activity. Figure 19.2 refers to process steps in QSM.10.003, a fictitious SOP covering the maintenance of the approved supplier list and supplier files. It also has provision for corrective action (exception reporting) if there are discrepancies. Escalation processes can also be built in this way.

Either of these tools can be used internally or externally. They work equally well with supplier–customer roles and responsibilities and within functions, as illustrated above, as well as between functions (e.g., quality and procurement).

In summary, relationship management is an important aspect of managing the supply chain. The customer–supplier relationship is voluntary, so managing it effectively requires a set of people-oriented skills, including communication,

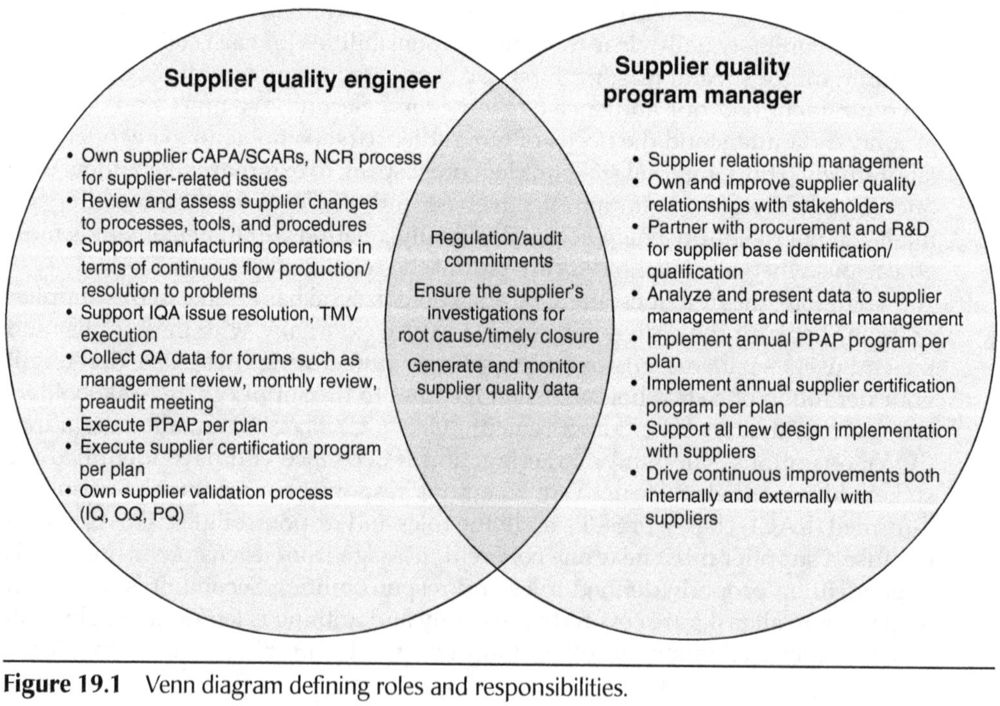

Figure 19.1 Venn diagram defining roles and responsibilities.

RACI chart template				
Activity / Role	PM leadership	PM team	SQE leadership	SQE team
QMS.10.003				
Initiate and submit form QSM 10.003.01 for required changes	R	A	R	R
Quarterly review of the site supplier list against the global ASL	N/A	C	R	A
Approve supplier changes as required (QSM 10.003.01)	A	R	R	C
Maintain the supplier documentation on record file (SharePoint)	A	R	R	R
Generate exception report (ER), as needed	I	C	A	R

Figure 19.2 RACI chart defining lines of authority and lines of communication.

leadership, collaboration, persuasion, conflict resolution and negotiation. These skills are necessary because SPQs often find themselves in situations where influencing without authority is required. This is as true with internal stakeholders as it is with suppliers. Furthermore, all business relationships are just that—relationships. There is no relationship that cannot be made better with these skill sets.

Part VII

Business Governance, Ethics, and Compliance

Chapter 20
ASQ Code of Ethics

> Determine appropriate behavior in situations requiring ethical decisions, including identifying conflicts of interest, and recognizing and resolving ethical issues. (Apply)
>
> **Body of Knowledge VII.A**

Trustworthy people acquire no worthy enemies as they go about their labors.

—Philip Crosby, *Quality Is Free: The Art of Making Quality Certain*

The CSQP Body of Knowledge includes the requirement that the SQP be able to determine appropriate behavior in situations requiring ethical decisions, including identifying conflicts of interest and recognizing and resolving ethical issues. Knowing when and how to apply a code of behavior such as the ASQ Code of Ethics or a corporate-level ethics policy is a crucial part of building and maintaining the perceived competence and prestige of the quality profession. If this is not done, the supplier quality profession risks becoming irrelevant or even detrimental to business and society as whole.

ASQ CODE OF ETHICS

Fundamental Principles

ASQ requires its representatives to be honest and transparent. Avoid conflicts of interest and plagiarism. Do not harm others. Treat them with respect, dignity, and fairness. Be professional and socially responsible. Advance the role and perception of the Quality professional.

Expectations of a Quality Professional

A. Act with Integrity and Honesty

1. Strive to uphold and advance the integrity, honor, and dignity of the Quality profession.
2. Be truthful and transparent in all professional interactions and activities.
3. Execute professional responsibilities and make decisions in an objective, factual, and fully informed manner.
4. Accurately represent and do not mislead others regarding professional qualifications, including education, titles, affiliations, and certifications.
5. Offer services, provide advice, and undertake assignments only in your areas of competence, expertise, and training.

B. Demonstrate Responsibility, Respect, and Fairness

1. Hold paramount the safety, health, and welfare of individuals, the public, and the environment.
2. Avoid conduct that unjustly harms or threatens the reputation of the Society, its members, or the Quality profession.
3. Do not intentionally cause harm to others through words or deeds. Treat others fairly, courteously, with dignity, and without prejudice or discrimination.
4. Act and conduct business in a professional and socially responsible manner.
5. Allow diversity in the opinions and personal lives of others.

C. Safeguard Proprietary Information and Avoid Conflicts of Interest

1. Ensure the protection and integrity of confidential information.
2. Do not use confidential information for personal gain.
3. Fully disclose and avoid any real or perceived conflicts of interest that could reasonably impair objectivity or independence in the service of clients, customers, employers, or the Society.
4. Give credit where it is due.
5. Do not plagiarize. Do not use the intellectual property of others without permission. Document the permission as it is obtained.

Ethical behavior that is driven solely by a desire or demand to conform to requirements (as in required review and approval of and adherence to a company ethics policy or standard of conduct) can be a slippery slope where decisions are made from the standpoint of, "It must be OK because the requirements don't say *not* to do it."

Although controls and legal ramifications have their place, the ASQ Code of Ethics strives to set a higher bar, calling for the SQP to internalize the fundamental principles and articles. This is consistent with the teachings of quality gurus such as W. Edwards Deming and Philip Crosby.

Quality and how it is perceived have been evolving since the days of mostly agricultural societies, where quality was determined by inspecting and rejecting bad product and variation was almost completely out of our hands, up to current modern businesses and products, where more focus rests on every individual having control and responsibility over quality using advanced SPC tools. However, as Deming put it, "The customer (or public) is (always) the most important part of the production line" (Deming 1982, 174), which aligns amazingly well with article 1.

Crosby wrote: "The integrity system concept is meant to produce an organization that, while producing a nice return on investment, doesn't make problems for itself or for others." Crosby went on to clarify: "I do not know of a single product safety problem where the basic cause was something other than a lack of integrity judgment on the part of some management individual. Product safety is not a legal problem; it is an ethical one" (Crosby 1979, 84). Crosby also wrote that integrity and compassion are vital and fundamental facets of character that cannot be learned or vastly improved from without, but we cannot succeed in management without them (Crosby 1979, 146).

In his 14 Points for Transformation (Deming 1982, 23, 59–62, 264–68), Deming listed point 8 as: "Drive out fear so that everyone may work effectively for the company." In the quality professional's environment, fear can be a result of failing to personally follow the ASQ Code of Ethics. Fear can also result from specific management actions, as Deming clearly pointed out.

While it may be true that behavior that is considered unethical in Europe and North America may be acceptable in some areas of the globe, implying that a single approach for the entire world may not be the most effective business practice, the ASQ Code of Ethics remains, and most confidently shall remain, fixed and stalwart. It is incumbent on any ASQ member to abide by its precepts, even when interacting and working with business partners and associates familiar with alternative interpretations of ethics.

The following are examples of decision making in ethically gray areas:

1. *Public at risk.* Your company has been using a supplier for years with little or no trouble and is happy with their cost and quality. Through a contact in the professional society you participate in, you become aware of a major safety recall that was initiated by another company due to defects in material from this same supplier you use. The identity of the supplier is not public knowledge as part of the recall, only the final manufacturer. Do you inform your management of this important information?

2. *Conflict of interest—supplier.* You are asked to audit and evaluate a potential supplier for a possible strategic partnership for a product your company has just developed. You own stock in the supplier that you purchased more than five years ago, prior to any discussion about the new product development. Should you disclose your financial interest in the supplier and recuse yourself from the evaluation?

3. *Conflict of interest—appearance of conflict.* Your spouse works in accounting for a company that your employer would like to evaluate as a supplier. The supplier has an excellent reputation for quality and price. Your spouse does not have any responsibility for delivered product quality or pricing. Should you disclose this potential conflict of interest?

4. *Work outside an area of competence.* You are a very accomplished and experienced supplier quality engineer working with printed circuit board assemblies. You are asked by your employer to perform a supplier audit on one of your current suppliers, a large multinational contract manufacturer that has been supplying PCBs. Your company wants to start purchasing large mechanical items from them that are cast metal components. Should you go ahead with the audit of the mechanical casting capabilities of the supplier since you already have knowledge of the supplier's quality system, or should you add an SME in casting to the audit team?

Chapter 21
Compliance and Sustainability

COMPLIANCE

> Understand issues of compliance and their applicable policies, laws, and regulations (e.g., conflict of interest, confidentiality, and bribery). (Apply)
>
> **Body of Knowledge VII.B.1**

Compliance and sustainability are two important concepts that are crucial for businesses and organizations to operate in a responsible and ethical manner.

Compliance refers to the adherence to laws, regulations, and standards that are set by government agencies, industry associations, and other relevant organizations. It involves following rules and guidelines related to various aspects of business operations such as health and safety, labor practices, environmental protection, and data privacy. Organizations must comply with these regulations to avoid penalties, legal action, and reputational damage.

Sustainability, on the other hand, refers to the practices and processes that allow an organization to operate in a manner that meets the needs of the present generation without compromising the ability of future generations to meet their own needs. This encompasses environmental, social, and economic aspects of a business and can include reducing waste and emissions, promoting diversity and equality, and implementing ethical business practices.

By combining compliance and sustainability, organizations can ensure that they are not only following rules and regulations, but also taking a responsible approach to their operations that protects the environment, promotes social responsibility, and supports the long-term viability of the business.

Vendor compliance is rapidly growing in importance as organizations look to outsourcing to reduce costs and increase focus on their core competencies. One key issue revolves around risk management and regulatory compliance.

As we look at compliance, we will address the following:

- Issues of compliance
- Laws and regulations

Issues of Compliance

In regulated industries such as medical devices, pharmaceutical, biotechnology, and food and beverages, compliance plays a key role in how we evaluate, assess, and report on design, development, and production activities.

Even in industries such as automotive, aerospace, and defense, issues of compliance play a crucial role. It is often challenging for companies to keep employees up to date due to the changing compliance landscape.

Additionally, compliance to the EU's REACH directive, RoHS, or the waste electrical and electronic equipment (WEEE) directive impacts companies differently depending on the industry, which adds to the complexity of selling in a global market.

There are several issues that organizations face when it comes to compliance:

1. Keeping up with regulations: The laws and regulations related to compliance are constantly changing, making it challenging for organizations to stay up to date and ensure they are following the latest guidelines.
2. Cost: Complying with regulations can be expensive, especially for small businesses and organizations. They may have to invest in new technology, training, and other resources to meet compliance requirements.
3. Lack of standardization: Different industries and regions have different regulations, making it difficult for organizations to implement a consistent compliance strategy.
4. Risk management: Organizations need to assess and manage the risks associated with noncompliance, such as fines, legal action, and reputational damage.
5. Employee awareness: Ensuring that all employees are aware of compliance requirements and understand their role in maintaining compliance can be a challenge.
6. Data privacy: With the increasing amount of data that organizations collect and store, ensuring that this information is protected and secure is a major compliance issue.
7. Global compliance: Organizations with a global presence need to comply with different regulations in different countries, making it challenging to maintain a consistent compliance strategy.

Laws and Regulations

There are numerous laws and regulations that impact the supplier process. In today's world of consumer affairs, with product liability litigation and an endless listing of government and industry regulations, businesses must develop a quality control system that satisfies customer requirements and remains cost effective. To this end, the business manager must rely heavily on customer requirements. Initial contact with potential customers should aim at understanding these requirements and regulations.

Laws and regulations are important components of compliance, and they vary by industry, location, and jurisdiction. Some of the most common laws and regulations that organizations must comply with include:

1. Health and safety regulations: Organizations must comply with regulations related to workplace health and safety, including requirements for protective equipment, training, and emergency response plans.
2. Environmental regulations: These regulations cover a range of environmental issues, including air and water pollution, waste management, and the use of natural resources.
3. Labor laws: Organizations must comply with labor laws that cover issues such as minimum wage, overtime pay, health benefits, and workplace discrimination.
4. Data protection laws: Organizations must comply with data protection laws, such as the General Data Protection Regulation (GDPR) in the European Union and the California Consumer Privacy Act (CCPA) in California, which govern how organizations collect, store, and use personal data.
5. Anti-corruption laws: Organizations must comply with anti-corruption laws, such as the Foreign Corrupt Practices Act (FCPA) in the United States and the United Kingdom's Bribery Act, which prohibit bribery and corruption in business transactions.
6. Trade laws: Organizations must comply with trade laws, such as import and export regulations, tariffs, and sanctions.

Food Safety

Food safety concerns are growing. Today's consumers are increasingly aware of the diverse health hazards associated with the production, processing, and marketing of foods. Although the United States has one of the safest food supplies in the world, our food still makes millions of people ill each year and causes numerous deaths. To help protect consumers, there are numerous local, state, and national government agencies that monitor and regulate the origin, composition, quality, safety, weight, labeling, packaging, marketing, and distribution of food sold in the United States. Most food industries are aware of the need to provide safe and wholesome food products to consumers. Adverse publicity about product safety or a company's negligence in complying with government safety regulations can result in large monetary losses, not to mention product liability lawsuits.

In 1938, the Federal Food, Drug, and Cosmetic (FD&C) Act was signed into law. This law gave FDA the responsibility for ensuring the safety of consumer products. Section 402(a)(3) of the FD&C Act defines a food as adulterated if "it consists in whole or in part of any filthy, putrid, or decomposed substance or if it is otherwise unfit for food." Section 402(a)(4) elaborates on the definition: "A food shall be deemed to be adulterated if it has been prepared, packed, or held under unsanitary conditions whereby it may have been contaminated with filth or whereby it may have been rendered injurious to health."

In 1979, the FD&C Act was amended to make it illegal to receive contaminated food. Any company involved with foods sold in interstate commerce, except for meat and poultry products, falls under FDA jurisdiction. Meat and poultry products are under the jurisdiction of the US Department of Agriculture (USDA).

The procurement department is the first line of defense against unsafe food products entering a company's facility. Verifying that suppliers can provide safe products requires both time and effort early in negotiations but can prevent unsafe product from ever being shipped to your facility. Relying on inspectors or lab analyses to identify and reject unsafe product after it has entered your facility is not only a risky way to ensure a safe food product but also a costly one.

The process of selecting a food product supplier includes addressing food safety issues. At a minimum, a supplier should demonstrate a management commitment to food safety by:

- Being aware of, and complying with, applicable government regulations
- Following good manufacturing practices (GMPs)
- Having a product identification and traceability system
- Having a process in place to control and prevent food contamination

How does one determine whether there is a management commitment to these basic food safety practices? Initially, ask, then follow up with a visit to verify that what was said actually is being practiced.

Determining if a company complies with the appropriate government regulations requires a basic knowledge of the regulations pertaining to the product that is being procured. After becoming familiar with the regulations pertaining to the product you are procuring, you can ask specific questions to ascertain whether the supplier is complying with the law. Suppliers that do not comply with food laws are a risk to do business with, both from a safety standpoint and because of the possibility of being implicated with them in the event of legal action.

To prevent the possibility of being criminally implicated with a supplier, a general requirement for doing business is to obtain a continuing food and drug guaranty. The guaranty is designed to exempt the buyer from criminal prosecution provided the buyer does not do anything to the product. This is a guaranty by the seller/supplier of a product subject to the FD&C Act that the product is not adulterated or misbranded within the meaning of the act. The guaranty should clearly state that it is the supplier's responsibility to ensure that products are in conformance with all the requirements of the applicable laws.

For all practical purposes, a request for a continuing food and drug guaranty should prompt a supplier to evaluate food safety practices to verify compliance with the FD&C Act. By having a signed guaranty on file, you have taken the first step in establishing that your supplier is in compliance with federal regulations.

GMPs are FDA regulations. The intent of GMP regulations is to interpret Section 402(a)(4) of the FD&C Act and establish criteria to aid industries in complying with the law. The GMPs summarize the correct procedures in a food production or processing facility. For this reason, FDA's GMPs are important in maintaining safe food products. They are contained in the Code of Federal

Regulations, Title 21, Part 110 (21 CFR 110). Any food supplier should have an internal set of rules that addresses every element of the GMPs, and every employee of the supplier should be following those rules. Ask a potential supplier to provide a copy of the company's GMPs and ask how it informs the employees of their roles in producing a safe food product.

A food supplier should have a coding system in place that identifies product by lot or batch and records the ingredients used with its individual lot identities. A system of this nature is necessary in the event that an individual ingredient is found to be unsafe. Traceability of ingredients into products and through distribution allows questionable product to be rapidly identified and completely quarantined or retrieved. A coding system should be reliable and understandable. Coding on product must contain the required information and must be legible. Verification of product identification and traceability is typically done at the supplier's facility by reverse tracing. This is conducted by choosing one lot of your product and tracing back through the manufacturing process records to determine exactly what lots or batches of ingredients were used in the manufacture of the product. A well-documented paper trail indicates that the supplier maintains good control of the identity of the product and its components during the manufacturing process. To provide a safe food product, a supplier must fundamentally prevent dangerous microorganisms and hazardous foreign matter from reaching the customer. The most cost-effective way to accomplish this is to prevent the occurrence of adulteration or contamination.

Prevention focuses on identifying and controlling the numerous variables that affect food product safety. These include the building (both the interior and exterior), pest control, ingredients, packaging material, employees, equipment, operating procedures, sanitation procedures, storage, and shipping. To control the myriad of variables that affect the safety of food, a systematic and comprehensive approach is required by the supplier.

HACCP (typically pronounced "has-sip") is a system that identifies and monitors safety criteria in a process to prevent hazards from occurring. HACCP is regarded as the most effective and economical way to prevent microbial, chemical, and physical contamination of food products. HACCP originated in the chemical processing industry over 40 years ago. In the 1950s, the Atomic Energy Commission began using HACCP principles to design nuclear power plants. The National Aeronautics and Space Administration (NASA) suggested that the HACCP approach be used in the production of space rations to minimize the possibility of foodborne illness among space crews. At that time, one of the contractors manufacturing space food was the Pillsbury Company. Pillsbury realized that by applying HACCP principles to the manufacture of food, it would minimize its liability. In 1971, Pillsbury adopted a HACCP process to maintain food safety.

A HACCP system:

- Identifies those control points in the food processing operation that are important in the prevention of adulteration
- Identifies the hazards associated with each control point
- Establishes adequate controls at each control point

- Establishes adequate monitoring of the controls at each control point
- Establishes corrective actions with documentation

The hazard analysis begins by flowcharting the entire manufacturing process—from raw material procurement to consumption by the ultimate customer. With the entire process diagrammed, the critical control points can be identified. Critical control points are those variables that (if not maintained) can result in the production of an unsafe food product. Once identified, the control limits, monitoring procedures and frequencies, and corrective actions are determined.

Several food safety regulatory agencies recognize the benefits of the HACCP approach and are recommending that food manufacturers adopt it. A HACCP program is voluntary; the regulatory agencies only recommend it. If you are responsible for procuring safe food products for your company, however, it is up to you to mandate that your suppliers have a hazard prevention and control system in place. Without a system, your supplier's problems become yours.

The Effects of Outsourcing on Compliance

Outsourcing of noncore activities is by now a fact of life in business. Services as diverse as data processing, building maintenance, gas compression, and power generation are being transferred out of many organizations. The reasons for this trend include, among others, cost reduction, improvement of company focus, and freeing assets from noncore activities.

Outsourcing requires constant monitoring to ensure that service levels meet contract requirements. This inevitably implies organizing a supplier management program that can involve consultants/contractors. One key area of focus that is important in the written contract is the contractor/consultant will provide on the basis of his or her experience, competence, and availability within the industry.

Bribery and Corruption

Bribery and corruption can significantly impact compliance. One of the key considerations for any anti-bribery and corruption (ABC) compliance program is how to think about the risk of third parties that may enter into a contractual relationship with your company via the supply chain. In many companies the supply chain function is a standalone discipline that is siloed from other functions. Typically, the greatest divide found in companies is between the sales side and supply side of the business. However, this divide also comes into play with a company's compliance function and the supply chain.

Even if the supply chain performs extensive due diligence or other background checks on a potential vendor, it does not do so in conjunction with the company's compliance function or even integrate an ABC compliance program approach into its calculation. If any ABC compliance analysis is involved, it is usually a "check-the-box" approach. Such an approach has potentially disastrous consequences for any ABC compliance program. The reason is that if you employ a check-the-box approach it may be not only inefficient but, more importantly, ineffective. This is because each ABC compliance program should be tailored to an organization's specific needs, risks, and challenges. This means that a company should assess its needs and, based on that assessment, design the third-party component of its ABC compliance program to be appropriate for its particular business. In the end, if designed carefully, implemented earnestly, and enforced fairly, a company's compliance program—no matter how large or small the organization—will allow the company, generally, to prevent violations, detect those that do occur, and remediate them promptly and appropriately.

Risk Ranking

The determination of the level of due diligence and categorization of a supplier depends on a variety of factors, including, but not limited to, whether the supplier is (1) located, or will operate, in a high-risk country; (2) associated with, recommended by, or required by a government official or his or her representative; (3) currently under investigation, the subject of criminal charges, or was recently convicted of criminal violations, including any form of corruption; (4) a multinational publicly traded corporation with a recognized exemplary system of compliance and internal controls, that has not been recently investigated or convicted of any corruption offense or that has taken appropriate corrective action to remedy such conduct; or (5) a provider of widely available services and products that are not industry specific, are offered to the public at large, and do not fall under the definition of low-risk supplier, detailed below.

A high-risk supplier is an individual or entity that is engaged to provide nonproject-specific goods or services to a company. It presents a higher level of compliance risk because of the presence of one or more of the following factors:

- It is based or operates in a country (including the supply of goods or services to a company) that poses a high risk for corruption, money laundering, or commercial bribery;
- It supplies goods or services to a company from a high-risk country;
- It has a reputation in the business community for questionable business practices or ethics; or
- It has been convicted of, or is alleged to have been involved in, illegal conduct and has failed to undertake effective remedial actions.

Finally, it presents one or more of the following factors:

- It is located in a country that has inadequate regulatory oversight of its activities;
- It is in an unregulated business;
- Its ultimate or beneficial ownership is difficult to determine;
- The company has an annual spend of more than $100,000 with the supplier;
- It was established or registered in a jurisdiction where ownership is not transparent or that permits ownership in the form of bearer shares;

- It is registered or conducts business in a jurisdiction that does not have anticorruption, anti–money laundering, and antiterrorism laws comparable to those of the United States and the United Kingdom; or
- It lacks a discernable and substantial business history.

A low-risk supplier is an individual or a non-publicly-held entity that conducts a business, such as a sole proprietorship, partnership, or privately held corporation, located in a low-risk country. Some indicia include that

- It supplies goods, equipment, or services directly to a company in a low-risk country;
- A company has an annual spend of less than $100,000 with the supplier; and
- It has no involvement with any foreign government, government entity, or government official.

However, if the supplier has other indicia of lower risk, such as that it is a publicly held company, it may be considered a low-risk supplier because it is subject to the highest disclosure and auditing and reporting standards, such as those under the US Securities Exchange Act of 1934, including those publicly traded on a reputable and highly regulated stock exchange, such as the New York or London exchanges, and are therefore subject to oversight by highly regarded regulatory agencies.

Assessing Supplier Risk

Supplier auditing is a key tool in managing the risk from the supply chain. An example of a company that handles this issue with a high profile is Apple, which annually publicizes vendors through its Supplier Responsibility Report. Apple conducts interviews with employees, contract workers, and senior management in relevant functional areas. Apple also conducts a physical inspection of manufacturing facilities and factory-managed dormitories and dining areas, as well as a review of records and relevant policies and procedures. Apple believes there may be cases where its audit reveals compliance in actual practice but the underlying management system may not be strong enough to prevent violations. For this reason, the Apple audits include examination of the management systems, such as policies and procedures, roles and responsibilities, and training programs, underlying every category in its compliance program.

Whistle-Blowers and the Supply Chain

Companies must be aware of whistle-blowers in today's interconnected world of suppliers. The smartphone is ubiquitous across the globe. Even where people do not have enough money, training, or skills to use a computer, they know how to use a smartphone. Digital pictures and videos can cause reputational damage. More importantly, these common modern-day tools can be used to facilitate a worldwide base of whistle-blowing. There are financial incentives for employees or anyone else to blow the whistle on companies that may be breaking the law.

In the United States, the Dodd-Frank Act has whistle-blower provisions that can pay a bounty of up to 30% of fines and penalties of companies that break laws such as the US Foreign Corrupt Practices Act. The US agency that administers this law, the Securities and Exchange Commission, has announced that even though Dodd-Frank is a US law, it will pay whistle-blowers if they are citizens of countries outside the United States. This means there is no territorial limit to the whistle-blower provisions, and any company subject to the Dodd-Frank law can be subject to paying whistle-blowers who provide information from outside the United States. These initiatives make it clear that the supply chain is becoming one of the highest risks in any business organization. The management of the supply chain will be a critical element for businesses going forward.

SUSTAINABILITY

> Understand and recognize the importance of environmental, social, and governance factors and adhere to applicable sustainability policies. (Understand)
>
> **Body of Knowledge VII.B.2**

Environmental, Social, and Governance Factors

Environmental, social, and governance (ESG) factors are a set of criteria that investors use to evaluate companies' sustainability and ethical practices. These factors are used to assess a company's impact on society and the environment, as well as the effectiveness of its management and governance practice.

1. Environmental factors: These factors assess a company's impact on the environment, including its carbon footprint, water and energy use, and waste management. Investors evaluate a company's environmental practices, such as its use of renewable energy sources, sustainability of its supply chain, and compliance with environmental regulations.
2. Social factors: These factors assess a company's impact on society, including its labor practices, human rights, community engagement, and product safety. Investors evaluate a company's social practices, such as its employee diversity and inclusion, labor standards, community involvement, and product safety.
3. Governance factors: These factors assess a company's management and governance practices, including its board composition, executive compensation, and shareholder rights. Investors evaluate a company's governance practices, such as its transparency and accountability, risk management, and compliance with regulatory requirements.

Investors use ESG factors to evaluate a company's sustainability and ethical practices and make investment decisions based on these criteria. Companies that perform well on ESG factors are seen as more attractive to investors who prioritize sustainable and responsible investing. By incorporating ESG factors into their operations, companies can improve their reputation, attract investors, and ensure long-term success.

Sustainability Policies

Sustainability refers to the practices and processes that allow an organization to operate in a manner that meets the needs of the present generation without compromising the ability of future generations to meet their own needs. This encompasses three main pillars: economic sustainability, social sustainability, and environmental sustainability.

1. Economic sustainability: This involves ensuring that an organization's operations are economically viable and sustainable over the long term. This includes managing finances, reducing costs, and investing in technology and infrastructure.
2. Social sustainability: This involves promoting social responsibility and ethical business practices. This includes promoting diversity and inclusion, implementing fair labor practices, and supporting local communities.
3. Environmental sustainability: This involves reducing the impact of an organization's operations on the environment. This includes reducing waste and emissions, implementing environmentally friendly practices, and using renewable energy sources.

Organizations can implement sustainability practices in various ways, such as reducing waste and emissions through recycling programs, using renewable energy sources, and implementing sustainable procurement practices. By integrating sustainability into their operations, organizations can improve their reputation, attract customers and employees who value sustainability, and ensure their long-term viability.

Chapter 22
Confidentiality

Information is readily available. It is no longer limited to books, journals, news publications, and the classroom. The digital age has made the search for any subject limitless. With so much data available, today's quality professional must learn not only to collect and assess data but also to comprehend it. In trying to understand the information, one must ask the following questions:

- Where did it come from?
- Is it factual or opinion based?
- Does someone own it?
- Do you need permission to use it?
- What are the consequences for not obtaining rightful use of it?
- Is this information legally protected?

In the supplier management relationship, the quality professional is exposed to information that must be handled with sensitivity and caution.

ORGANIZATIONAL POLICIES

> Apply organizational policies for executing appropriate agreements such as non-disclosure, quality, and change notification agreements. (Apply)
>
> **Body of Knowledge VII.C.1**

Businesses have put in place organizational policies to protect corporate information. Besides having employees sign confidentiality agreements to maintain and keep company information, the same requirement is extended to external partners. At initial onboarding, the appropriate agreements should be entered into with the supplier before they are assessed (Bossert 2004, p. 43). Such contractual agreements will provide supplier requirements that capture regulatory, legal, company, and quality expectations.

The model supplier quality agreement (SQA) covers the quality requirements under applicable quality system elements. Applicability may vary depending on the type of material or service a supplier provides, the quality standard they are certified to (if any), and the supplier's commitment level. Terms should also include nondisclosure of proprietary information and processes, management of intellectual property (if applicable), and timely change notifications.

Purchasing terms must be kept separate from the SQA. A supply or purchasing agreement will establish the supplier's responsibilities to (1) receive, inspect, account for, and store purchased materials; (2) contact the customer regarding disposition of any nonconforming material prior to using it; and (3) determine the total cost of such material that may be unacceptable after the supplier has performed work on it (Bossert 2004, p. 53). This type of agreement may include some quality requirements but not to the extent of an SQA.

In some cases, typically for service-oriented suppliers, a statement of work or SOW is also created to provide further details on the operational expectations. Key points of an SOW include material/service definition and scope, deliverables and their respective timing, provided inputs from the company and contractor, performance evaluation, price and payment, warranties, contract changes,

Parties
Company: ACME Manufacturing
Contractor: AAA Consulting Services

Service Definition and Scope
Provision of consulting services for determination of compliance to applicable standards.

Services will be carried out in ACME Manufacturing, Detroit, MI, and AAA Consulting Services headquarters in Ann Arbor, MI.

Work will be carried out following schedule presented in Exhibit A.

Deliverables
All deliverables shall be handed to ACME Manufacturing by June 30, 2017.

Deliverables include all the documents listed in Exhibit B.

Company-Provided Inputs
ACME Manufacturing shall provide:

- Access to documentation and records for review
- Appropriate subject matter experts (SMEs) to answer questions

ACME Manufacturing will pay for airfare and all other travel-related expenses, insofar as they are reasonable.

Performance Evaluation—Bonus/Penalty Clauses
AAA Consulting Services will pay US$5,000 for each week the termination date is not met.

Price and Payment
All work will be invoiced monthly, following the time reports approved by ACME Manufacturing's representative.

Parties agree that, unless any changes in scope are agreed, the contract price will in no case exceed US$100,000.

Figure 22.1 SOW example—consulting services.
Source: J. L. Bossert, *The Supplier Management Handbook*, 6th ed. (Milwaukee, WI: ASQ Quality Press, 2004).

confidentiality, key personnel, software and patents, and termination clauses. Not all of these elements are required in an SOW. Figure 22.1 gives of an example of an SOW developed for a consulting service.

Suppliers may also provide a copy of their own SOW template to their customers to simplify the contract negotiation process. There are pros and cons in doing so. The pros include clear identification of the supplier's commitments, an expeditious contract signing process, and timely onboarding for the supplier. The con is that the customer's requirements are not taken into consideration and may be in conflict with the supplier's existing SOW.

Before signing off on any contractual agreements with the supplier, one should always have the company's legal counsel review and approve them.

INTELLECTUAL PROPERTY

> Apply procedures for protecting the intellectual property of an organization and its suppliers. (Apply)
>
> Body of Knowledge VII.C.2

A number of businesses rely on intellectual property (IP) for their success and continued income flow. For this reason, the quality professional must recognize how organizations are able to protect their IP.

IP, depending on the type, can be protected through the following:

- Patents
- Copyrights
- Trademarks

The organization should develop a documented process in identifying what is considered IP and how IP will be managed with its suppliers and vice versa. Details of this process should be included in the contractual agreement with each supplier. IP owners may define limited terms to include how the IP will be used, where (i.e., within the United States or not), and for how long. It is very important that this information be written into the agreement and agreed on by both parties and their legal representatives.

ILLEGAL ACTIVITY

> Understand and interpret policies for reporting observations and deviations that could be perceived as illegal activity. (Apply)
>
> Body of Knowledge VII.C.3

Infringement or violation of one's IP rights may be subject to civil or criminal law, depending on the IP type, jurisdiction, and violations.

The Copyright Act of 1976, the Patent Act of 1952, and the Trademark Act of 1946 (also known as the Lanham Act) provide legal protection for IP against unauthorized use, theft, and other violations of the rights granted by those statutes to the IP owner. The federal Economic Espionage Act of 1996 provides criminal penalties for the theft or misappropriation of trade secrets, an alternative to patent law protection (Yeh 2012).

This should not be confused with those actions covered under a "safe harbor" for the use of IP (e.g., patented inventions) for research in the United States for philosophical purposes or to gather data in preparation for a drug approval.

An IP owner may file a lawsuit against an alleged infringer. In order to properly enforce IP rights, the owner needs to be aware of the statute of limitations and any federal agencies, in addition to the US Department of Justice, that may be involved in the prosecution. Penalties imposed will vary depending on the cause of action that has led to the violation.

In the United States, violations of IP rights may be reported to the National Intellectual Property Rights Coordination Center or electronically submitted using a form on the website of US Immigration and Customs Enforcement (https://help. cbp.gov/app/answers/detail/a_id/108).

The World Intellectual Property Organization (WIPO) Convention, the constituent instrument of WIPO, was signed in Stockholm on July 14, 1967, entered into force in 1970, and was amended in 1979. WIPO is an intergovernmental organization that in 1974 became one of the specialized agencies of the United Nations system. The origins of WIPO go back to 1883 and 1886, when the Paris Convention for the Protection of Industrial Property and the Berne Convention for the Protection of Literary and Artistic Works provided for the establishment of an "International Bureau." The two bureaus were united in 1893 and, in 1970, were replaced by WIPO, by virtue of the WIPO Convention.

WIPO administers 26 treaties, including the WIPO Convention, divided into three groups. The first general group of treaties defines internationally agreed basic standards of IP protection in each country:

- Beijing Treaty on Audiovisual Performances
- Berne Convention
- Brussels Convention
- Madrid Agreement (Indications of Source)
- Marrakesh VIP Treaty
- Nairobi Treaty
- Paris Convention
- Patent Law Treaty
- Phonograms Convention
- Rome Convention
- Singapore Treaty on the Law of Trademarks
- Trademark Law Treaty

- Washington Treaty
- WIPO Copyright Treaty (WCT)
- WIPO Performances and Phonograms Treaty (WPPT)

The second general group, known as the global protection system treaties, ensures that one international registration or filing will have effect in any of the relevant signatory states. The services provided by WIPO under these treaties simplify and reduce the cost of making individual applications or filings in all the countries in which protection is sought for a given IP right:

- Budapest Treaty
- Hague Agreement
- Lisbon Agreement
- Madrid Agreement
- Madrid Protocol
- Patent Cooperation Treaty (PCT)

The third and final general group is the classification treaties, which create classification systems that organize information concerning inventions, trademarks, and industrial designs into indexed, manageable structures for easy retrieval:

- Locarno Agreement
- Nice Agreement
- Strasbourg Agreement
- Vienna Agreement

Part VIII

Appendices

Appendix A
ASQ Certified Supplier Quality Professional Body of Knowledge

The topics in this Body of Knowledge include subtext explanations and the cognitive level at which the questions will be written. This information will provide useful guidance for both the Exam Development Committee and the candidate preparing to take the exam. The subtext is not intended to limit the subject matter or be all-inclusive of that material that will be covered in the exam. It is meant to clarify the type of content that will be included on the exam. The descriptor in parentheses at the end of each entry refers to the maximum cognitive level at which the topic will be tested. A complete description of cognitive levels is provided at the end of this document.

I. Supplier Strategy (20 Questions)

A. Supply Chain Vision/Mission

Assist in the development and communication of the supply chain vision/mission statement. (Apply)

B. Supplier Lifecycle Management

1. Supplier Selection

Develop the process for supplier selection and qualification, including the identification of sub-tier suppliers using tools such as SIPOC, decision analysis, and total risk factor analysis. (Create)

2. Performance Monitoring

Develop the supplier performance monitoring system, including expected levels of performance, process reviews, performance evaluations, improvement plans, and exit strategies. (Create)

3. Supplier Classification System

Define and develop a supplier classification system (e.g., non-approved, conditionally approved, approved, preferred, certified, partnership, and disqualified). (Create)

4. Partnerships and Alliances

Identify and analyze strategies for developing customer-supplier partnerships and alliances. (Analyze)

C. **Supply Chain Cost Analysis**

1. **Cost Reduction**

Identify and apply relevant inputs to prioritize cost reduction opportunities. (Analyze)

2. **Supply Chain Rationalization**

Interpret and analyze the optimization of a supply base to improve spending and leverage investments into supplier quality or risk reduction. (Analyze)

3. **Make/Buy Decisions**

Provide input on make/buy decisions using internal and external capability analysis. Apply tools such as SWOT analysis and use historical performance to analyze requirements. (Analyze)

D. **Supplier Agreements or Contracts**

1. **Terms and Conditions**

Review and provide input for developing terms and conditions that govern supplier relationships to ensure quality considerations are addressed. (Apply)

2. **Supplier Agreements**

Identify elements of supplier agreements (e.g., business and legal approach/requirements). (Understand)

3. **Quality Agreements**

Analyze the elements of quality agreements/requirements (e.g., other levels of approval/review). (Analyze)

4. **Finalization Controls**

Describe controls used to finalize terms and conditions that govern supplier relationships (e.g., agreements, contracts, and purchase orders). (Understand)

E. **Deployment of Strategy and Expectations**

Communicate strategy internally and communicate expectations to suppliers externally. (Apply)

II. **Risk Management (19 Questions)**

A. **Strategy**

1. **System**

Develop a risk-based approach to manage the supply base, including business continuity, contingency planning, and supply chain resilience. (Create)

2. **Product/Service Risk Mitigation**

Develop and implement a risk mitigation plan for predicting, minimizing, monitoring, and/or controlling risks. (Create)

3. **Prevention Strategies**

Identify and evaluate strategies and techniques such as supply chain mapping, avoidance, detection, and mitigation used to prevent the introduction of counterfeit parts, materials, and services. (Evaluate)

4. **Supplier Risk Identification and Categorization**

Identify supplier risks and develop categorization (e.g., organizational, business, security, and product) using tools and models, such as the Kraljic portfolio segmentation model. (Create)

B. **Analysis and Mitigation**

1. **Analysis**

Identify, assess, and prioritize risks to supplier quality using tools such as decision analysis (DA), failure mode and effects analysis (FMEA), fault tree analysis (FTA), and process auditing. (Evaluate).

2. **Mitigation Control**

Develop and deploy controls such as inspection and test plans. Prioritize mitigation activities and sustain a risk mitigation plan appropriate to the risk of the product/service. (Create)

3. **Mitigation Effectiveness**

Verify the effectiveness of the control plan and improve, if necessary, using continuous improvement methods such as plan-do-check-act (PDCA), lean, and product auditing tools. (Create)

III. **Supplier Selection and Part Qualification (27 Questions)**

A. **Product/Service Requirements Definition**

1. **Internal Design Reviews**

Identify and apply common elements of the design review process, including roles and responsibilities of the participants. (Apply)

2. **Identifying Requirements**

Identify and apply internal requirements (e.g., interrelated functional business units) for product or service in collaboration with stakeholders, including the requirements for supply chain, sub-tier suppliers, and manufacturability evaluation. (Evaluate)

B. **Supplier Selection Planning**

1. **Supplier Comparison**

Evaluate existing suppliers' and distributors' capabilities, capacities, past quality, delivery, price, lead times, and responsiveness against identified requirements. (Evaluate)

2. **Potential Suppliers Evaluation**

Assess potential new suppliers against identified requirements using tools such as self-assessments, audits, financial analysis, and quality function deployment. Verify third-party certification status and regulatory compliance and analyze and report on results of assessments to support the supplier selection process. (Evaluate)

3. Supplier Selection

Evaluate and select suppliers based on analysis of assessment reports and existing supplier evaluations using decision analysis tools such as weighted decision matrices and selection matrices. (Evaluate)

C. Part, Process, and Service Qualification

1. Technical Review

Interpret and evaluate technical specification requirements and characteristics such as views, title blocks, dimensioning and tolerancing, and apply GD&T symbols as they relate to the product and process. (Evaluate)

2. Supplier Relations

Collaborate with suppliers to define, interpret, and classify quality characteristics for the part/process/service. (Evaluate)

3. Process and Service Qualification Planning

Develop a part/process/service (e.g., calibration, laboratory, software, and design) qualification plan with supplier and internal team that includes service provider audit, calibration requirements, sample size, first article inspection, measurement system analysis (MSA), process flow diagram (PFD), failure mode and effects analysis (FMEA), control plans, critical to quality (CTQ), inspection planning, capability studies, material and performance testing, appearance approval, and internal process validation. (Analyze)

4. Part Approval

Understand production part approval process (PPAP) requirements and ensure suppliers understand the processes required to produce parts with consistent quality during an actual production run at production rates. (Understand)

5. Validate Requirements

Collaborate with internal team to interpret the results of the executed qualification plan for the part/process/service, including reviewing Certificate of Compliance (CoC), Certificate of Analysis (CoA), and production readiness reviews (PRR). (Evaluate)

IV. Supplier Performance Monitoring, and Improvement (29 Questions)

A. Supplier Performance Monitoring

1. Supplier Metrics

Define, implement, and monitor supplier performance metrics such as quality, delivery (e.g., on-time delivery [OTD] and on-time in full delivery [OTIF]), cost, and responsiveness. (Evaluate)

2. Supplier Performance

Analyze supplier performance data (e.g., warranty analysis/field returns and defect rates) and develop periodic reports (e.g., scorecard and dashboards). (Analyze)

3. Supplier Process Performance

Define and implement lean principles and applications such as 5S, kaizen, value stream mapping, supplier process capabilities and controls, 8 wastes, single minute exchange of dies (SMED), kanban, muda, standardized work, takt time, and error-proofing to reduce waste and increase performance. (Evaluate)

B. Assess Nonconforming Product/Process/Service

Segregate, control, and evaluate nonconforming materials to determine whether a material review board (MRB) requires disposition. Conduct risk assessments to prevent future discrepancies. (Evaluate)

C. Supplier Corrective and Preventive Action (CAPA)

1. Root Cause Analysis Tools and Methods

Evaluate the root cause analysis of a problem using tools such as cause and effect diagrams, Pareto analysis, 5 Why's, fault tree analysis, design of experiments (DOE), brainstorming, check sheets, measurement system analysis (MSA), production records, and review of process flow. (Evaluate)

2. Collaboration with Supplier

Evaluate and implement supplier corrective/preventive action and review its effectiveness and robustness with supplier. Understand the process of updating failure mode and effects analysis (FMEA) and process control plan, and understand statistical process control (SPC), 8D, and product/process design change. (Evaluate)

V. Supplier Quality Management (26 Questions)

A. Supplier Quality Monitoring

1. Supplier Audit

Apply the stages of a quality audit, including audit planning, conducting the initial audit, and executing periodic reevaluation. Understand and apply the various types of quality audits (e.g., product, process, and management system) and audit methods (e.g., virtual, on-site, and desktop). (Apply)

2. Audit Reporting and Follow-up

Apply and analyze audit reporting and follow up, including verification of the effectiveness of corrective action. (Analyze)

3. Supplier Communication

Evaluate various communication techniques such as periodic reviews, metric and performance indices, change management, notifications, recalls, change requests, and business updates. Maintain active communication with suppliers to assess risk and take appropriate action. (Evaluate)

4. Supplier Development and Remediation

Identify and analyze present and future training needs and gaps, using quality methods and tools such as kaizen and benchmarking. Use process improvement tools such as DMAIC, cycle time reduction, defect rate, and cost reduction. Evaluate supplier remediation to develop and manage improvement plans. (Evaluate)

5. **Project Management Basics**

Understand and apply various types of project reviews, such as phase-end, management, and retrospectives or post-project reviews to assess project performance and status, to review issues and risks, and discover and capture lessons learned from the project. Apply forecasts, resources, schedules, and task and cost estimates to develop and monitor project plans. (Apply)

B. **Teams and Team Processes**

1. **Team Development**

Identify and describe the various types of teams and the classic stages of team development: forming, storming, norming, performing, and adjourning. (Apply)

2. **Team Roles**

Define and describe various team roles and responsibilities for leader, facilitator, coach, and individual member. (Understand)

3. **Performance and Evaluation**

Describe various techniques to evaluate training, including evaluation planning, feedback surveys, pre-training testing, and post-training testing. (Understand)

C. **Compliance with Requirement and Supplier Categorization**

Understand and evaluate compliance with regulations and industry standards (e.g., RoHS, Governmental regulatory authorities, and ISO), specifications, contracts, agreements, and certification authority. Evaluate and categorize suppliers based on risk and performance. (Evaluate)

VI. **Relationship Management (16 Questions)**

A. **Supplier Onboarding**

Understand and apply processes for orientation of suppliers such as providing overview of company, vision, mission, guiding principles, overall requirements, expectations, and criticality of product, service, and delivery requirements. (Apply)

B. **Communication**

1. **Techniques and Mediation**

Identify and apply communication techniques (e.g., oral, written, and presentation) specifically for internal stakeholders and suppliers to resolve issues. Apply different techniques when working in multi-cultural environments. Identify and describe the impact that culture, communications, and Diversity, Equity, and Inclusion (DEI) can have on an organization. (Evaluate)

2. **Reporting Using Quality Tools**

Use appropriate technical and managerial reporting techniques for effective presentation and reporting, including the seven classic quality tools: Pareto charts, cause and effect diagrams, flowcharts, control charts, check sheets, scatter diagrams, and histograms. (Analyze)

C. Leadership and Collaboration

Understand and apply techniques for coaching suppliers through regular communications, influencing without authority, negotiation techniques, conflict resolution techniques, and establish clear roles and responsibilities of internal stakeholders and suppliers using tools such as a RACI matrix (responsible, accountable, consulted, and informed). (Evaluate)

VII. Business Governance, Ethics, and Compliance (13 Questions)

A. ASQ Code of Ethics

Determine appropriate behavior in situations requiring ethical decisions, including identifying conflicts of interest, and recognizing and resolving ethical issues. (Apply)

B. Compliance and Sustainability

1. Compliance

Understand issues of compliance and their applicable policies, laws, and regulations (e.g., conflict of interest, confidentiality, and bribery). (Apply)

2. Sustainability

Understand and recognize the importance of environmental, social, and governance factors and adhere to applicable sustainability policies. (Understand)

C. Confidentiality

1. Organizational Policies

Apply organizational policies for executing appropriate agreements such as non-disclosure, quality, and change notification agreements. (Apply)

2. Intellectual Property

Apply procedures for protecting the intellectual property of an organization and its suppliers. (Apply)

3. Illegal Activity

Understand and interpret policies for reporting observations and deviations that could be perceived as illegal activity. (Apply)

LEVELS OF COGNITION BASED ON BLOOM'S TAXONOMY—REVISED (2001)

In addition to **content** specifics, the subtext for each topic in this BOK also indicates the intended **complexity level** of the test questions for that topic. These levels are based on "Levels of Cognition" (from Bloom's Taxonomy—Revised, 2001) and are presented below in rank order, from least complex to most complex.

Remember

Recall or recognize terms, definitions, facts, ideas, materials, patterns, sequences, methods, principles, etc.

Understand

Read and understand descriptions, communications, reports, tables, diagrams, directions, regulations, etc.

Apply

Know when and how to use ideas, procedures, methods, formulas, principles, theories, etc.

Analyze

Break down information into its constituent parts and recognize their relationship to one another and how they are organized; identify sublevel factors or salient data from a complex scenario.

Evaluate

Make judgments about the value of proposed ideas, solutions, etc., by comparing the proposal to specific criteria or standards.

Create

Put parts or elements together in such a way as to reveal a pattern or structure not clearly there before; identify which data or information from a complex set is appropriate to examine further or from which supported conclusions can be drawn.

Appendix B
Sample Risk Assessment Template

Step 1: Risk identification	Step 2: Risk assessment		Step 3: Managing risks					
Possible risks	Impact (H/M/L)	Likelihood (HML)	What are we already doing about it?	What more can we do about it?	When will it be done?	Who will do it?	How will we review progress?	Reviewed level of risk

Person/group responsible for review	Date to be reviewed

Appendix C
Acronym List

AAR—appearance approval report
ABC—anti-bribery and corruption
AIAG—Automotive Industry Action Group
ALARP—as low as reasonably possible
ALI—American Law Institute
ANOVA—analysis of variance
ANSI—American National Standards Institute
AP—action priority
API—active pharmaceutical ingredient
APQP—Advanced Product Quality Planning
AQL—acceptable quality level
ASAP—as soon as possible
ASL—approved supplier list
ASME—American Society of Mechanical Engineers
ASN—advanced shipping notice
ASQ—American Society for Quality
ASTM—American Society for Testing and Materials
AVL—approved vendor list
BOK—body of knowledge
BoM—bill of materials
CA—confidentiality agreement
CAPA—corrective and preventive action
CCP—critical control point
CDA—confidential disclosure agreement
CFR—Code of Federal Regulations
cGMP—current good manufacturing practice
CISG—Contracts for International Sale of Goods

CLEAR—collaborative-limited-emotional-appreciable-refinable
CM—corrective maintenance
CMDCAS—Canadian Medical Devices Conformity Assessment System
CMDR—Canadian Medical Device Regulation
CO—change order
COA—certificate of analysis
COC—certificate of compliance
COGS—cost of goods sold
COPQ—cost of poor quality
COPQA—cost of poor quality analysis
COTS—commercial off-the-shelf
CP—control plan
C_{pk}—process capability index
CPP—critical process parameters
CQ—component qualification
CQA—critical quality attribute
CQI—continuous quality improvement
CR—capability ratio
CSD—Customer-Supplier Division
CSQP—certified supplier quality professional
CSR—corporate social responsibility
CTQ—critical to quality
CTS—critical to safety
DA—decision analysis
DCO—document change order
DEI—diversity, equity, and inclusion
DFSS—Design for Six Sigma
DHR—device history record
DMAIC—define-measure-analyze-improve-control
DMF—drug master file
DoD—Department of Defense
DOE—design of experiments
DPMO—defects per million opportunities
DQ—design qualification
ECHA—European Chemical Agency
ECO—engineering change order

EDMS—electronic document management system
ERP—enterprise resource planning
ESI—early supplier involvement
EU—European Union
FAI—first article inspection
FAL—first article layout
FAT—factory acceptance testing
FCPA—Foreign Corrupt Practices Act
FCR—facility change request
FDA—US Food and Drug Administration
FMA—failure mode analysis
FMEA—failure modes and effects analysis
FMECA—failure modes, effects, and criticality analysis
FRP—Full-Rate Production
FTA—fault tree analysis
GD&T—geometric dimensioning and tolerancing
GIP—good importer practice
GMP—good manufacturing practice
HACCP—hazard analysis and critical control points
HAZOP—hazard operability analysis
IATF—International Automotive Task Force
ICH—International Conference on Harmonization
IMDS—international material data sheet
IM&TE—inspection, measuring, and test equipment
IP—intellectual property
IPC—in-process control
IQ—installation qualification
IQA—incoming quality assurance
ISIR—initial sample inspection report
ISO—International Organization for Standardization
ITAR—International Traffic in Arms Regulations
IVDR—in-vitro device regulation
JIT—just-in-time
KPI—key performance indicator
KPRI—key process indicator
LCL—lower control limit

LPA—layered process audit
LRIP—Low-Rate Initial Production
LSL—lower specification limit
MBNQA—Malcolm Baldrige National Quality Award
MBR—master batch record
MDD—Medical Device Directive
MDR—Medical Device Regulation
MRB—material review board
MRP—material requirements planning
MRP II—manufacturing resource planning
MSA—measurement system analysis
MSDS—material safety data sheet
MTBF—mean time before failure
MVP—master validation plan
NCCUSL—National Conference of Commissioners on Uniform State Laws
NCM—nonconforming materials
NCR—nonconformance report
NDA—nondisclosure agreement
OEE—overall equipment effectiveness
OEM—original equipment manufacturer
OFAT—one factor at a time
OIL—open issues list
OOC—out of calibration
OOS—out of specification
OOT—out of tolerance
OQ—operational qualification
OTD—On-time delivery
OTIF—On-Time In Full
OTS—off the shelf
PA—preventive action
PCI—process capability index
PCP—process control plan
PCT—Patent Cooperation Treaty
PDCA—plan-do-check-act
PEST—political, economic, social, and technological
PESTLE—political, economic, social, technological, legal, and environmental

PHA—preliminary hazard analysis
PIA—proprietary information agreement
PID—project initiation document
PM—preventive maintenance
PMBOK—Project Management Body of Knowledge
PMI—Project Management Institute
PO—purchase order
PPAP—production part approval process
PPQ—process performance qualification
PPV—purchase price variance
PQ—performance qualification
PQR—product quality review
PRR—production readiness review
PSW—Part Submission Warrant
QA—quality assurance
QbD—quality by design
QC—quality control
QCU—quality control unit
QFD—quality function deployment
QMS—quality management system
QSR—Quality System Regulation
QU—quality unit
R&D—research and development
R&R—roles and responsibilities
RACI—responsible-accountable-consulted-informed
REACH—registration, evaluation, authorization, and restriction of chemicals
RFI—request for information
RFP—request for proposal
RFQ—Request for quote
RMP—risk management plan
RoHS—Restriction of Hazardous Substances
ROI—return on investment
RPN—risk priority number
SA—secrecy agreement
SAE—Society of Automotive Engineers
SC—supply chain

SCAR—supplier corrective action request
SCCAF—special characteristics communication and agreement form
SDS—safety data sheet
SCM—supply chain management
SEC—Securities and Exchange Commission
SIM—supplier information management
SIPOC—suppliers-inputs-process-outputs-customers
SKU—stock keeping unit
SLM—supplier lifecycle management
SMART—specific-measurable-attainable-realistic-timely
SME—subject matter expert
SMED—single-minute exchange of dies
SOP—standard operating procedure
SOW—statement of work
SPC—statistical process control
SQA—supplier quality agreement
SQAM—Supplier Quality Assurance Manual
SQE—supplier quality engineer
SQP—supplier quality professional
SRM—supplier relationship management
STS—ship-to-stock
SWOT—strengths, weaknesses, opportunities, and threats
TBE—technical bid evaluation
TCO—total cost of ownership
TE—technical expert
TMV—test method validation
TRF—total risk factor
UCC—Uniform Commercial Code
UDI—Unique Device Identifier
UL—Underwriters Laboratories
USDA—US Department of Agriculture
USML—United States Munitions List
VDA—Verband der Automobilindustrie
VMI—vendor-managed inventory
VMP—validation master plan
VOE—verification of effectiveness

VOI—verification of implementation
WCT—WIPO Copyright Treaty
WEEE—waste electrical and electronic equipment
WIPO—World Intellectual Property Organization
WPPT—WIPO Performances and Phonograms Treaty
WTO—World Trade Organization

Glossary

A

acceptance sampling—Sampling inspection in which decisions are made to accept or not accept product or service; the methodology that deals with procedures by which decisions to accept or not accept are based on the results of the inspection of samples.

accreditation—The formal recognition by an independent body, generally known as an accreditation body, that a certification body operates according to international standards.

accuracy—The degree to which the result of a measurement, calculation, or specification conforms to the correct value or a standard.

action level—Specification or limit established in published regulation (or a guidance, guideline, or standard that is regarded by competent authorities as de facto regulation) that, when exceeded, requires immediate intervention, including investigation of cause, immediate remediation, and/or corrective action.

actual yield—The quantity that is actually produced at any appropriate phase of manufacture, processing, or packing of a particular product.

Advanced Product Quality Planning (APQP)—Quality process for developing new products that uses up-front quality planning and evaluates the output to determine whether customers are satisfied.

alpha (α) risk—The probability of erroneously claiming a difference in two averages or two variances; the risk that the decision will be made that a part is defective when it really is not.

analysis of variance (ANOVA)—Basic statistical technique for analyzing experimental data. It subdivides the total variation of a data set into meaningful component parts associated with specific sources of variation, including interactions.

appraisal cost—Cost associated with measuring, evaluating, or auditing products, components, and purchased material to ensure conformance with quality standards and performance requirements.

attribute measurement—Qualitative measurement that typically shows only the number of parts or the number of defects per part failing to conform to specified criteria.

audit—The on-site verification activity, such as inspection or examination, of a process or quality system to ensure compliance to requirements. An audit can apply to an entire organization or might be specific to a function, process, or production step.

B

batch number—Any distinctive combination of letters, numbers, or symbols from which the complete history of the manufacture, processing, packing, holding, and distribution of a batch or lot of product or other material can be determined.

beta (β) risk—The probability of erroneously not claiming a difference in two averages or two variances; the risk that the decision will be made that a part is not defective when it really is.

bias—The difference between a measured value and a known or accepted reference value.

bilateral tolerance—Splitting of a tolerance by a median axis so that each side is identical.

business continuity plan—A document specifying tasks or activities needed to ensure that critical business functions in an organization will continue to be available to customers, suppliers, regulators, and other entities that must have access to those functions.

C

calibration—Comparison of two instruments or measuring devices, one of which is a standard of known accuracy traceable to national standards, to detect, correlate, report, or eliminate by adjustment any inaccuracy of the instrument or measuring device, as compared to the standard.

can—Indicates a possibility or a capability in regard to a standard.

capability ratio (Cr)—Measurement of the proportion of specification width that is consumed by process variation.

certification—The provision by an independent body of written assurance (a certificate) that the product, service, or system in question meets specific requirements.

change control—A written procedure that describes the action to be taken if a change is proposed to facilities, materials, equipment, and/or processes used in the fabrication, packaging, and testing, or that may affect the operation of the quality or support system.

cleanroom—A room that is maintained to be virtually free of contaminants, such as dust or bacteria, used in laboratory work and in the production of precision parts for electronic or aerospace equipment.

common cause variation—Causes of variation that are inherent in a process over time. They affect every outcome of the process and everyone working in the process.

confidence interval—Range within which a parameter of a population (e.g., mean and standard deviation) may be expected to fall, on the basis of measurement, with some specified confidence level.

consumer's risk—In sampling, the potential risk that bad products will be accepted and shipped to the consumer.

continual improvement—Ongoing activities to evaluate and positively change products, processes, and the quality system to increase effectiveness.

contract—An agreement entered into voluntarily by two or more parties with the intention of creating a legal obligation. It may have elements in writing, although contracts can be made orally.

control chart—A statistical tool including upper and lower control limits and the process average, which visually displays process performance.

control limits—The area three standard deviations on either side of the centerline, or mean, of data plotted on a control chart.

control number. See *batch number*.

control plan—Plan that establishes and maintains adequate written procedures covering all critical characteristics and key processes to ensure a consistent and acceptable quality product.

controlled area—A nonclassified room or operating area designed to control or minimize the presence, proliferation, and/or ingress of particulates, and in which specific environmental conditions (e.g., temperature, humidity, directional airflow, and viable and nonviable particulate limits) are defined and monitored to prevent contamination of exposed products.

correction—Repair, rework, or adjustment relating to the disposition of an existing discrepancy (also called *remedy* or *remediation*); usually the first step in a corrective action.

corrective action—Resolution of problems between user and producer arising due to product nonconformance.

critical process parameter (CPP)—Quantifiable equipment setting whose variability has an impact on a critical quality attribute and therefore should be monitored or controlled to ensure that the process produces the desired quality, purity, potency, and safety.

critical quality attribute (CQA)—Physical, chemical, biological, or microbiological property or characteristic that should be within an appropriate limit, range, or distribution to ensure the desired product quality.

customer—Person or organization (internal or external) that receives a product or service at any point in the product's lifecycle.

customer complaint—Formal or informal allegation by the customer due to failure in meeting previously agreed-upon requirements by the supplier.

D

dashboard—Visuals used in manufacturing to identify good or poor performance of a process.

decision maker(s)—Person(s) with the competence and authority to make appropriate and timely decisions.

design of experiments (DOE) or **experimental design**—Statistical technique used for planning, conducting, analyzing, and interpreting sets of experiments, aimed at making sound decisions most efficiently.

design space—Multidimensional combination and interaction of input variables (e.g., material attributes) and process parameters that have been demonstrated to provide assurance of quality.

destructive testing—Measurement, testing, and inspection of product or material that damages or destroys the product or material so it is not usable. Contrast with non-destructive testing.

detectability—Ability to discover or determine the existence, presence, or fact of a hazard.

deviation—Process through which a producer is authorized to ship nonconforming products to the user with the user's concurrence.

discrepancy—Datum or result outside of the expected range; an unfulfilled requirement; may be called nonconformity, defect, deviation, out-of-specification, out-of-limit, or out-of-trend.

discrimination—Discrimination, or *resolution*, of a measurement system is its capability to detect and indicate even small changes in the measured characteristic. The main consideration when selecting or analyzing a measurement system.

disposition—Final arrangement or settlement of nonconforming product in an orderly way.

distributor—Nonmanufacturing source of product, usually where no transformation of product takes place; an intermediary between a manufacturer and retailer or customer.

drawing (blueprint)—Sketch of the product being produced with specified tolerances of each characteristic.

E

effectiveness—The state of having produced a decided-on or desired effect.

engineering support—Essential part of a good quality system that encompasses product design and development as well as reliability testing of new or revised products.

external failure cost—Cost generated by defective products in the field after having been shipped to customers.

F

failure costs—Costs resulting from materials, products, or services failing to conform to requirements or customer/user needs, or not manufactured in compliance with applicable regulations.

failure modes analysis (FMA)—A procedure used to determine which malfunction symptoms appear immediately before or after the failure of a critical parameter in a system. After all possible causes are listed for each symptom, the product is designed to eliminate the problems.

failure modes and effects analysis (FMEA)—A systematized group of activities performed to recognize and evaluate the potential failure of a product or process and its effects, identify actions that could eliminate or reduce the occurrence of the potential failure, and document the process.

failure modes, effects, and criticality analysis (FMECA)—A procedure performed after a failure modes and effects analysis to classify each potential failure effect according to its severity and probability of occurrence.

final inspection—Examination of a product to ensure that it conforms to all applicable specifications and requirements before it is packaged and shipped to the customer.

financial analysis—An assessment used to evaluate financial stability for continued business through the contract phase or financial stability to support the manufacturing process and resources.

finished product—Batch of product in its final pack for release to the market.

first article inspection (FAI)—The practice of inspecting the first item, part, unit, group, or batch produced from a production run. A complete, independent, and documented physical and functional inspection process to verify that prescribed production methods have produced an acceptable item as specified by engineering drawings, planning, purchase order, engineering specifications, and/or other applicable design documents.

flowchart—Diagram that shows step-by-step progression of a product through a manufacturing system showing all factors that could adversely affect quality at the point where they occur.

flow-down requirement—A supplier requirement that must be passed down or delegated to sub-tier suppliers. Material or service specifications are often flowed down or passed down to the supplier's suppliers.

frequency distribution—Tabulation of the number of times a given outcome has occurred within the sample of products being checked.

G

gage repeatability—Measurement of the consistency obtained with one gage when used several times by one operator while measuring the identical characteristic on the same parts.

gage repeatability and reproducibility (gage R&R)—The evaluation of a gauging instrument's accuracy by determining whether its measurements are repeatable and reproducible.

gage reproducibility—Measurement of the consistency of different operators using the same gage while measuring the identical characteristic on the same parts.

gage variability—Measurement of the consistency of at least two sets of measurements obtained with a gage on the same parts as a result of time.

guard banding—The practice of utilizing the gage R&R results to reduce the range of acceptance to ensure that features are within the prescribed specifications.

H

harm—Damage to health, including the damage that can occur from loss of product quality or availability.

hazard—Potential source of harm.

hazard analysis and critical control points (HACCP)—A QMS for effectively and efficiently ensuring farm-to-table food safety in the United States. HACCP regulations for various sectors are established by the Department of Agriculture and FDA.

housekeeping—General cleaning (including sweeping) and removal of accumulated process waste, dirty equipment, utensils, and other nonproduct.

I

incoming inspection—Inspection of purchased parts at the customer's facility after the shipment of parts from the supplier to ensure supplier compliance with specifications and contractual agreements.

in-process control—One of various control strategies that measures a critical quality attribute (CQA) at a particular point in a manufacturing process.

inspection—Process of measuring, examining, testing, gauging, or otherwise comparing the unit with the applicable requirements.

installation qualification (IQ)—Establishes, by objective evidence, that all key aspects of the process equipment and ancillary system installations adhere to the manufacturer's approved specification and that the recommendations of the supplier of the equipment are suitably considered.

internal failure costs—Costs generated by a producer in making defective and nonconforming materials and products that do not meet company quality specifications.

K

key process characteristics—Manufacturing processes deemed crucial in producing a product to meet its design intent.

key product characteristics—Properties deemed crucial by the user to satisfy the design intent.

Kraljic Portfolio Purchasing Model—is a strategic purchasing model used to help purchasers reduce costs by maximizing purchasing power.

L

laboratory—A test facility that can include chemical, metallurgical, dimensional, physical, electrical, and reliability testing or test validation.

lead time—The time between the initiation and completion of a process.

line clearance—A procedure used to prevent mix-ups and/or comingling involving product, packaging, and labeling.

lot—A batch, or a specific identified portion of a batch, having uniform character and quality within specified limits.

lot control—System that provides the means to trace pertinent information about the materials and/or components comprising a given product.

lot number. See *batch number*.

M

material certification—Process through which documented evidence establishes that a product is in compliance with designated specifications.

material review board (MRB)—Quality control committee or team, usually employed in manufacturing or other materials-processing installations, that has the responsibility and authority to deal with items or materials that do not conform to fitness-for-use specifications.

may—Indicates a permission/recommendation in regard to a standard.

measurement system analysis (MSA)—An analysis that considers operations, procedures, devices, and other equipment or personnel used to assign a value to the characteristic being measured.

measuring and testing devices—Equipment used to evaluate a product's conformance to its specifications.

must—Indicates a requirement in regard to a standard.

N

nonconforming material—Material that is not in compliance with specifications.

nonconforming report (NCR)—A permanent record—made in writing—for accounting and preserving the knowledge of a nonconforming condition for the purposes of documenting facts or events.

nonconformity—Deficiency in a characteristic, product specification, process parameter, record, or procedure that renders the quality of a product unacceptable, indeterminate, or not according to specified requirements.

normal distribution—Condition where measured variation is symmetric about a central value and has a bell-shaped form.

O

operational—Occupancy state in which the installation is functioning in a defined manner, with a specified number of personnel present and working in a defined manner.

operational qualification (OQ)—Establishes by objective evidence that the equipment process control limits meet all predetermined requirements.

outliers—Observed points that are distant from other observations; abnormal responses resulting from special causes or uncontrolled influences that occur during an experiment.

out-of-control process—A process in which the statistical measure being evaluated is not in a state of statistical control. In other words, the variations between the observed sampling results cannot be attributed to a constant system of chance causes.

out of specification (OOS)—All in-process laboratory tests that are outside of established specifications.

out of tolerance (OOT)—During calibration, an item that is not within specification limits.

P

Pareto chart—A graphical tool for prioritizing effects.

performance qualification (PQ)—Establishes by objective evidence that a process consistently produces a result and/or product that meets the predetermined requirements (reproducible and repeatable).

pest control—System for preventing, evaluating, and eliminating infestation by rodents, insects, birds, and other vermin.

precision—The number of significant digits to which a value has been reliably measured; the ability of an instrument to repeat the same reading when making the same measurement in the same manner and under identical conditions.

prevention costs—The cost of all activities specifically designed to prevent poor quality in materials, products, or services (e.g., validation, PPAP, capability analysis, quality planning, training and quality education, design review, FMEA, DFSS, quality by design).

preventive action—Action taken to eliminate the cause of a potential discrepancy or other potential undesirable situation to make such an occurrence less probable.

preventive maintenance—Service, cleaning, lubrication of parts, alignment, adjustment, functional test, repair, modification, and overhaul, as required, to ensure there is no deterioration of equipment performance.

procedure—The steps in a process and how these steps are to be performed in order for the process to fulfill a customer's requirements; usually documented.

process—Combination of people, equipment, materials, methods, and environment that produces output to a planned effect.

process audit—Analysis of elements of a process and appraisal of completeness, correctness, or conditions.

process capability—A statistical measure of the inherent process variability of a given characteristic. The most widely accepted formula for process capability is Six Sigma.

process capability index—The value of the tolerance specified for the characteristic, divided by the process capability. The several types of process capability indexes include the widely used C_{pk} and C_p.

process control—The method for keeping a process within boundaries; the act of minimizing the variation of a process.

process development studies—Experiments that help rule in or rule out the choice and sequence of specific unit operations and unit processes and their associated detailed choices of equipment models and critical process parameters for manufacturing.

process performance qualification (PPQ)—The collection and evaluation of data, from the process-design stage through commercialization, that establishes scientific evidence that a process is capable of consistently delivering quality products.

process survey—Survey used to evaluate whether a supplier has process controls in place to ensure that the supplier's process will manufacture quality products. Process controls include proper tooling, equipment, and inspection.

process validation—Establishing by objective evidence that a process consistently produces a result or product meeting its predetermined requirements.

procurement quality—Any and all aspects dealing with the purchasing of products.

producer's risk—For a given sampling plan, the probability of not accepting a lot, the quality of which has a designated numerical value representing a generally desirable level. Usually, the designated value will be the acceptable risk and quality level.

product audit—Quantitative assessment of conformance to required product characteristics.

product lifecycle—All phases in the life of the product, from the initial development through marketing and the product's discontinuation.

product/process characteristic—Any given attribute of a product or process.

product/service—Intended results of activities or processes; can be tangible or intangible.

purchase order—A commercial document issued by a buyer to a seller that indicates types, quantities, and agreed prices for products or services the seller will provide to the buyer.

Q

qualitative analysis—Testing performed to establish the composition of natural or synthetic substances.

quality—A measure of a product's or service's ability to satisfy the customer's stated or implied needs; also, the degree to which a set of inherent properties of a product, system, or process fulfills requirements.

quality agreement—Detailed quality requirements that establish and maintain the systematic business relationship and expectations of the customer and supplier relationship.

quality assurance (QA)—Proactive and retrospective activities that provide confidence that product requirements are fulfilled.

quality by design (QbD)—A systematic approach to development that begins with predefined objectives and emphasizes product and process understanding and process control based on sound science and quality risk management.

quality control (QC)—Steps taken during the generation of a product or service to ensure that it meets requirements and that the product or service is reproducible.

quality function deployment (QFD)—A focused methodology for carefully listening to the voice of the customer and then effectively responding to those needs and expectations.

quality management—Accountability for the successful implementation of the quality system.

quality objectives—Specific, measurable activities or processes designed to meet the organization's intentions and directions as defined in the quality policy.

quality plan—Documented result of quality planning that is disseminated to all relevant levels of the organization.

quality planning—Management activity that sets quality objectives and defines the operational and/or quality system processes and the resources needed to fulfill the objectives.

quality policy—Statement of intentions and direction issued by the highest level of the organization, related to satisfying customer needs. It is similar to a strategic direction, which communicates quality expectations that the organization is striving to achieve.

quality system—Formalized business practices that define management responsibilities for organizational structure, processes, procedures, and resources needed to fulfill product/service requirements, customer satisfaction, and continual improvement.

quality unit (QU)—Group organized within an organization to promote quality in general practice.

quantitative analysis—Analytical techniques used to identify the amount or concentration of an analyte and quantify any compound or substance in a sample.

R

rating method—Quantitative method of evaluating systems and performance.

reliability—The susceptibility of a measurement device with a visual display to having its indications converted to a meaningful number, also expressed as the legibility of a visual display, normally defined as the minimum measure and increment that can be discriminated in the terms of the display.

request for information—An inquiry to a potential supplier about that supplier's product or service for potential use in the organization. The inquiry can provide certain organization requirements or be of a more general exploratory nature.

request for proposal (RFP)—A method of soliciting ideas from potential suppliers that may be incorporated into a final design of a product or service for a later quote; a document used to solicit vendor responses when the functional requirements and features are known but no specific product is in mind.

request for quote (RFQ)—An invitation to suppliers to bid on specific products or services.

resolution—The smallest change in input necessary to produce the smallest detectable change in output of the instrument under test.

risk—Combination of the probability of occurrence of harm and the severity of that harm.

risk acceptance—Decision to accept risk.

risk analysis—Estimation of the risk associated with the identified hazards.

risk assessment—Systematic process for organizing information to support a risk decision that is made within a risk management process (the process consists of the identification of hazards and the analysis and evaluation of risks associated with exposure to those hazards).

risk communication—The sharing of information about risk and risk management between the decision maker and other stakeholders.

risk control—Actions implementing risk management decisions.

risk evaluation—Comparison of the estimated risk to given risk criteria using a quantitative or qualitative scale to determine the significance of the risk.

risk identification—Systematic use of information to identify potential sources of harm (hazards) referring to the risk question or problem description.

risk management—Systematic application of quality management policies, procedures, and practices to the tasks of assessing, controlling, communicating, and reviewing risk.

risk reduction—Actions taken to lessen the probability of occurrence of harm and the severity of that harm.

risk review—Review or monitoring of output/results of the risk management process considering (if appropriate) new knowledge and experience about the risk.

rounding—Used to shorten numbers by either increasing or decreasing a number to the next digit.

run chart—Plot of process output over time without superimposed control limits.

S

sample—In acceptance sampling, one or more units of product (or a quantity of material) drawn from a lot for purposes of inspection to reach a decision regarding acceptance of the lot.

self-assessment—Typically a questionnaire sent to a new supplier to gather information on capability, capacity, staffing, and resources available to support the potential procurement or contract.

senior management—Top management officials in a firm who have the authority and responsibility to mobilize resources.

sensitivity—The ability of a measuring device to detect small differences in a quantity being measured.

severity—A measure of the possible consequences of a hazard.

shall—Indicates a requirement in regard to a standard.

ship-to-stock (STS)—Program in which the supplier and customer work together for improved quality and conformance of manufactured parts to eliminate the need for incoming or source inspection of purchased parts or products. Under this program, individual products or processes are qualified as opposed to an overall supplier certification. Maintenance of this program is provided through audits.

should—Indicates a recommendation in regard to a standard.

significant digit—Any of the digits of a number, beginning with the digit farthest to the left, that is not zero, and ending with the last digit farthest to the right that is either not zero or that is a zero but is considered to be exact.

skip-lot—Plan in acceptance sampling in which some lots in a series are accepted without inspection when the sampling results for a stated number of immediately preceding lots meet stated criteria.

source inspection—Inspection of purchased parts at the supplier's facility by a customer representative to ensure supplier compliance with specifications and contractual agreements.

special cause—Source of variation that is not inherent in the system and can be prevented.

specifications—Specific limits or parameters that are required to ensure the success of a product to perform as designed.

spend—Cost to an organization for a given purchased product or service for a specified period of time.

stakeholder—An individual or organization having an ownership or interest in the delivery, results, and metrics of the quality system framework or business process improvements; also, any individual, group, or organization that can affect, be affected by, or perceive itself to be affected by a risk. Decision makers might also be stakeholders.

standard deviation/sigma—Measurement of the spread of dispersion of a set of values about their average value.

statistical control—A process is considered to be in a state of statistical control if variations among the observed sampling results can be attributed to a constant system of chance causes.

statistical process control (SPC)—Use of statistical techniques to analyze a process or its output to take required actions to achieve and maintain a state of statistical control and improve the process capability.

statistical process control chart—Plot of process output over time with a superimposed central tendency line and upper and lower control limit lines.

sterilization—The act or process, physical or chemical, of destruction or elimination of all viable organisms (including bacterial and fungal spores, viruses, protozoa) in the inanimate environment.

stratified sampling—The process of collecting a representative sample by selecting units deliberately from various identified locations.

sub-tier supplier—A supplier for the main supplier of a product of service.

supplier—A source of materials, service, or information input provided to a process.

supplier certification—Program aimed at qualifying suppliers already on an approved status to a higher level of approval called certification. This usually encompasses review of the supplier's past delivered product history and an in-depth quality system survey. Certification of a supplier usually is all-encompassing and covers all products. Once certification is granted to a supplier, the customer may institute reduced sampling at incoming inspection.

survey—Broad overview of a supplier's system or process used to evaluate the adequacy of that system or process to produce quality products.

sustainability—A process by which organizations manage their financial, social, and environmental risks, obligations, and opportunities.

system audit—Documented activity performed to verify, by examination and evaluation of objective evidence, that applicable elements of the quality system are suitable and have been developed, documented, and effectively implemented in accordance with specified requirements.

system survey—Survey conducted to assess whether the supplier has appropriately controlled systems that will adequately prevent the manufacture of nonconforming products.

T

test method validation (TMV)—See *measurement system analysis.*

theoretical yield—The quantity that would be produced at any appropriate phase of manufacture, processing, or packing of a product, based on the quantity of components to be used, in the absence of any loss or error in actual production.

tolerance—The maximum and minimum limit values a product can have and still meet customer requirements.

total quality costs—The sum of costs, representing the difference (delta value) between the actual (although not commonly captured in the accounting systems) cost of material, product, or service and what the potential reduced cost would be in the absence of poor quality. This "delta" value is where cost reduction opportunities will be found. This is a crucial input to the determination of total cost of ownership.

traceability—The ability to track the history, application, or location of what is under consideration from the origin of materials and parts, the processing history, and distribution, and location of the product after delivery.

trend—The tendency of a variable, attribute, or characteristic to increase, decrease, or remain unchanged over time.

U

unclassified area—A room or area not designated by grades but that needs to be designed and maintained such that its environment does not adversely impact the quality, purity, and integrity of the products. Unclassified areas may or may not be controlled areas.

V

validation—Confirmation, through the provision of objective evidence, that the requirements for a specific intended use or application have been fulfilled.

validation lifecycle—Process, equipment, and facility assurance that depends on the following stages (separated in time) being executed properly:

Stage 1—Process, equipment, and/or facility design

Stage 2—Process, equipment, and/or facility qualification

Stage 3—Process, equipment, and/or facility continued verification/monitoring

Stage 4—Process, equipment, and/or facility retirement

variable measurement—Quantitative data in which physical properties are measured, such as hole diameters or coating thickness.

variables data—Measurement quantities that are measured on a continuous and infinite scale.

variance—The square of the standard deviation.

variation—A change in data, characteristic, or function caused by one of four factors: special causes, common causes, tampering, or structural variation. Sources of variation include time to time, part to part, and within-sample.

vendor rating—System of measurement of vendor or supplier performance against set goals or standards.

Venn diagram—A diagram that shows all possible logical relations between a finite collection of different sets.

verification—Confirmation, through the provision of objective evidence, that specified requirements have been fulfilled.

Bibliography

ANSI/ASQC. 1987. Standard Q94-1987. *Quality Management and Quality Systems Elements Guidelines.* Milwaukee, WI: ASQ Quality Press.

Arter, D. 2003. *Quality Audits for Improved Performance.* Milwaukee, WI: ASQ Quality Press. Autry, J. A. 2001. *The Servant Leader.* Danvers, MA: Three Rivers Press.

Blanchard, K. 1999. *Heart of a Leader.* Colorado Springs: David C. Cook.

Blanchard, K., and J. L. Stoner. 2011. *Full Steam Ahead!* Oakland, CA: Berrett-Khoeler Publishers.

Blanchard K., and T. Waghorn. 1999. *Mission Possible: Becoming a World-Class Organization While There's Still Time.* New York: McGraw-Hill.

Blanchard, K., P. Zigarmi, and D. Zigarmi. 1985. *Leadership and the One Minute Manager: Increasing Effectiveness through Situational Leadership.* New York: Morrow.

Bossert, J. L. 2004. *The Supplier Management Handbook.* 6th ed. Milwaukee, WI: ASQ Quality Press.

Cavinato, J. L. 2006. *The Supply Management Handbook.* 7th ed. New York: McGraw-Hill.

Chibba, A. 2007. "Measuring Supply Chain Performance Measures." Licentiate thesis, Luleå University of Technology.

Crosby, P. B. 1979. *Quality Is Free: The Art of Making Quality Certain.* New York: McGraw-Hill.

Daskal, Lolly. 2015. "The 100 Best Leadership Quotes of All Time." *Inc.,* April 3. http://www.inc.com/lolly-daskal/the-100-best-leadership-quotes-of-all-time.html.

Davis, Shirley. 2022. "Diversity, Equity and Inclusion for dummies", Hoboken, NJ, Wiley & Sons.

Deming, W. E. 1982. *Out of the Crisis.* Cambridge, MA: MIT CAES.

DeYong, C., and K. Case. 1996. "Operationalizing Customer Satisfaction Dimensions." *51st Annual Quality Congress Proceedings.* Milwaukee, WI: ASQC.

Durivage, M. A. 2022. *Practical Engineering, Process, and Reliability Statistics.* 2nd ed. Milwaukee, WI: ASQ Quality Press.

———. 2015. *Practical Attribute and Variable Measurement Systems Analysis (MSA).* Milwaukee, WI: ASQ Quality Press.

———. 2016. *Practical Design of Experiments (DOE).* Milwaukee, WI: ASQ Quality Press.

Durivage, M. A., and R. Mehta. 2016. *Practical Process Validation.* Milwaukee, WI: ASQ Quality Press.

Fisher, R., and W. Ury. 2011. *Getting to Yes: Negotiating Agreement without Giving In.* New York: Penguin.

Frost, Stephen. Ed. 2022. "The Key to Inclusion: A practical guide to diversity, equity and belonging for you, your team and your organization," New York, NY, Kogan Page.

Garrison, D., F. Khalil, and P. O'Reilly. 2001. "Purchasing Professional Services." *NAPM Info Edge*, May.

Global Harmonization Task Force (GHTF) Study Group 3. 2004. *Quality Management Systems—Process Validation Guidance*. January.

Gorman, C. K. 2011. "Communicating Across Cultures." American Management Association reprint, March.

Griffith, G. K. 2012. *The Quality Technician's Handbook*. 6th ed. Milwaukee, WI: ASQ Quality Press.

Heinrichs, J. 2013. *Thank You for Arguing*. New York: Crown Publishing.

Hoerl, R. W., and R. Snee. 2012. *Statistical Thinking: Improving Business Performance*. 2nd ed. Hoboken, NJ: John Wiley & Sons.

Hughes, J. 2008. "From Vendor to Partner: Why and How Leading Companies Collaborate with Suppliers for Competitive Advantage." *Global Business and Organizational Excellence* 27, no. 3: 21–37.

"Introduction to Kaizen,"Creative Safety Supply, 2015.

Juran, J. M., and J. A. De Feo. 2010. *Juran's Quality Handbook: The Complete Guide to Performance Excellence*. 6th ed. New York: McGraw-Hill.

Kang, C. W., and P. H. Kvam. 2011. *Basic Statistical Tools for Improving Quality*. Hoboken, NJ: John Wiley & Sons.

Kraljic, P. 1983. "Purchasing Must Become Supply Management." *Harvard Business Review*, September.

McKeller, J. M. 2014. *Supply Chain Management Demystified*. Columbus, OH: McGraw-Hill Education.

Mills, C. A. 1989. *The Quality Audit: A Management Evaluation Tool*. Boston: McGraw-Hill. Myers, M. B. 2010. "The Many Benefits of Supply Chain Collaboration." *Supply Chain Management Review*, November 12.

Patterson, K., J. Grenny, R. McMillan, and A. Switzler. 2002. *Crucial Conversations: Tools for Talking When Stakes Are High*. New York: McGraw-Hill.

Pohlig, H. 2002a. "Drafting Consulting Agreements." *Inside Supply Management*, November.

———. 2002b. "Legal Issues of Contracting for Services." *Inside Supply Management*, September.

Porter, Kathey. 2022. "Supplier Diversity for dummies", Hoboken, NJ, Wiley & Sons.

Reece, B. L., and R. Brandt. 1990. *Effective Human Relations in Organizations*. 4th ed. Boston: Houghton Mifflin.

Rodríguez-Pérez, J. 2011. *Quality Risk Management in the FDA-Regulated Industry*. Milwaukee, WI: ASQ Quality Press.

Rodríguez-Pérez, J., and M. Peña-Rodríguez. 2012. "Fail-Safe FMEA." *Quality Progress*, January, 31–36.

Russell, J. P. 2010. *The Process Auditing Techniques Guide*. 2nd ed. Milwaukee, WI: ASQ Quality Press.

Russell, J. P. 2013. *The ASQ Auditing Handbook*. 4th ed. Milwaukee, WI: ASQ Quality Press.

Saxton, B. W. 2001. "Reasons, Regulations, and Rules: A Guide to the Validation Master Plan." *Pharmaceutical Engineering* 21, no. 3: 1–6.

Sayle, A. J. 1988. *Management Audits: The Assessment of Quality Management Systems*. 2nd ed. Columbus, OH: McGraw-Hill Education.

Shore, J., and J. Freije. 2016. *Proactive Supplier Management in the Medical Device Industry*. Milwaukee, WI: ASQ Quality Press.

Skhmot, N., 2017. "The 8 Wastes of Lean." *The Lean Way*, August 5. https://theleanway.net/The-8-Wastes-of-Lean.

Sonya Krakoff, Senior Content Marketing Specialist, Champlain College Online. https://online.champlain.edu/blog/top-conflict-resolution-strategies.

Thompson, Jason R. 2022. "Diversity and Inclusion Matters: Tactics and Tools to Inspire Equity and Game-Changing Performance", Hoboken, New Jersey, Wiley & Sons.

Toyota. 2006. "Ask 'Why' Five Times About Every Matter." Toyota Traditions, March. http://www.toyota-global.com/company/toyota_traditions/quality/mar_apr_2006.html.

Westcott, R. T., ed. 2014. *The Certified Manager of Quality/Organizational Excellence Handbook*. 4th ed. Milwaukee, WI: ASQ Quality Press.

"What is Mistake Proofing?", ASQ, https://asq.org/quality-resources/mistake-proofing.

"What is Value Stream Mapping (VSM)?," ASQ, https://asq.org/quality-resources/lean/value-stream-mapping.

Womack, J. P., and D. T. Jones. 2003. *Lean Thinking*. 2nd ed. New York: Simon & Schuster.

Wood, D. C., ed. 2013. *Principles of Quality Costs: Financial Measures for Strategic Implementation of Quality Management*. Milwaukee, WI: ASQ Quality Press.

Yeh, B. T. 2012. *Intellectual Property Rights Violations*. Congressional Research Service, December 13.

ISO STANDARDS

ISO 9001:2015 *Quality management systems—Requirements.*

ISO 13485:2016 *Medical devices—Quality management systems—Requirements for regulatory purposes.*

ISO 14971:2019 *Medical devices—Application of risk management to medical devices*

ISO 17025:2017 *General requirements for the competence of testing and calibration laboratories.*

ISO 19011:2018 *Guidelines for auditing management systems.*

ISO 31000:2018 *Risk management—Principles and guidelines.*

Index

Note: Figures and tables are indicated by *f* and *t* following the page numbers.

B

C

D

E

F

www.ingramcontent.com/pod-product-compliance
Lightning Source LLC
LaVergne TN
LVHW080644210926
848485LV00003B/321